Advances and Technologies in High Voltage Power Systems Operation, Control, Protection and Security

Advances and Technologies in High Voltage Power Systems Operation, Control, Protection and Security

Editors

Hassan Haes Alhelou
Pierluigi Siano

MDPI • Basel • Beijing • Wuhan • Barcelona • Belgrade • Manchester • Tokyo • Cluj • Tianjin

Editors
Hassan Haes Alhelou
University College Dublin
Ireland

Pierluigi Siano
University of Salerno
Italy

The work of Hassan Haes Alhelou was supported in part by the Science Foundation Ireland (SFI) through the SFI Strategic Partnership Programme under Grant SFI/15/SPP/E3125, and in part by the University College Dublin (UCD) Energy Institute. The opinions, findings, and conclusions or recommendations expressed in this material are those of the authors and do not necessarily reflect the views of the Science Foundation Ireland. For the purpose of open access the author has applied a CC BY public copyright license to any author accepted manuscript version arising from this submission.

Editorial Office
MDPI
St. Alban-Anlage 66
4052 Basel, Switzerland

This is a reprint of articles from the Special Issue published online in the open access journal *Applied Sciences* (ISSN 2076-3417) (available at: https://www.mdpi.com/journal/applsci/special_issues/power_security).

For citation purposes, cite each article independently as indicated on the article page online and as indicated below:

LastName, A.A.; LastName, B.B.; LastName, C.C. Article Title. *Journal Name* **Year**, *Volume Number*, Page Range.

ISBN 978-3-0365-1140-5 (Hbk)
ISBN 978-3-0365-1141-2 (PDF)

Contents

About the Editors

Hassan Haes Alhelou (Senior Member, IEEE) is currently with University College Dublin, Dublin 4, Ireland. He has published more than 100 research papers in high-quality peer-reviewed journals and international conferences. He has participated in more than 15 industrial projects. His major research interests include power systems, power system dynamics, power system operation and control, dynamic state estimation, frequency control, smart grids, micro-grids, demand response, load shedding, and power system protection. Dr. Alhelou is included in the 2018 and 2019 Publons list of the top 1% Best Reviewer and Researchers in the field of engineering. He was a recipient of the Outstanding Reviewer Award from *Energy Conversion and Management* journal in 2016, *ISA Transactions* journal in 2018, *Applied Energy* journal in 2019, and many other Awards. He was a recipient of the Best Young Researcher in the Arab Student Forum Creative, among 61 researchers from 16 countries, at Alexandria University, Egypt, in 2011. He has also performed reviews for highly prestigious journals, including *IEEE Transactions on Power Systems, IEEE Transactions on Smart Grid, IEEE Transactions on Industrial Informatics, IEEE Transactions on Industrial Electronics, Energy Conversion and Management, Applied Energy*, and *International Journal of Electrical Power and Energy Systems*. He has participated in more than 15 industrial projects. His major research interests are power systems, power system dynamics, power system operation and control, dynamic state estimation, frequency control, smart grids, micro-grids, demand response, and load shedding.

Pierluigi Siano (M'09–SM'14) received an MS.c. degree in electronic engineering and a Ph.D. in information and electrical engineering from the University of Salerno, Salerno, Italy, in 2001 and 2006, respectively. He is a Professor and Scientific Director of the Smart Grids and Smart Cities Laboratory with the Department of Management and Innovation Systems, University of Salerno. His research activities are centered on demand response, on energy management, on the integration of distributed energy resources in smart grids, on electricity markets and on the planning and management of power systems. In these research fields, he has co-authored more than 500 articles, including more than 300 international journal papers that received, in Scopus, more than 10,100 citations with an H-index equal to 49. In 2019 and 2020, he received the award of Highly Cited Researcher by ISI Web of Science Group. He has been the Chair of the IES TC on Smart Grids. He is an Editor for the Power and Energy Society Section of *IEEE Access, IEEE Transactions on Industrial Informatics, IEEE Transactions on Industrial Electronics*, the open journals of *IEEE IES, IET Smart Grid* and *IET Renewable Power Generation*.

applied sciences

MDPI

Editorial

Special Issue on Advances and Technologies in High Voltage Power Systems Operation, Control, Protection, and Security

Hassan Haes Alhelou [1,*] and Pierluigi Siano [2,*]

[1] Department of Electrical Power Engineering, Tishreen University, Lattakia 2230, Syria
[2] Department of Management and Innovation Systems, University of Salerno, 84084 Salerno, Italy
* Correspondence: alhelou@tishreen.edu.sy (H.H.A.); psiano@unisa.it (P.S.)

1. Introduction

The electrical demands in several countries around the world are increasing due to the huge energy requirements of prosperous economies and the human activities of modern life. In order to economically transfer electrical powers from generation-side to demand-side, these powers need to be transferred at high-voltage levels through suitable transmission systems and power substations. To this end, high-voltage transmission systems and power substations are in demand. Actually, they are at the heart of interconnected power systems, in which any faults might lead to unsuitable consequences, abnormal operation situations, security issues, and even power cuts and blackouts. In order to cope with the ever-increasing operation and control complexity and security in interconnected high-voltage power systems, new architectures, concepts, algorithms, and procedures are essential. This special issue aims at encouraging researchers to address the technical issues and research gaps in high-voltage transmission systems and power substations in modern energy systems.

2. Published Papers

This special issue contains 10 highly qualified papers related to the latest advances and technologies in Power Systems Operation, Control, Protection, and Security. In what follows, each paper published in this special issue is introduced briefly by highlighting its main contributions. Authors in [1] have proposed a novel very fast transient overvoltage (VFTO) suppression method with great prospects in engineering, called the spiral tube damping busbar. The paper results showed that the improved damping busbar has a significant suppressing effect on the amplitude and the frequency of very fast transient overvoltage.

In [2], an adaptive ensemble square root Kalman filter (AEnSRF) has been proposed, in which the ensemble square root filter (EnSRF) and Sage–Husa algorithm are utilized to estimate measurement noise online. Simulation results obtained by applying the proposed method showed that the estimation accuracy of AEnSRF is better than that of ensemble Kalman filter (EnKF), and AEnSRF can track the measurement noise when the measurement noise changes.

Authors in [3] have presented a transient electromagnetic method (TEM) to determine grounding grid orientation without excavation. Unlike the existing pathological solutions, TEM does not enhance the surrounding electromagnetic environment. A secondary magnetic field as a consequence of induced eddy currents is subjected to inversion calculation. A digital simulator has been proposed in [4] that enables the dynamic testing of protective relays without using any steady test and expensive real-time simulators. This simulator includes both external waveform imports and internal waveform generation functions.

In [5], authors presented a comparative study on mother wavelets using a fault type classification algorithm in a power system. The study aims to evaluate the performance of the protection algorithm by implementing different mother wavelets for signal analysis and

Citation: Haes Alhelou, H.; Siano, P. Special Issue on Advances and Technologies in High Voltage Power Systems Operation, Control, Protection, and Security. *Appl. Sci.* **2021**, *11*, 274. https://doi.org/10.3390/app11010274

Received: 18 December 2020
Accepted: 28 December 2020
Published: 30 December 2020

determines a suitable mother wavelet for power system protection applications. An online multiprocessor scheduling multiprocessor with bounded speed (MBS) has been proposed in [6]. The comparative analysis shows that the multiprocessor with bounded speed (MBS) outperforms other algorithms. The competitiveness of MBS is the least to date.

In [7], a new hybrid control algorithm for the cascaded modular multilevel converter has been presented. The Harris hawk's optimization (HHO) and atom search optimization (ASO) are used to optimally design the controller of the hybrid modular multilevel converters (MMC). This study in [8] has focused on one such application. In this proposed work, a direct current (DC)-based intermediate DC-DC power converter (i.e., a modified LUO (M-LUO) converter) is used to extricate the availability of power in the high range from the PV array.

Further to the above mentioned research papers, this special issues contains a case study. The case study in [9] applied laser-induced breakdown spectroscopy (LIBS) to analyze contamination on insulator surfaces. Moreover, a review paper [10] has also been published in this special issue. The review paper in [10] has addressed the existing approaches attending to cyber-physical security in power systems from a microgrid-oriented perspective.

Funding: This research received no external funding.

Acknowledgments: This issue would not have been possible without the help of a variety of talented authors, professional reviewers, and the dedicated editorial team of Applied Sciences. Thank you to all the authors and reviewers for this opportunity. Finally, thanks to the Applied Sciences editorial team.

Conflicts of Interest: The authors declare no conflict of interest.

References

1. Almenweer, R.A.; Su, Y.-X.; Xixiu, W. Numerical Analysis of a Spiral Tube Damping Busbar to Suppress VFTO in 1000 kV GIS. *Appl. Sci.* **2019**, *9*, 5076. [CrossRef]
2. Nan, D.; Wang, W.; Wang, K.; Mahfoud, R.J.; Haes Alhelou, H.; Siano, P. Dynamic State Estimation for Synchronous Machines Based on Adaptive Ensemble Square Root Kalman Filter. *Appl. Sci.* **2019**, *9*, 5200. [CrossRef]
3. Qamar, A.; Ul Haq, I.; Alhaisoni, M.; Qadri, N.N. Detecting Grounding Grid Orientation: Transient Electromagnetic Approach. *Appl. Sci.* **2019**, *9*, 5270. [CrossRef]
4. Kim, J.; Cho, G.-J.; Kim, J. Development of Railway Protective Relay Simulator for Real-Time Applications. *Appl. Sci.* **2020**, *10*, 191. [CrossRef]
5. Pothisarn, C.; Klomjit, J.; Ngaopitakkul, A.; Jettanasen, C.; Asfani, D.A.; Negara, I.M.Y. Comparison of Various Mother Wavelets for Fault Classification in Electrical Systems. *Appl. Sci.* **2020**, *10*, 1203. [CrossRef]
6. Singh, P.; Khan, B.; Mahela, O.P.; Haes Alhelou, H.; Hayek, G. Managing Energy Plus Performance in Data Centers and Battery-Based Devices Using an Online Non-Clairvoyant Speed-Bounded Multiprocessor Scheduling. *Appl. Sci.* **2020**, *10*, 2459. [CrossRef]
7. Diab, A.A.Z.; Ebraheem, T.; Aljendy, R.; Sultan, H.M.; Ali, Z.M. Optimal Design and Control of MMC STATCOM for Improving Power Quality Indicators. *Appl. Sci.* **2020**, *10*, 2490. [CrossRef]
8. Darcy Gnana Jegha, A.; Subathra, M.S.P.; Manoj Kumar, N.; Subramaniam, U.; Padmanaban, S. A High Gain DC-DC Converter with Grey Wolf Optimizer Based MPPT Algorithm for PV Fed BLDC Motor Drive. *Appl. Sci.* **2020**, *10*, 2797. [CrossRef]
9. Lu, S.; Wang, X.; Wang, T.; Qin, X.; Wang, X.; Jia, Z. Analysis of Salt Mixture Contamination on Insulators via Laser-Induced Breakdown Spectroscopy. *Appl. Sci.* **2020**, *10*, 2617. [CrossRef]
10. Canaan, B.; Colicchio, B.; Ould Abdeslam, D. Microgrid Cyber-Security: Review and Challenges toward Resilience. *Appl. Sci.* **2020**, *10*, 5649. [CrossRef]

applied sciences

MDPI

Article

Dynamic State Estimation for Synchronous Machines Based on Adaptive Ensemble Square Root Kalman Filter

Dongliang Nan [1,2], Weiqing Wang [1], Kaike Wang [2], Rabea Jamil Mahfoud [3], Hassan Haes Alhelou [4,*] and Pierluigi Siano [5,*]

1 School of Electrical Engineering, Xinjiang University, Urumqi 830047, China; ndlxju@126.com (D.N.); wwq59@xju.edu.cn (W.W.)
2 Electric Power Research Institute, State Grid Xinjiang Electric Power Co., Ltd., Urumqi 830011, China; wangkaike@dky.xj.sgcc.com.cn
3 College of Energy and Electrical Engineering, Hohai University, Nanjing 210098, China; Rabea7mahfoud@hotmail.com
4 Department of Electrical Power Engineering, Faculty of Mechanical and Electrical Engineering, Tishreen University, Lattakia 2230, Syria
5 Department of Management & Innovation Systems, University of Salerno, 84084 Salerno, Italy
* Correspondence: h.haesalhelou@gmail.com (H.H.A.); psiano@unisa.it (P.S.); Tel.: +39-320-464-6454 (P.S.)

Received: 13 November 2019; Accepted: 26 November 2019; Published: 29 November 2019

Abstract: Dynamic state estimation (DSE) for generators plays an important role in power system monitoring and control. Phasor measurement unit (PMU) has been widely utilized in DSE since it can acquire real-time synchronous data with high sampling frequency. However, random noise is unavoidable in PMU data, which cannot be directly used as the reference data for power grid dispatching and control. Therefore, the data measured by PMU need to be processed. In this paper, an adaptive ensemble square root Kalman filter (AEnSRF) is proposed, in which the ensemble square root filter (EnSRF) and Sage–Husa algorithm are utilized to estimate measurement noise online. Simulation results obtained by applying the proposed method show that the estimation accuracy of AEnSRF is better than that of ensemble Kalman filter (EnKF), and AEnSRF can track the measurement noise when the measurement noise changes.

Keywords: dynamic state estimation (DSE); synchronous machine; ensemble square root filter (EnSRF); Sage–Husa algorithm

1. Introduction

In order to obtain the optimal control strategy of generator in power system, phasor measurement unit (PMU) is required to observe the generator data (power angle, electric angular velocity, etc.) to determine its state. However, there are random noises and measurement errors in PMU measurements, which cannot be directly used as the true state value [1–3]. Statistical method is used to calculate the estimated value of generator state variables and predict the future state. This method is called dynamic state estimation (DSE). Therefore, the prediction of generator's future state by DSE is helpful for power system regulation and has important significance to the stability of power network.

In recent years, the problem of electromechanical transient state estimation of generators has attracted wide attentions. Generator is a complex nonlinear system, and its DSE needs corresponding nonlinear filtering algorithm. In the field of generator DSE algorithms, many studies mainly focus on extended Kalman filter (EKF), unscented Kalman filter (UKF) and cubature Kalman filter (CKF) [4,5]. In [6], EKF was implemented to estimate the dynamic state of generator, where the fourth-order model of generator with unknown input was utilized. However, the linearization process of the generator model by EKF will lead to large truncation errors, which cannot be applied to the strongly nonlinear systems. In [7,8], the deterministic sampling filter-UKF algorithm was applied to estimate the dynamic state of generator. Due to the poor flexibility of UKF in parameter selection, the CKF method was proposed by S. Haykin [9,10]. In [11], the CKF was used for the DSE of generator, which could achieve much better performance than both EKF and UKF.

On the other hand, considering the system uncertainties, in [12], based on the second-order model of generator, adaptive interpolation and H_∞ theory was introduced into EKF to estimate the dynamic state of generator, by which the robustness was improved. In [13], an online state estimation method for synchronous generators based on UKF was proposed, and the fourth-order nonlinear model of generators was used to verify that the algorithm has not only high filtering speed and accuracy but also strong robustness. In [14], square root filtering was introduced into CKF to ensure the nonnegative qualitative and numerical stability of covariance matrix, by which, the asymmetric or nonnegative covariance of CKF in the iteration of DSE could be avoided and the estimation accuracy of CKF could also be improved. To the best of the authors' knowledge, there have been few results concerning DSE for generator by using ensemble Kalman filtering algorithm [15], which has been widely utilized in meteorology.

In this paper, based on the improved EnKF-EnSRF, which can approximate the nonlinear system by random sampling method, an adaptive ensemble root mean square Kalman filter algorithm is proposed. The proposed method does not need to interfere with the measured values, avoiding the problem of underestimating the analysis error covariance in EnKF and improving the filtering precision. Meanwhile, the Sage–Husa algorithm is introduced to estimate measurement noise online. When there is a deviation in measurement noise, the method can revise the covariance of measurement noise dynamically and filter it, which reduces the influence of noise error on filtering. Extensive simulation results show that the proposed method can not only achieve a higher filtering precision than the conventional EnKF but also with a stronger robustness to noise. Hence, the main contributions of this paper are:

- Proposing a novel, robust AEnSRF method applicable for measurement noise estimation.
- The proposed robust AEnSRF method is based on the combination between the EnSRF and the Sage–Husa algorithm.
- The proposed AEnSRF does not need to interfere with the measured values, avoiding the problem of underestimating the analysis error covariance in EnKF and improving the filtering precision.
- The suggested technique utilizes Sage–Husa algorithm to adjust the covariance matrices of measurement noise dynamically and filter it when there is a deviation in measurement noise.
- The proposed method mitigates the adverse influence of noise error on the filtering result.
- The proposed method has a higher filtering precision than the conventional EnKF and a stronger robustness to noise.

The rest of the paper is organized as follows: In Section 2, the second-order generator DSE model is presented. Section 3 describes the main algorithm in detail. IEEE 9-bus system and the real power grid system are used to test and illustrate the usefulness of the proposed method, and some necessary comparison results with ENKF and ENSRF are also performed in Section 4. Finally, the conclusions are provided in Section 5.

2. Generator Dynamic State Estimation Model

2.1. Mathematical Model of Dynamic State Estimation

State space model can be used to describe the dynamic system as

$$\begin{cases} X_{k+1} = \Phi X_k + \Gamma U_k + W_k \\ Z_k = C X_k + V_k \end{cases} \tag{1}$$

where k indicates the time instant, W_k represents system noise, V_k represents measurement noise, Φ denotes the state transition matrix, Γ is the input matrix, and C is the observation matrix.

The statistical characteristics of noise W_k and V_k are given as follows

$$\begin{cases} E[W_k] = E[V_k] = 0 \\ E[W_k W_k^T] = Q\delta_{kj} \\ E[V_k V_k^T] = R\delta_{kj} \end{cases} \tag{2}$$

where Q is the variance matrix of system process noise, R is the variance matrix of system measurement noise, $\delta_{kj} = \begin{cases} 1, k = j \\ 0, k \neq j \end{cases}$.

2.2. Equation of Measurement and State Estimation

In the electromechanical transient process of power system, the system topology and bus voltage cannot be acquired in real time. Thus, the rotor power angle and angular velocity, which do not abruptly change and are constrained by the rotor differential equation, are selected as the state variables of the generator DSE.

The second-order differential equations of synchronous generators can be expressed as follows

$$\begin{cases} \frac{d\delta}{dt} = (\omega - 1)\omega_0 \\ \frac{d\omega}{dt} = \frac{1}{T_J}(T_m - T_e - D\omega) = \frac{1}{T_J}(\frac{P_m}{\omega} - \frac{P_e}{\omega} - D\omega) \end{cases} \tag{3}$$

where δ indicates the power angle, ω represents the electric angular velocity; T_m denotes the mechanical torque, T_e is the electromagnetic torque; P_m is the mechanical power, P_e is the electromagnetic power, T_J represents the generator's rotor's inertia time constant, and D is the damping coefficient.

The differential equation of generator model in (3) can be rewritten as

$$\begin{bmatrix} \dot{\delta} \\ \dot{\omega} \end{bmatrix} = \begin{bmatrix} 0 & \frac{180}{\pi}\omega_0 \\ 0 & -\frac{D}{T_J} \end{bmatrix} \begin{bmatrix} \delta \\ \omega \end{bmatrix} + \begin{bmatrix} 1 & 0 \\ 0 & \frac{1}{T_J} \end{bmatrix} \begin{bmatrix} -\frac{180}{\pi}\omega_0 \\ (P_m - P_e)/\omega \end{bmatrix} \tag{4}$$

where the unit of δ is degree. For convenience, one has

$$\begin{cases} X = \begin{bmatrix} \delta \\ \omega \end{bmatrix} \\ A = \begin{bmatrix} 0 & \frac{180}{\pi}\omega_0 \\ 0 & -D/T_J \end{bmatrix} \\ B = \begin{bmatrix} 1 & 0 \\ 0 & 1/T_J \end{bmatrix} \\ U = \begin{bmatrix} -\frac{180}{\pi}\omega_0 \\ (P_m - P_e)/\omega \end{bmatrix} \end{cases} \tag{5}$$

In general, the first three terms of Taylor expansion of general state transition matrix can meet the precision requirements

$$\Phi(T) \approx I + AT + \frac{A^2T^2}{2} \tag{6}$$

$$\Gamma(T) = \left[\int_0^T \Phi(\alpha)d(\alpha)\right]B \tag{7}$$

where T represents the sampling period.

The power angle and angular velocity of generator can be measured directly by PMU, which are usually selected as the measurements

$$\begin{cases} Z = \begin{bmatrix} \delta \\ \omega \end{bmatrix} \\ C = \begin{bmatrix} 1 & 0 \\ 0 & 1 \end{bmatrix} \end{cases} \tag{8}$$

According to (4) to (7), the discrete form of generator DSE model can be obtained

$$\begin{cases} X_{k+1} = \Phi X_k + \Gamma U_k = f(X_k, U_k) \\ Z_k = CX_k = h(X_k, U_k) \end{cases} \tag{9}$$

2.3. Error Analysis

Equations (4) to (9) show that the imprecision of the model parameters T_J, D and the errors in the measurement of P_e and P_m will lead to the imprecision of the model. The obtained value of T_J is generally accurate. In general, the value of D is very small, indicating the influence of mechanical friction and wind resistance on the operation of generators. Therefore, the measurement errors of P_e and P_m are the main components of generator process noise. Actually, it is difficult to obtain precise P_m value, assuming that P_m value is constant when there is only governor but no shut-off and fast shut-off valve; the change of P_m caused by governor action is treated as the process noise of the system.

According to [16,17], the measurement error is generally between 1% and 2%. Considering the influence, the process noise variance matrix is set as

$$Q = diag(0, 0.0004P_{e0} + 0.0001) \tag{10}$$

where P_{e0} is the standard unit of output electromagnetic power in steady state.

The measurement errors of δ and ω are mainly caused by PMU measurement errors. Ideally, the measurement error variance of δ is 2^2, ω is 10^{-6}, and the measurement noise variance matrix of the system is set as

$$R = diag(2^2, 10^{-6}) \tag{11}$$

3. Adaptive Ensemble Square Root Kalman Filter

3.1. Ensemble Kalman Filter

Regarding the nonlinear systems and in order to track their state trajectories, the extended Kalman filter and the nonlinear filtering method which combines the techniques of unscented transformation and numerical difference without computing Jacobian matrix are widely used. However, when the system's nonlinearity is strong and the noise is non-Gaussian, the performance of these algorithms will be degraded significantly, resulting in unreliable state estimation. Therefore, ensemble Kalman filter, a sub-optimal filtering method, has attracted more and more attention. Based on Kalman filtering, ensemble prediction is introduced into ensemble Kalman filtering. Sequential Monte Carlo simulation method is employed to create the initial set of samples representing the state statistics. The state transition function is used to calculate the new set of states at the current time for each sample in the set, then the samples of the new set of states are calculated. Mean and covariance are used to get the optimal value of the current time estimate [18].

In the traditional Kalman filter, the calculation formulas of state prediction error covariance and analysis error covariance are presented as

$$\begin{aligned} P^f &= \overline{(X^f - X^t)(X^f - X^t)^T} \\ P^a &= \overline{(X^a - X^t)(X^a - X^t)^T} \end{aligned} \tag{12}$$

where X denotes the state vector, P represents the state covariance, T is the matrix transposition. The superscript f indicates the prediction state, the superscript a is the analysis state, and the superscript t is the real state.

In the EnKF, the posterior distribution of the state function is approximated by the ensemble elements, and the degree of approximation depends on the number of elements in the ensemble. Assuming that the number of set elements is N, XE^f represents the set storing all predicted states, and XE^a represents the set storing all analyzed states. When N tends to infinity, the real value of the state can be replaced by the mean value of the set of state estimators. It follows from [19] that the prediction error covariance and the analysis error covariance can be calculated by

$$\begin{aligned} P^f &\approx \tfrac{1}{N-1}(XE^f - \overline{XE^f})(XE^f - \overline{XE^f})^T \\ P^a &\approx \tfrac{1}{N-1}(XE^a - \overline{XE^a})(XE^a - \overline{XE^a})^T \end{aligned} \tag{13}$$

The EnKF uses the finite element in the ensemble for the estimation of the state error covariance, thus avoiding the prediction of the covariance matrix.

In the process of EnKF, firstly, according to the known state priori information, an initial state set of N elements is obtained by SMC method, then a prediction set XE^f is obtained by predicting each element in the initial state set through Kalman filter, in which the elements are expressed by X^{f_i}. Secondly, the observation set of state by SMC method as well, and the prediction state set is modified by covariance of XE^f to get the analysis set XE^a of state. The elements are expressed by X^{a_i}, and the average value of the state analysis set XE^a is the optimal state estimation value. The analysis set of states is used for prediction [20,21].

The iteration formula of ensemble Kalman filter algorithm is given as follows

(1) State prediction:

$$X_{k+1}^{f_i} = f(X_k^{a_i}, U_k) + W_k^i \tag{14}$$

$$\overline{X}_{k+1}^f = \frac{1}{N}\sum_{i=1}^{N} X_{k+1}^{f_i} \tag{15}$$

$$E_{X,k+1}^f = \begin{pmatrix} X_{k+1}^{f_1} - \overline{X}_{k+1}^f & \cdots & X_{k+1}^{f_N} - \overline{X}_{k+1}^f \end{pmatrix} \tag{16}$$

$$E_{Z,k+1}^f = \begin{pmatrix} Z_{k+1}^{f_1} - \overline{Z}_{k+1}^f & \cdots & Z_{k+1}^{f_N} - \overline{Z}_{k+1}^f \end{pmatrix} \tag{17}$$

$$P_{XZ,k+1}^f = P_{k+1}^f H_{k+1}^T = \frac{1}{N-1} E_{X,k+1}^f (E_{Z,k+1}^f)^T \tag{18}$$

$$P_{ZZ,k+1}^f = H_{k+1} P_{k+1}^f H_{k+1}^T = \frac{1}{N-1} E_{Z,k+1}^f (E_{Z,k+1}^f)^T \tag{19}$$

(2) State update:

$$K_{k+1} = P_{XZ,k+1}^f (P_{ZZ,k+1}^f)^{-1} \tag{20}$$

$$X_{k+1}^{a_i} = X_{k+1}^{f_i} + K_{k+1}(Z_{k+1}^{f_i} - h(X_{k+1}^{f_i})) \tag{21}$$

$$\overline{X}_{k+1}^a = \frac{1}{N}\sum_{i=1}^{N} X_{k+1}^{a_i} \tag{22}$$

where $X_{k+1}^{f_i}$ and $Z_{k+1}^{f_i}$ are the i elements of prediction state set and measurement state set, respectively; $X_k^{a_i}$ are the i elements of analysis state set, $\overline{X}_{k+1}^f$, $\overline{X}_{k+1}^a$ and $\overline{Z}_{k+1}^f$ are the average values of predicted state set, analyzed state set and measured state set at $k+1$ time instant, respectively, $E_{X,k+1}^f$ and $E_{Z,k+1}^f$ are the deviation matrices of elements and average values in predicted state set and measured state set, respectively, $P^f_{XZ,k+1}$ is the covariance of predictive state error and observation state error at $k+1$ time, $P^f_{ZZ,k+1}$ is the covariance of observation state error at $k+1$ time.

The EnKF algorithm not only overcomes the problem that Kalman filter is limited to linear systems, but also avoids the problem of large amount of computing resources when calculating prediction error covariance. Tangent linear model and adjoint model are no longer needed. The method also gives the optimal value of the estimated result and the confidence interval of the estimated value at the same time.

The EnKF is a sub-optimal estimation method which uses the ensemble to describe the traditional Kalman filter algorithm. A prediction set estimation is used to calculate the state prediction error covariance matrix of the Kalman gain matrix. If each element of the set is updated with the same observation value and the same gain, the set will systematically underestimate the state update error covariance matrix. It even can lead to subsequent analysis degradation and filtering divergence. This problem can be alleviated by adding random perturbations to the measurements with correct statistical data. However, the introduction of random perturbation will inevitably increase the additional error of sampling error related to measurement error covariance. This additional error will reduce the covariance accuracy of the observation state error as well as the estimation accuracy.

3.2. Ensemble Square Root Kalman Filter

In EnSRF, the calculation of the analytic state variables is divided into two parts [22], one is the update of the mean value of the analytic state set, the other is the update of the mean deviation of the analytic state set sample

$$\overline{X}^a = \overline{X}^f + K(\overline{Z} - H\overline{X}^f) \tag{23}$$

$$X'^a = X'^f + \widetilde{K}(Z' - HX'^f) \tag{24}$$

where $\overline{X}^a$ represents the mean of analysis state set, X'^a indicates the mean deviation of analysis state set samples, $\overline{X}^f$ denotes the mean deviation of prediction state set, X'^f is the mean deviation of prediction state set samples, $\overline{Z}$ is the observation value obtained by PMU, Z' is the random observation deviation obeying the probability distribution of observation error, K is the gain of traditional Kalman filter, $\widetilde{K}$ is the mean deviation gain of the update set.

In EnKF, when all sets of samples are updated with the same observation value and the same gain [23], i.e., $Z' = 0$ and $\widetilde{K} = K$, the covariance of the analysis state set can be expressed as (25). In the traditional Kalman filter, the covariance expression of state analysis error is shown in (26). As can be seen from (25) and (26), the value of P^a in EnKF lacks the term KRK^T, so EnKF has the problem of systematic underestimation

$$P^a = (I - KH)P^f(I - KH)^T \tag{25}$$

$$P^a = (I - KH)P^f(I - KH)^T + KRK^T = (I - KH)P^f \tag{26}$$

EnSRF eliminates the problem of systematic underestimation of EnKF without disturbing the observed values [22–24]. In EnSRF, (24) can be reformulated as

$$X'^a = X'^f - \widetilde{K}HX'^f = \left(I - \widetilde{K}H\right)X'^f$$

$\widetilde{K}$ is substituted for (25) and the analysis error covariance of the set $\widetilde{K}$ needs to satisfy (26). Furthermore, if $\widetilde{K} = \alpha K$ and α is a constant, one can further derive

$$(I - \widetilde{K}H)P^f(I - \widetilde{K}H)^T = (I - \widetilde{K}H)P^f \tag{27}$$

$$\frac{HP^fH^T}{HP^fH^T + R}\widetilde{K}\widetilde{K}^T - K\widetilde{K}^T - \widetilde{K}K^T + KK^T = 0 \tag{28}$$

$$\frac{HP^fH^T}{HP^fH^T+R}\alpha^2-2\alpha+1=0 \tag{29}$$

$$\alpha=(1+\sqrt{\frac{R}{HP^fH^T+R}})^{-1} \tag{30}$$

The traditional Kalman filtering algorithm is a linear unbiased minimum variance estimation algorithm under standard conditions. Kalman filter can obtain better estimation results when the dynamic system model and noise statistics are known. However, in the practical application, it is difficult to obtain accurate system model as well as accurate noise statistical characteristics, which will affect the accuracy of Kalman filter estimation results.

3.3. *Sage–Husa Algorithm*

In 1969, Sage and Husa presented an improved noise filtering algorithm. The algorithm can compute both matrices of process noise variance Q and measurement noise variance R [25,26] online and in real time when the system noise mean and covariance are unknown. Comparing with the adaptive filtering algorithm based on maximum likelihood criterion, this algorithm adds a forgetting factor, which strengthens the influence of newer observations on filtering and weakens the influence of older observations on the results. This algorithm effectively reduces the influence of model error on filtering, restrains filter divergence, and improves estimation accuracy [27].

The steps of Sage–Husa algorithm are given as:

(1) Filtering equation:

$$\hat{X}_{k+1|k}=F_k\hat{X}_{k|k}+\hat{q}_k \tag{31}$$

$$P_{k+1|k}=F_kP_{k|k}F_k^T+\hat{Q}_k \tag{32}$$

$$K_{k+1}=P_{k+1|k}H_{k+1}^T(H_{k+1}P_{k+1|k}H_{k+1}^T+\hat{R}_{k+1})^{-1} \tag{33}$$

$$e_{k+1}=Z_{k+1}-H_{k+1}\hat{X}_{k+1|k}-\hat{r}_{k+1} \tag{34}$$

$$\hat{X}_{k+1|k+1}=\hat{X}_{k+1|k}+K_{k+1}e_{k+1} \tag{35}$$

$$P_{k+1|k+1}=(I-K_{k+1}H_{k+1})P_{k+1|k} \tag{36}$$

(2) Time-varying noise estimator:

$$\hat{r}_{k+1}=(1-d_{k+1})\hat{r}_k+d_{k+1}(Z_{k+1}-H_{k+1}\hat{X}_{k+1|k}) \tag{37}$$

$$\hat{R}_{k+1}=(1-d_{k+1})\hat{R}_k+d_{k+1}(e_{k+1}e_{k+1}^T-H_{k+1}P_{k+1|k}H_{k+1}^T) \tag{38}$$

$$\hat{q}_{k+1}=(1-d_{k+1})\hat{q}_k+d_{k+1}(\hat{X}_{k+1|k+1}-F_{k+1}\hat{X}_{k|k}) \tag{39}$$

$$\begin{aligned}\hat{Q}_{k+1}=&(1-d_{k+1})\hat{Q}_k+d_{k+1}(K_{k+1}e_{k+1}e_{k+1}^TK_{k+1}^T\\&+P_{k+1|k+1}\ -F_{k+1}P_{k+1|k+1}F_{k+1}^T)\end{aligned} \tag{40}$$

where $\hat{q}_{k+1}$ represents the mean of process noise at $k+1$; $\hat{r}_{k+1}$ is the mean of $k+1$, e_{k+1} is the residual at time instant $k+1$; $d_{k+1}=(1-b)/(1-b^{k+2})$, b is the forgetting factor and $0<b<1$.

Generally, Sage–Husa algorithm can simultaneously estimate process noise $\hat{Q}_{k+1}$ and measurement noise $\hat{R}_{k+1}$. In fact, the algorithm cannot estimate both $\hat{Q}_{k+1}$ and $\hat{R}_{k+1}$ at the same time. It can be seen from (37) and (39) that the calculation of $\hat{Q}_{k+1}$ and $\hat{R}_{k+1}$ depends on e_{k+1}. The change of e_{k+1} will affect the calculation of $\hat{Q}_{k+1}$ and $\hat{R}_{k+1}$ at the same time, which cannot guarantee the accuracy of the estimation results. In addition, there are minus signs in the calculation of $\hat{Q}_{k+1}$ and $\hat{R}_{k+1}$, which may affect the semi-positive and positive definiteness of $\hat{Q}_{k+1}$ and $\hat{R}_{k+1}$, leading to filter divergence.

In the electromechanical transient process of generator, the influence of measurement noise $\hat{R}_{k+1}$ on the estimation is more important. Generally speaking, process noise Q is a constant. Assuming that both process noise as well as measurement noise are Gauss, white noise with mean value equals to zero, i.e., $r = 0$ and $q = 0$.

Sage–Husa algorithm can be simplified as follows.

(1) Filtering equation:

$$\hat{X}_{k+1|k} = F_k \hat{X}_{k|k} \tag{41}$$

$$P_{k+1|k} = F_k P_{k|k} F_k^T + Q \tag{42}$$

$$K_{k+1} = P_{k+1|k} H_{k+1}^T (H_{k+1} P_{k+1|k} H_{k+1}^T + \hat{R}_{k+1})^{-1} \tag{43}$$

$$e_{k+1} = Z_{k+1} - H_{k+1} \hat{X}_{k+1|k} \tag{44}$$

$$\hat{X}_{k+1|k+1} = \hat{X}_{k+1|k} + K_{k+1} e_{k+1} \tag{45}$$

$$P_{k+1|k+1} = (I - K_{k+1} H_{k+1}) P_{k+1|k} \tag{46}$$

(2) Measurement noise estimator:

$$\hat{R}_{k+1} = (1 - d_{k+1}) \hat{R}_k + d_{k+1} (e_{k+1} e_{k+1}^T - H_{k+1} P_{k+1|k} H_{k+1}^T) \tag{47}$$

In order to ensure the positive covariance of measurement noise, (47) is replaced by (48). It can be seen from the formula that a certain filtering-precision is sacrificed to ensure the stability of the filter.

$$\hat{R}_{k+1} = (1 - d_{k+1}) \hat{R}_k + d_{k+1} (e_{k+1} e_{k+1}^T) \tag{48}$$

3.4. Adaptive Ensemble Square Root Kalman Filter

The proposed AEnSRF combines ensemble root mean square Kalman filter with Sage–Husa algorithm. Comparing with ensemble Kalman filter, the proposed AEnSRF not only has a higher filtering-precision but also can be effectively corrected while measurement noise deviates, so as to suppress filter divergence and improve estimation accuracy.

The specific steps are as follows:

(1) **Initialization**: The initial value ω_0 of generator rotor angular velocity is 1. The initial value δ of generator power angle, electromagnetic power P_e and mechanical power P_m is its steady-state operation value.

(2) The initial value of the state set with the number of set elements is generated by Monte Carlo method, and the initial estimation error covariance matrix is taken as the unit matrix.

$$X^0 = \begin{bmatrix} x_{11} & x_{12} & x_{13} & \cdots & x_{1N} \\ x_{21} & x_{22} & x_{23} & \cdots & x_{2N} \\ \vdots & \vdots & \vdots & \vdots & \vdots \\ x_{31} & x_{32} & x_{33} & \cdots & x_{3N} \\ x_{n1} & x_{n2} & x_{n3} & \cdots & x_{nN} \end{bmatrix} \tag{49}$$

(3) **State prediction:**

$$X_{k+1}^{f_i} = f(X_k^{a_i}, U_k) + W_k^i \tag{50}$$

$$\overline{X}_{k+1}^f = \frac{1}{N} \sum_{i=1}^{N} X_{k+1}^{f_i} \tag{51}$$

$$E_{X,k+1}^f = \begin{pmatrix} X_{k+1}^{f_1} - \overline{X}_{k+1}^f & \cdots & X_{k+1}^{f_N} - \overline{X}_{k+1}^f \end{pmatrix} \tag{52}$$

$$E_{Z,k+1}^f = \begin{pmatrix} Z_{k+1}^{f_1} - \overline{Z}_{k+1}^f & \cdots & Z_{k+1}^{f_N} - \overline{Z}_{k+1}^f \end{pmatrix} \tag{53}$$

$$P^f_{XZ,k+1} = P^f_{k+1}H^T_{k+1} = \frac{1}{N-1}E^f_{X,k+1}(E^f_{Z,k+1})^T \tag{54}$$

$$P^f_{ZZ,k+1} = H_{k+1}P^f_{k+1}H^T_{k+1} = \frac{1}{N-1}E^f_{Z,k+1}(E^f_{Z,k+1})^T \tag{55}$$

(4) **Calculating Kalman gain:**

$$K_{k+1} = P^f_{XZ,k+1}(P^f_{ZZ,k+1})^{-1} \tag{56}$$

(5) **Calculate the measurement noise covariance:**

$$e_{k+1} = Z_{k+1} - \overline{Z}^f_{k+1} \tag{57}$$

$$d_{k+1} = (1-b)/(1-b^{k+2}) \tag{58}$$

$$\hat{R}_{k+1} = (1-d_{k+1})\hat{R}_k + d_{k+1}(e_{k+1}e^T_{k+1}) \tag{59}$$

(6) **Calculate the mean deviation gain of the updated set:**

$$\alpha = [1 + \sqrt{\hat{R}_{k+1}(P^f_{ZZ,k+1})^{-1}}]^{-1} \tag{60}$$

$$\widetilde{K}_{k+1} = \alpha K_{k+1} \tag{61}$$

(7) **State update:**

$$\overline{X}^a_{k+1} = \overline{X}^f_{k+1} + K_{k+1}e_{k+1} \tag{62}$$

$$X'^{a_i}_{k+1} = X'^{f_i}_{k+1} - \widetilde{K}_{k+1}h(X'^{f_i}_{k+1}) \tag{63}$$

$$\overline{X}'^a_{k+1} = \frac{1}{N}\sum_{i=1}^{N} X'^{a_i}_{k+1} \tag{64}$$

$$X^a_{k+1} = \overline{X}^a_{k+1} + \overline{X}'^a_{k+1} \tag{65}$$

For convenience, the proposed AEnSRF method's implementation flow chart is presented in Figure 1.

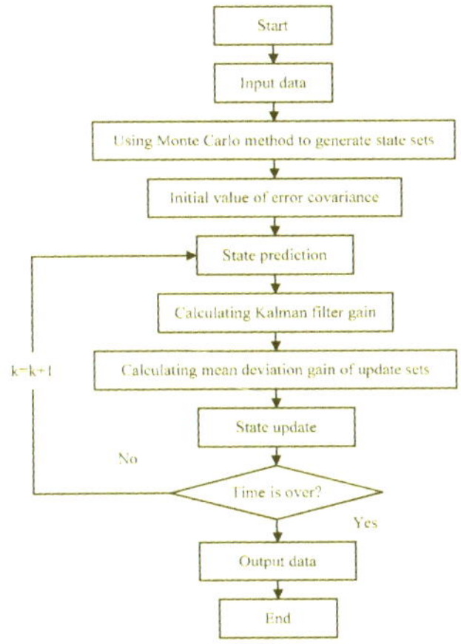

Figure 1. The flowchart of adaptive ensemble square root Kalman filter (AEnSRF) algorithm.

4. Case Study

4.1. Basic Test Analysis

The IEEE-9 bus system is firstly taken as the test system in this paper. The single-line diagram of it is given in Figure 2. The classical second-order model of generator considering governor is adopted in simulation. The inertia time constants of G1, G2, and G3 are respectively 47.28, 12.8, 6.02 s; and the damping coefficient is set to 2. The initial values of generator state variables and control variables are set to the values of the system's stable operation before fault occurred. Suppose that in the IEEE 9 bus system, three-phase short-circuit fault occurs at the beginning of branch 4–8 at 0.8 s. After 0.36 s, the switch on both sides of branch jumps off and the fault is removed. The whole simulation time is 6 s.

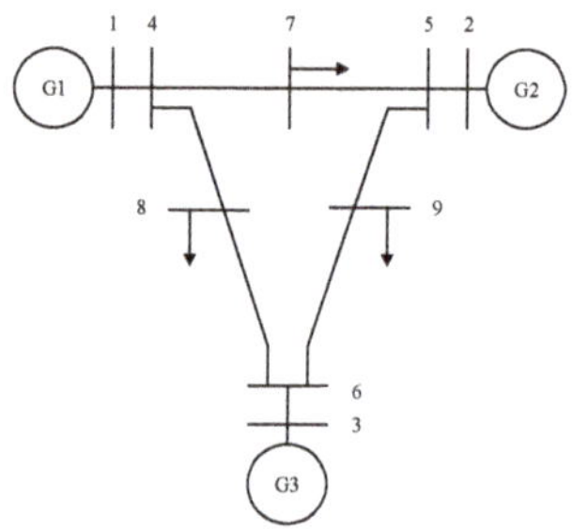

Figure 2. The single-line diagram of IEEE-9 bus test system.

The real values of power angle and angular velocity of generator are obtained by Bonneville Power Administration (BPA) simulation software, and the measured values are obtained by adding the real values to the random noise where its mean value obeys normal distribution with zero standard deviation. In this paper, the DSE algorithm is implemented in Matlab. The initial values of the estimated generator state variables are set to stable operation value, and the initial state error covariance matrix is set to unit matrix.

In order to further compare the difference of filtering results of each algorithm, the root mean square error (*RMSE*) is used

$$RMSE = \sqrt{\frac{1}{n}\sum_{k=1}^{n}\left(\hat{X}_{k|k} - X_k\right)^2} \tag{66}$$

where n is the sampling period.

The process noise variance matrix and the measurement noise variance matrix of the generator are selected as the corresponding values in [10,11]. Figure 3 is a comparison of the filtering results of G1 by utilizing the EnKF and EnSRF algorithms under normal noise. Table 1 gives the RMSE results of EnKF and EnSRF algorithms.

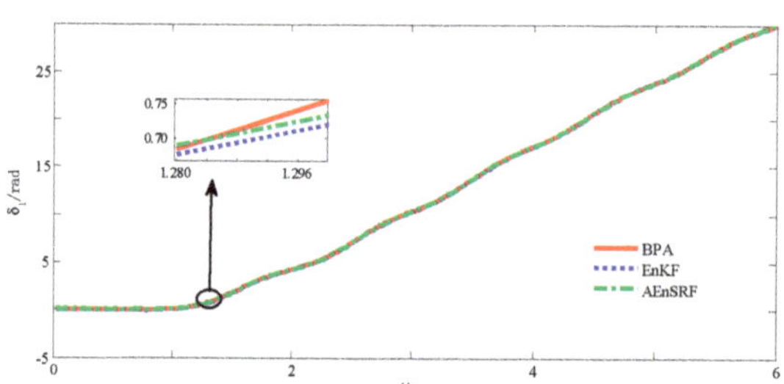

(a)

Figure 3. *Cont.*

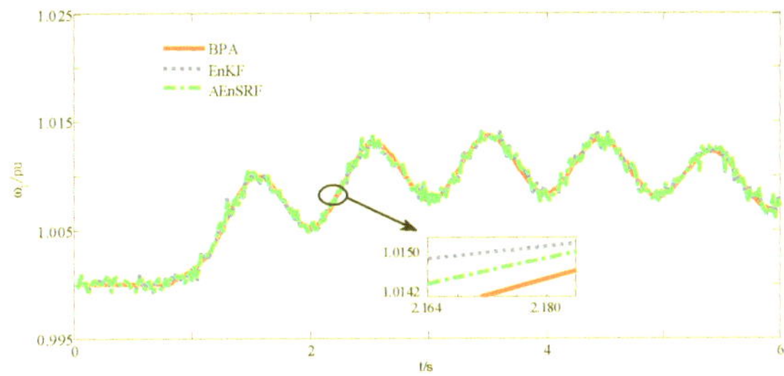

(b)

Figure 3. Comparison of G1 state filtering results under ensemble Kalman filter (EnKF) and AEnSRF: (**a**) Power angle dynamic estimation curve of G1; (**b**) electric angular velocity dynamic estimation curve of G1.

Table 1. The RMSE results of EnKF and EnSRF.

Generator Number	Algorithm	δ/rad	ω/pu
G1	EnKF	0.0345	9.7396×10^{-4}
	AEnSRF	0.0329	9.1950×10^{-4}
G2	EnKF	0.0346	10.000×10^{-4}
	AEnSRF	0.0330	9.6680×10^{-4}
G3	EnKF	0.0354	11.000×10^{-4}
	AEnSRF	0.0344	10.000×10^{-4}

As it might be noted from Figure 3 and Table 1, EnKF and AEnSRF both show good filtering effect without deviation of measurement noise. Because the observation value in AEnSRF does not introduce disturbance, the problem of underestimation in EnKF filtering is avoided. The filtering-precision of AEnSRF is improved comparing with EnKF.

Figure 4 is a sketch of the noise estimator tracking the measured noise in AEnSRF without deviation from the measured noise.

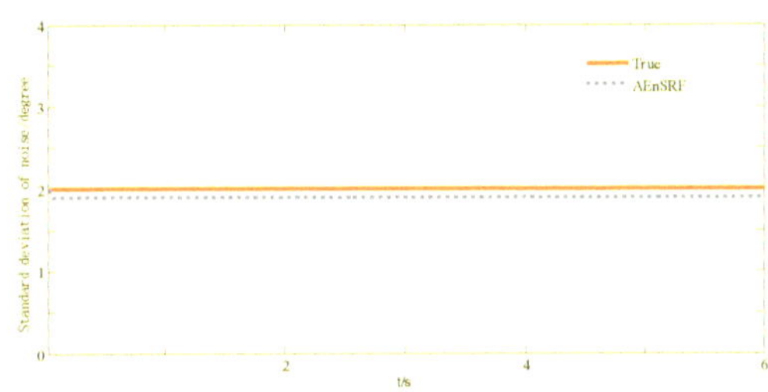

(a)

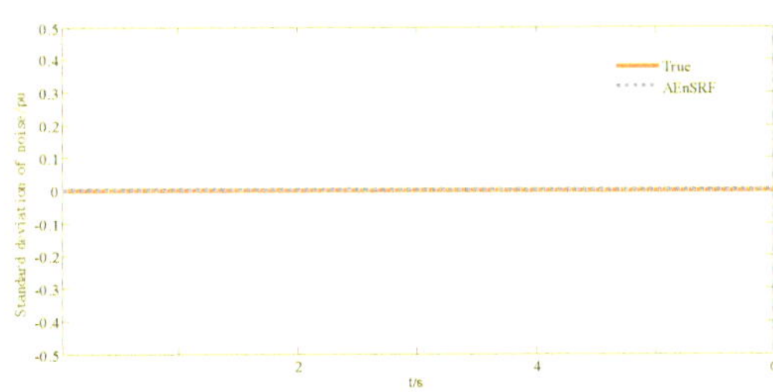

(b)

Figure 4. Measurement noise tracking curve of G1 under AEnSRF: (**a**) Power angle measurement noise tracking curve of G1; (**b**) angular velocity measurement noise tracking curve of G1.

From the simulation, we can see that in the process of tracking power angle measurement noise by AEnSRF, the standard deviation of 0.14 s is stable at 1.9 degrees, and the angular velocity is too small (almost 0). Combining with Figure 4, it can be remarked that AEnSRF can track the measured noise more accurately.

4.2. Noise Measurement and Analysis on the IEEE 9-Bus System

In this subsection, G2 is taken as an example to test the estimation effect of AEnSRF on generator state as well as the tracking of measurement noise when it is biased. Assuming that the actual value of the standard deviation of power angle measurement noise is 3 degrees, the value is set as 2 degrees in the filtering program to detect the tracking effect of AEnSRF on power angle measurement noise.

Figure 5 depicts the comparison results of G2 by utilizing EnKF and AEnSRF algorithms, respectively. It can be obviously noted from Figure 5 that both of them can show good estimation results under the consideration of measurement noise deviation condition.

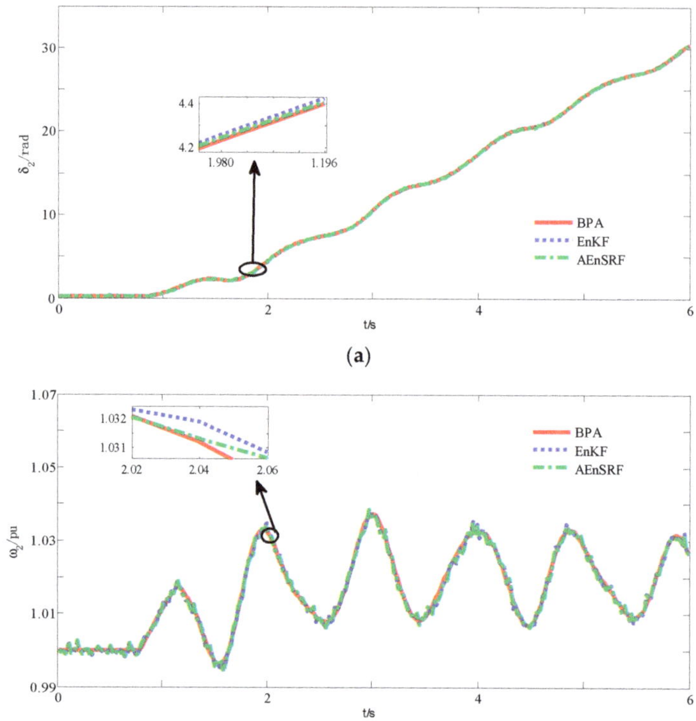

(**b**)

Figure 5. Comparisons of G2 state filtering results between EnKF and AEnSRF in the presence of measurement noise deviation: (**a**) Power angle dynamic estimation curve of G2; (**b**) electric angular velocity dynamic estimation curve of G2.

The simulation results of G2 show that under AEnSRF algorithm, the RMSE of power angle estimation is 0.0516, the RMSE of electric angular velocity estimation is 0.0011. Under EnKF algorithm, the RMSE of power angle estimation is 0.0536 and the RMSE of electric angular velocity estimation is 0.0012. By comparing both algorithms, AEnSRF has higher filtering-precision.

Figure 6 is the simulation diagram of AEnSRF tracking curve for measurement noise when there is a deviation in measurement noise. As it can be noticed from this figure, when the power angle measurement noise is biased, the Sage–Husa noise estimator with on-line real-time estimation noise is introduced by AEnSRF, which makes the estimated standard deviation of power angle measurement noise stabilize around 3 degrees at about 1.1 s.

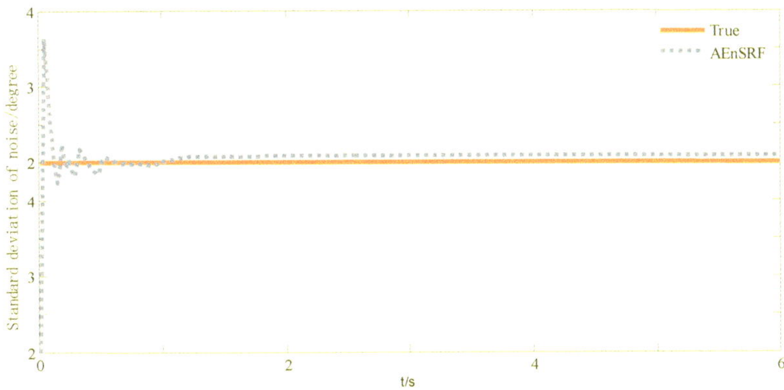

(a)

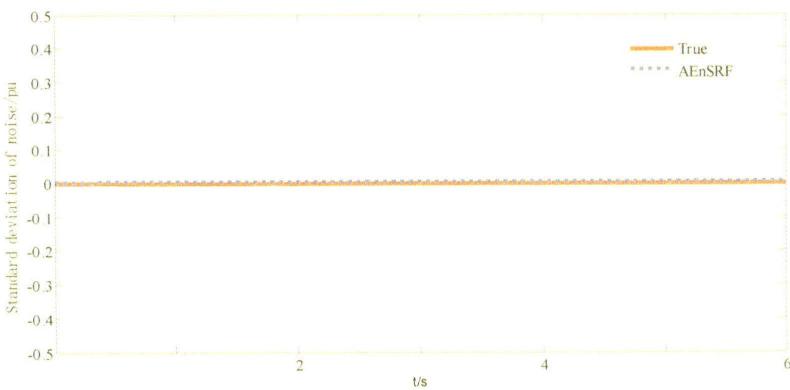

(b)

Figure 6. Tracking curve of G2 measurement noise by AEnSRF in the case of measurement noise deviation: (**a**) Power angle measurement noise tracking curve of G2; (**b**) angular velocity measurement noise tracking curve of G2.

4.3. Noise Measurement and Analysis on a Real Power System

For further verification of the proposed algorithm's effectiveness, simulation tests are also carried out on a real power grid system. One generator outgoing circuit has a short-circuit fault, and the generator parameters are based on the actual data. The initial value of the state variable is the steady value, and the initial error covariance matrix is the unit matrix. Assuming that the actual value of the standard deviation of power angle measurement noise is 2.5 degrees, the value is set to 2 degrees in the filtering program.

Comparing EnKF with the proposed method, Figures 7 and 8 show the simulation results of DSE of generator in the presence of measurement noise deviation. The simulation results of G2 illustrate that under the AEnSRF, the RMSE of power angle estimation is 0.0435, the RMSE of electric angular velocity estimation is 0.0010. Under the EnKF, the RMSE of power angle estimation is 0.0438 and the RMSE of electric angular velocity estimation is 0.0013.

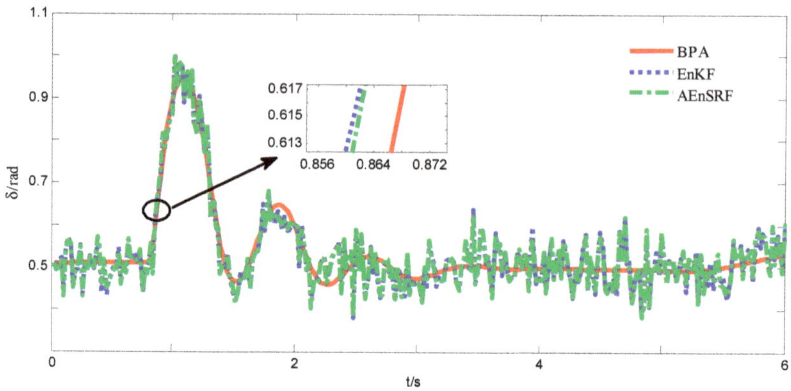

(a)

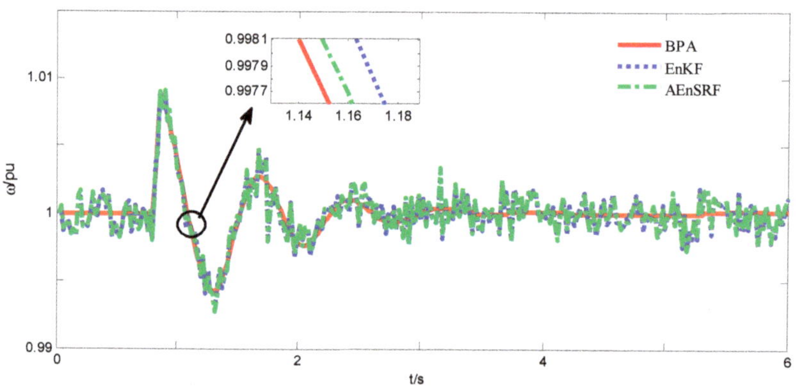

(b)

Figure 7. Contrast chart of EnKF and AEnSRF for generator state filtering with deviation of measurement noise: (**a**) Power angle dynamic estimation curve of generator; (**b**) dynamic angular velocity estimation curve of generator.

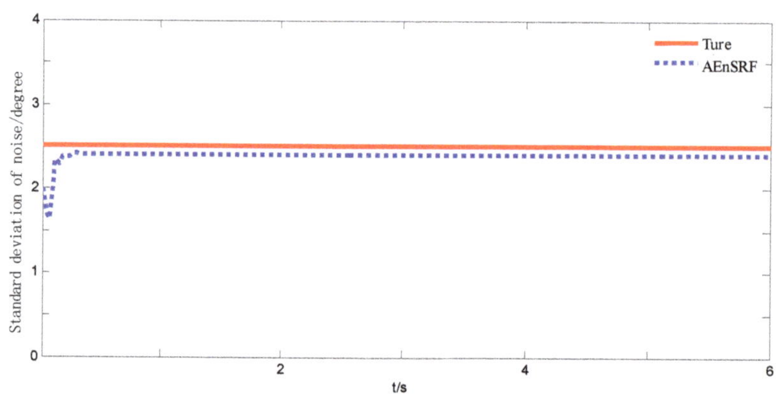

(a)

Figure 8. *Cont.*

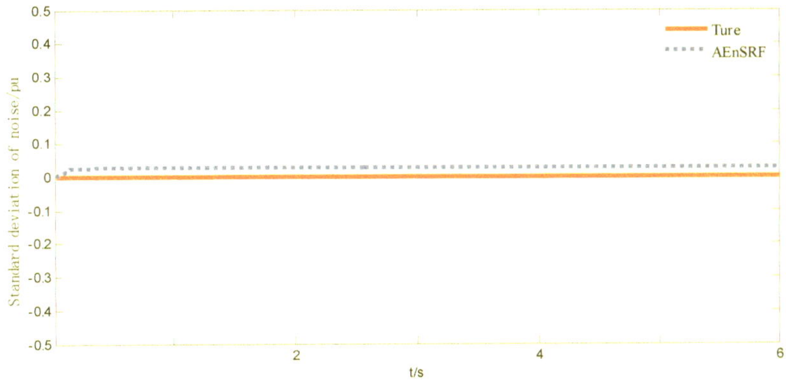

(b)

Figure 8. Tracking curve of AEnSRF for generator measurement noise with deviation of measurement noise: (**a**) Power angle measurement noise tracking curve of generator; (**b**) angular velocity measurement noise tracking curve of generator.

Comparing with EnKF, AEnSRF has higher filtering-precision. In AEnSRF, the noise estimator tracks the power angle measurement noise at 0.3 s and stabilizes at about 2.5 degrees. This shows that AEnSRF has a good tracking effect on measurement noise.

5. Conclusions

In this paper, a novel DSE method was proposed based on the EnSRF; a simplified Sage–Husa adaptive Kalman filter with relatively simple principle and good practicability was introduced. In the proposed AEnSRF method, set members were utilized to approximate the posterior distribution of the real state without random disturbance to the measured values. Comparing with EnKF, AEnSRF could enhance the accuracy of filtering. At the same time, Sage–Husa noise estimator was added between the prediction and correction steps to estimate the measurement noise online and in real time, which effectively avoided the problem that the filtering-precision was reduced or even divergent due to the deviation of the measurement noise.

Author Contributions: D.N. created models, developed methodology, wrote the initial draft, and designed computer programs; W.W. supervised and was responsible for leading the research activity planning and presented critical review; K.W. conducted research and investigation process and edited the initial draft. D.N., W.W., R.J.M., H.H.A., and P.S. reviewed the manuscript and synthesized study data. All authors read and approved the manuscript.

Funding: This work was supported by the National Natural Science Foundation of China under Grant 51667020.

Conflicts of Interest: The authors declare no conflict of interest.

References

1. Wang, Y.; Sun, Y.; Dinavahi, V.; Wang, K.; Nan, D. Robust dynamic state estimation of power system with model uncertainties based on adaptive unscented H infinity filter. *IET Gener. Transm. Distrib.* **2019**, *13*, 2455–2463. [CrossRef]
2. De La Ree, J.; Centeno, V.; Thorp, J.S.; Phadke, A.G. Synchronized phasor measurement applications in power systems. *IEEE Trans. Smart Grid* **2010**, *1*, 20–27. [CrossRef]
3. Wang, Y.; Sun, Y.; Dinavahi, V. Robust forecasting-aided state estimation for power system against uncertainties. *IEEE Trans. Power Syst.* **2019**. [CrossRef]
4. Singh, H.; Alvarado, L.; Liu, W. Constrained LAV state estimation using penalty functions. *IEEE Trans. Power Syst.* **2002**, *12*, 383–388. [CrossRef]

5. Mili, L.; Cheniae, M.; Rousseeuw, P. Robust state estimation of electric power systems. *IEEE Trans. Circuits Syst. I Fundam. Theory Appl.* **2002**, *41*, 349–358. [CrossRef]
6. Ghahremani, E.; Kamwa, I. Dynamic state estimation in power system by applying the extended Kalman filter with unknown inputs to phasor measurements. *IEEE Trans. Power Syst.* **2011**, *26*, 2556–2566. [CrossRef]
7. Wang, S.; Gao, W.; Meliopoulos, A. An alternative method for power system dynamic state estimation based on unscented transform. *IEEE Trans. Power Syst.* **2012**, *27*, 942–950. [CrossRef]
8. Qi, J.; Sun, K.; Wang, J.; Liu, H. Dynamic state estimation for multi-machine power system by unscented Kalman filter with enhanced numerical stability. *IEEE Trans. Smart Grid* **2018**, *9*, 1184–1196. [CrossRef]
9. Arasaratnam, I.; Haykin, S. Cubature Kalman filters. *IEEE Trans. Autom. Control* **2009**, *54*, 1251–1269. [CrossRef]
10. Arasaratnam, I.; Haykin, S. Cubature Kalman filtering for continuous-discrete systems: Theory and simulations. *IEEE Trans. Signal Process.* **2010**, *58*, 4977–4993. [CrossRef]
11. Li, Y.; Li, J.; Qi, J.; Chen, L. Robust cubature Kalman filter for dynamic state estimation of synchronous machines under unknown measurement noise statistics. *IEEE Access* **2019**, *7*, 29139–29148. [CrossRef]
12. Ai, M.; Sun, Y.; Lv, X. Dynamic state estimation for synchronous machines based on interpolation H∞ extended Kalman filter. In Proceedings of the IEEE 33rd Youth Academic Annual Conference of Chinese Association of Automation (YAC), Nanjing, China, 18–20 May 2018.
13. Esmaeil, G.; Innocent, K. Online state estimation of a synchronous generator using unscented Kalman filter from phasor measurements units. *IEEE Trans. Energy Convers.* **2011**, *26*, 1099–1108.
14. Gopinath, G.; Shyama, P. Sensorless control of permanent magnet synchronous motor using square-root cubature Kalman filter. In Proceedings of the IEEE 8th International Power Electronics and Motion Control Conference (IPEMC-ECCE Asia), Hefei, China, 22–26 May 2016.
15. Evensen, G. Sequential data assimilation with a nonlinear quasi-geostrophic model using Monte Carlo methods to forecast error statistics. *J. Geophys. Res. Ocean.* **1994**, *106*, 10143–10162. [CrossRef]
16. Huang, R.; Diao, R.; Li, Y.; Sanchez-Gasca, J.; Huang, Z.; Thomas, B.; Etingov, P.; Kincic, S.; Wang, S.; Fan, R.; et al. Calibrating parameters of power system stability models using advanced ensemble Kalman filter. *IEEE Trans. Power Syst.* **2018**, *33*, 2895–2905. [CrossRef]
17. Lee, W.; Gim, J.; Chen, M.; Wang, S.; Li, R. Development of a real-time power system dynamic performance monitoring system. *IEEE Trans. Ind. Appl.* **1997**, *33*, 1055–1060.
18. Gillijns, S.; Mendoza, O.B.; Chandrasekar, J.; De Moor, B.L.R.; Bernstein, D.S.; Ridley, A. What is the ensemble Kalman filter and how well does it work? In Proceedings of the IEEE American Control Conference, Minneapolis, MN, USA, 14–16 June 2006.
19. Kotecha, J.; Djuric, P. Gaussian particle filtering. *IEEE Trans. Signal Process.* **2003**, *51*, 2592–2601. [CrossRef]
20. Fan, R.; Huang, R.; Diao, R. Gaussian mixture model-based ensemble Kalman filter for machine parameter calibration. *IEEE Trans. Energy Convers.* **2018**, *33*, 1597–1599. [CrossRef]
21. Fan, R.; Liu, Y.; Huang, R.; Diao, R.; Wang, S. Precise fault location on transmission lines using ensemble Kalman filter. *IEEE Trans. Power Del.* **2018**, *33*, 3252–3255. [CrossRef]
22. Burgers, G.; Peter, J.; Evensen, G. Analysis scheme in the ensemble Kalman filter. *Mon. Weather Rev.* **1998**, *126*, 1719–1724. [CrossRef]
23. Whitaker, J.S.; Hamill, T.M. Ensemble data assimilation without perturbed observations. *Mon. Weather Rev.* **2003**, *130*, 1913–1924. [CrossRef]
24. Tippett, M.K.; Anderson, J.L.; Bishop, C.H.; Hamill, T.M.; Whitaker, J.S. Ensemble square root filters. *Mon. Weather Rev.* **2003**, *131*, 1485–1490. [CrossRef]
25. Zhu, Z.; Liu, S.; Zhang, B. An improved Sage-Husa adaptive filtering algorithm. In Proceedings of the IEEE 31st Chinese Control Conference, Hefei, China, 25–27 July 2012.
26. Wang., Y.; Sun, Y.; Dinavahi, V.; Cao, S.; Hou, D. Adaptive robust cubature Kalman filter for power system dynamic state estimation against outliers. *IEEE Access* **2019**, *7*, 105872–105881. [CrossRef]
27. Deng, F.; Chen, J.; Chen, C. Adaptive unscented Kalman filter for parameter and state estimation of nonlinear high-speed objects. *J. Syst. Eng. Electron.* **2013**, *24*, 655–665. [CrossRef]

applied sciences

MDPI

Article

Managing Energy Plus Performance in Data Centers and Battery-Based Devices Using an Online Non-Clairvoyant Speed-Bounded Multiprocessor Scheduling

Pawan Singh [1], Baseem Khan [2], Om Prakash Mahela [3], Hassan Haes Alhelou [4,5] and Ghassan Hayek [4,5,*]

1 Department of Computer Science and Engineering, Amity School of Engineering and Technology, Amity University Uttar Pradesh, Lucknow Campus, Lucknow 226010, India; pawansingh51279@gmail.com
2 Department of Electrical and Computer Engineering, Hawassa University, Hawassa P.O. Box 05, Ethiopia; baseem.khan04@gmail.com
3 Power System Planning Division, Rajasthan Rajya Vidhyut Prasaran Nigam Ltd., Jaipur 302005, India; opmahela@gmail.com
4 Department of Electrical Power Engineering, Faculty of Mechanical and Electrical Engineering, Tishreen University, 2230 Lattakia, Syria; h.haesalhelou@gmail.com
5 Department of Electrical Power Engineering, Dresden University, 01069 Dresden, Germany
* Correspondence: prof.hayek@gmail.com

Received: 5 March 2020; Accepted: 30 March 2020; Published: 3 April 2020

Abstract: An efficient scheduling reduces the time required to process the jobs, and energy management decreases the service cost as well as increases the lifetime of a battery. A balanced trade-off between the energy consumed and processing time gives an ideal objective for scheduling jobs in data centers and battery based devices. An online multiprocessor scheduling multiprocessor with bounded speed (MBS) is proposed in this paper. The objective of MBS is to minimize the importance-based flow time plus energy (IbFt+E), wherein the jobs arrive over time and the job's sizes are known only at completion time. Every processor can execute at a different speed, to reduce the energy consumption. MBS is using the tradition power function and bounded speed model. The functioning of MBS is evaluated by utilizing potential function analysis against an offline adversary. For processors $m \geq 2$, MBS is O(1)-competitive. The working of a set of jobs is simulated to compare MBS with the best known non-clairvoyant scheduling. The comparative analysis shows that the MBS outperforms other algorithms. The competitiveness of MBS is the least to date.

Keywords: multiprocessor system; online non-clairvoyant scheduling; weighted flow time; potential analysis; energy efficiency

1. Introduction

There are number of server farms equipped with hundreds of processors. The cost of energy used for cooling and running a machine for around three years surpasses the hardware cost of the machine [1]. Consequently, the major integrated chips manufacturers such as Intel and AMD are producing the dynamic speed scaling (DSS) enabled multiprocessor/multi-core machine and software such as Intel's SpeedStep [2], which support the operating system in managing the energy by varying the execution speed of processors. A founder chip maker Tilera forecasted that the numbers of processors/cores will be doubled every eighteen months [3], which will increase the energy demand to a great extent. Data centers consume 1.5% of total electricity usage in United States [4]. To avoid such critical circumstances, the current issue in the scheduling is to attain the good quality of service by generating an optimal schedule of jobs and to save the energy consumption, which is a conflicting and complicated problem [5].

The power P consumed by a processor running at speed s is sV^2, where V is a voltage [6]. The traditional power function is $P = s^\alpha$ ($\alpha \geq 2$ for CMOS based chips [7,8]). There are two types of speed models: the first unbounded speed model, in which the processor's speed range is, i.e., $[0, \infty)$; the second bounded speed model, in which the speed of a processor can range from zero to some maximum speed, i.e., $[0, \eta]$. This DSS plays a vital role in energy management, where in a processor can regulate its speed to save energy. A few qualities of service metrics are slowdown, throughput, makespan, flow time and weighted flow time. At low speed, the processor finishes jobs slower and save energy, whereas at high speed, the processor finishes jobs faster but consumes more energy, as shown in Figure 1. To get a better quality of service and low energy consumption the objective should be to minimize the sum of flow time and energy; in case, if the importance or priority is attached, the objective should be to minimize the sum of importance-based flow time and energy. The objective of minimizing the IbFt+Ehas a natural explanation, as it can be considered in monetary terms [9].

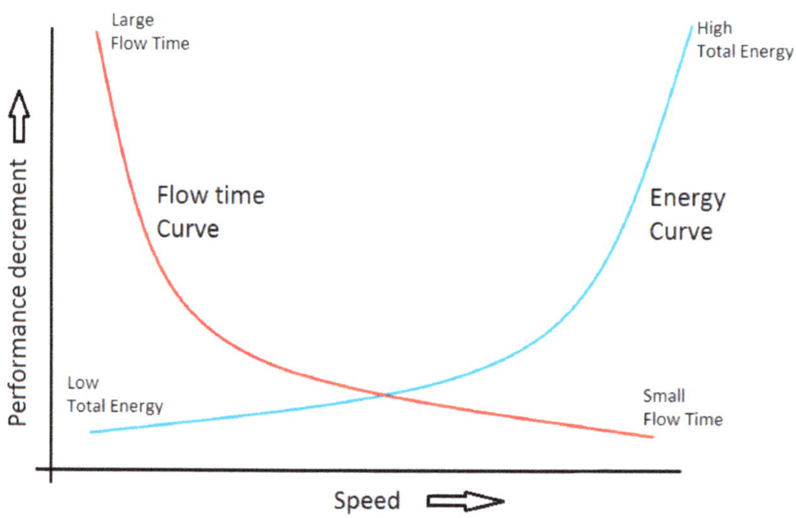

Figure 1. Performance and speed curve.

In the multiprocessor systems, there is a requirement of three different policies: the first policy is job selection, which decides the next job to be executed on every processor; the second policy is speed scaling, which decides every processor's execution speed at all time; the third policy is job assignment, which indicates that to which processor the new job should be assigned. In the c-competitive online scheduling algorithm, for each input the cost received is less than or equal to c times the cost of optimal offline algorithm [9]. Unlike non-clairvoyant scheduling, the size of job is unknown at arrival time, such as in UNIX operating system where jobs arrive with no information of processing requirement. Unlike online modes, in the offline mode, the whole job progression is known in advance. No online algorithm can attain a constant competitiveness with equal maximum speed to optimal offline algorithm [10].

Motwani et al. [10] commenced the study of the non-clairvoyant scheduling algorithms. Yao et al. inducted the theoretical study of speed scaling scheduling algorithm [11]. Yao et al. proposed an algorithm average rate heuristic (AVR) with a competitive ratio at most $2^{\alpha-1}\alpha^\alpha$ using the traditional power function. Koren et al. [12] presented an optimal online scheduling algorithm D^{over} for a overloaded uniprocessor system with competitive ratio-$\left(\frac{1}{(1+\sqrt{k})^2}\right)$ for the objective of minimizing the throughput, where k is the importance ratio. The competitiveness of shortest remaining processing time (SRPT) for multiprocessor system is $O\left(min\left(log\left(\frac{m}{n}\right), log\ \sigma\right)\right)$, where m is number of processors, n is total number of jobs and σ represents the ratio of minimum to maximum job size [13]. Kalyanasundaram et al. [14] presented the idea of resource augmentation. If the resources are augmented and, $(2+\Delta)$-speed p processors are used then the competitive ratio of Equi-partition lies between $\frac{2}{3}(1+\Delta)$ and $\left(2+\frac{4}{\Delta}\right)$ [15]. Multilevel feedback queue, a randomized algorithm with n jobs is $O(log\ n)$-competitive [16,17]. The first algorithm with non trivial guarantee is $O\left(\log^2 \sigma\right)$-competitive [18], where σ is the ratio of minimum to maximum job size. There are different algorithms proposed with different objectives over a span of time [19–27].

Chen et al. [19] proposed algorithms with different approximation bounds for processors with/without constraints on the maximum processor speed. The concept of merging dual objective of energy used and total flow time into single objective of energy used plus total flow time is proposed by Albers et al. [20]. Bansal et al. [21] proposed an algorithm, which uses highest density first (HDF) for the job selection with a traditional power function. Lam et al. [22] proposed a multiprocessor algorithm for homogeneous processors in which job assignment policy is a variant of round robin, the job selection. Random dispatching can provide $(1+\Delta)$-speed $O\left(\frac{1}{\Delta^3}\right)$-competitive non-migratory algorithm [23]. Chan et al. [24] proposed an $O(1)$-competitive algorithm using sleep management for the objective of minimizing the flow time plus energy. Albers et al. [25] studied an offline problem in polynomial time and proposed a fully combitorial algorithm that relies on repeated maximum flow computation. Gupta et al. [26] proved that highest density first, weighted shortest elapsed time first and weighted late arrival processor sharing are not $O(1)$-speed $O(1)$-competitive for the objective of minimizing the weighted flow time even in fixed variable speed processors for heterogeneous multiprocessor setting. Chan et al. [27] studied an online clairvoyant sleep management algorithm scheduling with arrival-time-alignment (SATA) which is $(1+\Delta)$-speed $O\left(\frac{1}{\Delta^2}\right)$-competitive for the objective of minimizing the flow time plus energy. For a detailed survey refer to [28–34].

In this paper, the problem of online non-clairvoyant (ON-C) DSS scheduling is studied and an algorithm multiprocessor with bounded speed (MBS) is proposed with an objective to minimize the IbFt+E. On the basis of potential function analysis MBS is O(1)- competitive. The notations used in this paper are mentioned in the Table 1.

Table 1. Notations used.

Notations	Meaning
t	Current time
j	A job
u	A processor
$r(j)$ or r_j	Release/arrival time of a job j
$p(j)$	Processing requirement (size) of a job j
m	Number of processors
lg_u	On a processor u, the count of lagging jobs, at time t
η	Maximum speed of a processor using Opt
P	Power of a processor at speed s
$s(t)$ or s	At time t, speed of some processor
α	A constant, commonly believed that its value is 2 or 3
Δ	A constant, its value depends on the value of α
I, S	A set of jobs and their schedule, respectively
$pwk(j,\ t)$, $pwk_a(j,\ t)$ and $pwk_o(j,\ t)$	Remaining/pending work of a job j at time t, using MBS and Opt, respectively
$F(j)$	Flow time of a job j
F	Total importance-based flow time
$imp_{uj}(t)$ or $imp_u(j)$	Importance/weight of a job j, at time t on a processor u
$imp_{ua}(t)$ or imp_{ua} and $imp_{uo}(t)$ or imp_{uo}	Importance of all active jobs using MBS and Opt at time t on a processor u, respectively
$imp_{lg}(t)$ or imp_{lg} and $imp_{lg_u}(t)$ or imp_{lg_u}	Total importance of lagging jobs, at time t on all m processors and on a processor u, respectively
$n_a(t)$ or n_a and $n_o(t)$ or n_o	Total number of active jobs (NoAJ) in MBS and Opt at time t on all m processors, respectively
$n_{ua}(t)$ or n_{ua} and $n_{uo}(t)$ or n_{uo}	NoAJ in MBS and Opt at time t on a processor u, respectively
$s_{ua}(t)$ or s_{ua} and $s_{uo}(t)$ or s_{uo}	Speed of a processor u for MBS and Opt at time t, respectively
$imp(t)$	Total importance of all active jobs n_a, at time t

Table 1. *Cont.*

Notations	Meaning
E	Energy consumed by processors
G	Total IbFt+E
c	Competitiveness
μ	A constant ($0 < \mu < 1$), its value depends on the value of Δ
$G_a(t)$ or G_a and $G_o(t)$ or G_o	IbFt+E acquired till time t by the MBS and Opt, respectively
$\frac{dG_a(t)}{dt}$ or $\frac{dG_a}{dt}$ and $\frac{dG_o(t)}{dt}$ or $\frac{dG_o}{dt}$	Rate of change (RoC) of G_a due to MBS and G_o due to Opt at time t, respectively
$G_{ua}(t)$ or G_{ua} and $G_{uo}(t)$ or G_{uo}	IbFt+E acquired on a processor u till time t by the MBS and Opt, respectively
$\frac{dG_{ua}(t)}{dt}$ or $\frac{dG_{ua}}{dt}$ and $\frac{dG_{uo}(t)}{dt}$ or $\frac{dG_{uo}}{dt}$	RoC of G_a due to MBS and Opt at time t on a processor u, respectively
γ	A constant (> 0)
c_i	Coefficient of a job j_i at time t
ω_i	Difference of pending work of a job j_i using MBS and Opt at time t
δ	A constant depends on α, its value is $\left(\frac{1}{2\alpha}\right)$
LG_u	A set of lagging jobs using MBS on a processor u
LG	A set of all lagging jobs using MBS on all m processors
$\Phi(t)$ or Φ	Total potential value of all m processors at time t
$\Phi_u(t)$ or Φ_u	Potential value of a processor u at time t
$\frac{d\Phi_o}{dt}$ and $\frac{d\Phi_a}{dt}$	RoC of Φ due to Opt and MBS, respectively
$\frac{d\Phi}{dt}$	RoC of Φ due to Opt and MBS
$\frac{d\Phi_{uo}}{dt}$ and $\frac{d\Phi_{ua}}{dt}$	RoC of Φ due to Opt and MBS on a processor u, respectively
$\frac{d\Phi_u}{dt}$	RoC of Φ due to Opt and MBS on a processor u

The organization of the paper is as follows. In Section 2, some related non-clairvoyant algorithms are explained and their competitive values are compared to the proposed algorithm MBS. Section 3 presents the preliminary definition and information for the proposed work. In Section 4, the proposed algorithm, its flow chart and potential function analysis is presented. The processing of a set of jobs are simulated using MBS and the best identified algorithm to observe the working of MBS. Section 6 provides the conclusion and future scope of the work.

2. Related Work

Gupta et al. [35] gave an online clairvoyant scheduling algorithm GKP (proposed by Gupta, Krishnaswamy and Pruhs) for the objective of minimizing the weighted flow time plus energy. Under the traditional power function, GKP is $O(\alpha^2)$-competitive without a resource augmentation for power heterogeneous processors. GKP uses highest density first (HDF) for the selection of jobs on each processor; the speed of any processor scales such that the power of a processor is the fractional weight of unfinished jobs; jobs are assigned in such a way that it gives the least increase in the projected future weighted flow time. Gupta et al. [35] used a local competitiveness analysis to prove their work. Fox et al. [36] considered the problem of scheduling the parallelizable jobs in the non-clairvoyant speed scaling settings for the objective of minimizing the weighted flow time plus energy and they used the potential function analysis to prove it. Fox et al. presented weighted latest arrival processor sharing with energy (WLAPS+E), which schedules the late arrival jobs and every job use the same number of machines proportioned by the job weight. WLAPS+E spares some machines to save the energy. WLAPS+E is $(1 + 6\Delta)$-speed $(5/\Delta^2)$-competitive, where $0 < \Delta \leq 1/6$. Thang [37] studied the online clairvoyant scheduling problem for the objective of minimizing the weighted flow time plus energy in the unbounded speed model and using the traditional power function. Thang gave an algorithm (ALG_{Thang}) on unrelated machines and proved that ALG_{Thang} is $8(1 + \alpha/ln\alpha)$-competitive. In AlG_{Thang}, the speed of any processor depends on the total weight of pending jobs on that machine, and any new job is assigned to a processor that minimizes the total weighted flow time.

Im et al. [38] proposed an ON-C scheduling algorithm SelfishMigrate-Energy (SM-E) for the objective of minimizing the weighted flow time plus energy for the unrelated machines. Using the traditional power function SM-E is $O(\alpha^2)$-competitive. In SM-E, a virtual queue is maintained on every processor where the new or migrated jobs are added at tail; the jobs migrate selfishly until equilibrium is gained. Im et al. simulates sequential best response (SBR) dynamics and they migrates each job to the machine that is provided by the Nash equilibrium. The scheduling policy applied on every processor is a variant of weighted round robin (WRR), wherein the larger speed is allotted to jobs residing at the tail of the queue (like Latest Arrival Processor Sharing (LAPS) and Weighted Latest Arrival Processor Sharing (WLAPS)). Bell et al. [39] proposed an online deterministic clairvoyant algorithm dual-classified round robin (DCRR) for the multiprocessor system using the traditional power function. The motive of $\left(2^{4\alpha}\left(log^{\alpha} P + \alpha^{\alpha} 2^{\alpha-1}\right)\right)$-competitive DCRR is to schedule the jobs so that they can be completed within deadlines using minimum energy, i.e., the objective is to maximize the throughput and energy consumption. In DCRR, the sizes and the maximum densities (= size/(deadline – release time)) of jobs are known and the classification of jobs depends on the size and the maximum density both. The competitive ratio of DCRR is high, as it considers the jobs with deadlines and using a variation of round robin with the speed scaling.

Azar et al. [40] gave an ON-C scheduling algorithm NC-PAR (Non-Clairvoyant for Parallel Machine) for the identical parallel machines, wherein the job migration is not permitted. Using traditional function NC-PAR is $\left(\alpha + \frac{1}{\alpha-1}\right)$-competitive for the objective of minimizing the weighted flow time plus energy in unbounded speed model. In NC-PAR a global queue of unassigned jobs is maintained in First In First Out (FIFO) order. A new job is assigned to a machine, when a machine becomes free. In NC-PAR jobs are having uniform density (i.e., $weight/size = 1$) and the jobs are not immediately allotted to the processors at release time. The speed of a processor using NC-PAR is based on the total remaining weight of the active jobs. In non-clairvoyant model with known arbitrary weights no results are known [40].

An ON-C multiprocessor speed scaling scheduling algorithm MBS is proposed and studied against an offline adversary with an objective of minimizing IbFt+E. The speed of a processor using MBS is proportional to the sum of importance of all active jobs on that processor. In MBS, the processor's maximum speed can be $(1 + {}^{\Delta}/_{3m})\eta$ (i.e., the range of speed is from zero to $(1 + {}^{\Delta}/_{3m})\eta$), whereas the processor's maximum speed using Opt (Optimal algorithm) is η, where m is number of processors and $0 < \Delta \leq (3\alpha)^{-1}$ a constant. In MBS, a new job is assigned to an idle processor (if available) or to a processor having the minimum sum of the ratio of importance and executed size for all jobs on that processor; the policy for job selection is weighted/importance-based round robin, and each active job receives the processor speed equal to the ratio of its importance to the total importance of jobs on that processor. In this paper, the performance of MBS is analysed using a competitive analysis, i.e., the worst-case comparison of MBS and optimal offline scheduling algorithm. MBS is $(1 + {}^{\Delta}/_{3m})$-speed, $\left(\frac{9}{8} + \frac{3\Delta}{8}\right)\cdot\left(1 + (1 + {}^{\Delta}/_{3m})^{\alpha}\right) = O(1)$ competitive, i.e., the value for competitive ratio c for $m = 2$, $\alpha = 2$ is 2.442; for $m = 2, \alpha = 3$ is 2.399; the detailed results for different values of m, $\Delta = (3\alpha)^{-1}$ and $\alpha = 2$ & 3 is shown in Table 2. The comparison of results is given along with the summary of results in Table 1.

Table 2. Results of multiprocessor with bounded speed (MBS).

Number of Processors (m)	α=2		α=3	
	Speed Ratio (s_r)	**Competitive Ratio (c)**	**Speed Ratio (s_r)**	**Competitive Ratio (c)**
2	1.02778	2.44189	1.01852	2.39936
4	1.01389	2.40822	1.00926	2.36604
8	1.00694	2.39155	1.00463	2.34961
16	1.00347	2.38326	1.00231	2.34145
64	1.00086	2.37706	1.00058	2.33536
128	1.00043	2.37603	1.00029	2.33435
512	1.00010	2.37526	1.00007	2.33359
1024	1.00005	2.37513	1.00003	2.37513
4096	1.00001	2.37503	1.00001	2.33336
11,264	1.000004	2.37501	1.000003	2.33334

Note: the speed ratio $s_r = \frac{Maximum\ speed\ of\ a\ processor\ using\ MBS}{Maximum\ speed\ of\ a\ processor\ using\ Opt}$.

On the basis of the values mentioned in the Table 2, it can be observed that in proposed algorithm MBS if the number of processor increases then the speed ratio and competitive ratio increases. The data mentioned in Table 3 describe the competitive values of different scheduling algorithm. Some clairvoyant and non-clairvoyant algorithms competitive ratio are considered at $\alpha = 2, \alpha = 3$. The lower competitive value represents the better algorithm. The value of competitiveness is least for the proposed algorithm MBS.

Table 3. Summary of Results.

Multiprocessor	Competitiveness for Weighted Flow Time + Energy			Modelling Criteria
Algorithms	**General**	α=2	α=3	**(Bounded (BS)/Unbounded Speed(US)) (Clairvoyant (C)/Non-Clairvoyant (NC)**
GKP [35]	α^2	4	9	US, C
WLAPS+E [36]	$5/\Delta^2$ (where $0 < \Delta \leq \frac{1}{6}$)	180	180	US, NC
ALG_{Thang} [37]	$8\left(1 + \frac{\alpha}{ln\alpha}\right)$	31.085	29.85	US, C
SM-E [38]	α^2	4	9	US, NC
DCRR [39]	$2^{4\alpha}\left(log^{\alpha}P + \alpha^{\alpha}2^{\alpha-1}\right)$ *where P is the ratio between the maximum and minimum job size*	>2048	>442368	US, C
NC-PAR [40]	$\left(\alpha + \frac{1}{\alpha-1}\right)$	3	3.5	US, NC
MBS [This Paper]	$\left(\frac{9}{8} + \frac{3\Delta}{8}\right)\left(1 + (1 + \Delta/3m)^{\alpha}\right)$ (where $0 < \Delta \leq (3\alpha)^{-1}$)	2.442	2.399	BS, NC

3. Definitions and Notations

An ON-C job scheduling on a multiprocessor using speed bounded setting is considered, where the jobs arrive over time, the job's importance/weight are known at release time and the size of a job is revealed only after the job's completion. Processor's speed using Opt can vary dynamically from 0 to the maximum speed η i.e., $[0, \eta]$. The nature of jobs is sequential as well as unrestricted pre-emption is permitted without penalty. The traditional power function $Power\ P = speed^{\alpha}$ is considered, where $\alpha > 1$ a fixed constant. If s is the processor's speed then a processor executes s unit of work per unit time. An active job j has release time lesser than the current time t, and it is not completely executed. The flow time $F(j)$ of job j is the time duration since j released and until it is completed. The total importance-based flow time F is $\sum_{j \in I} imp(j)F(j)$. Amortized analysis is used for algorithms where an occasional operation is very slow, but most of the other operations are faster. In amortized analysis, we analyse a sequence of operations and guarantee a worst case average time which is lower than the worst case time of a particular expensive operation.

4. Methodology

In this study, the amortized potential function analysis of the objective is used to examine the performance of the proposed algorithm. Amortized analysis is a worst-case analysis of a sequence of operations—to obtain a tighter bound on the overall or average cost per operation in the sequence than is obtained by separately analyzing each operation in the sequence. The amortized potential method, in which we derive a potential function characterizing the amount of extra work we can do in each step. This potential either increases or decreases with each successive operation, but cannot be negative. The objective of study is to minimize the total IbFt+E, denoted by $G = F + E$. It reflects that the target is to minimize the quality of service and energy consumed. The input to the problem is the set of jobs I. A scheduler generates the schedule S of jobs in I. The total energy consumption E for the scheduling is $\int_0^{\infty} s(t)^{\alpha}\, dt$. Let Opt be an optimal offline algorithm such that for any job sequence I, IbFt+E $F_{Opt(I)} + E_{Opt(I)}$ of Opt is minimized among all schedule of I. The notations used in MBS are mentioned in the Table 1. Any online algorithm ALG is said to be c-competitive for $c \geq 1$, if for all job sequences I and any input the cost incurred is never greater than c times the cost of optimal offline algorithm Opt, and the following inequality is satisfied:

$$\left(F_{ALG(I)} + E_{ALG(I)}\right) \le c \cdot \left(F_{Opt(I)} + E_{Opt(I)}\right)$$

The traditional power function is utilized to simulate the working of the proposed algorithm and compare the effectiveness by comparing with the available best known algorithm. The jobs are taken of different sizes and the arrival of jobs is considered in different scenario to critically examine the performance of the proposed algorithm. Different parameters (such as IbFt, IbFt+E, speed of processor and speed growth) are considered to evaluate the algorithm.

5. An $O(1)$-Competitive Algorithm

An ON-C multiprocessor scheduling algorithm multiprocessor with bounded speed (MBS) is explained in this section. The performance of MBS is observed by using potential function analysis, i.e., the worst-case comparison of MBS with an offline adversary Opt. The competitiveness of MBS is $O(1)$ with an objective to minimize the IbFt+E for m processors with the highest speed $(1 + \Delta/3m)\eta$.

5.1. Multiprocessor with Bounded Speed Algorithm: MBS

At time t, the processing speed of u adjusts to $s_{ua}(t) = (1 + \Delta/3m) \cdot min\left(\left(\frac{imp_{ua}(t)}{\Upsilon}\right)^{1/\alpha}, \eta\right)$, where $0 < \Delta \le \left(\frac{1}{3\alpha}\right)$, $\Upsilon \ge 1$ and $\alpha \ge 2$ are constants. The importance $imp(j)$ of a job is uninformed and acknowledged only at release time $r(j)$. The policies considered for the multiprocessor scheduling MBS are as follows:

Job selection policy: The importance-based/weighted round robin is used on every processor.

Job assignment policy: a newly arrived job is allotted to an idle processor (if available) or to a processor having the minimum sum of the ratio of importance to the executed size for all jobs on that processor (i.e., $min \sum_{f=1}^{n_{ua}} \left(\frac{imp_u(j_f)}{exs_u(j_f)}\right)$).

Speed scaling policy: The speed of every processor is scaled on the bases of the total importance of active jobs on that processor. Every active job j_i on u obtains the fraction of speed:

$$processor's\ speed\left(\frac{importance\ of\ j_i}{total\ importance\ of\ all\ active\ jobs\ on\ that\ processor}\right)$$

i.e., $s_{ua} \cdot \left(\frac{imp_u(j_i)}{\sum_{k=1}^{n_{ua}} imp_u(j_k)}\right)$ or $s_{ua} \cdot \left(\frac{imp_u(j_i)}{imp_{ua}}\right)$. The speed of any processor gets adjusted (re-evaluated) on alteration in total importance of active jobs on that processor. MBS is compared against an optimal offline algorithm Opt, using potential function analysis. The principal result of this study is stated in Theorem 1. The Algorithm 1 of MBS is given next and the flow chart for MBS is given in Figure 2.

Theorem 1. *When using more than two processors (i.e., $m \ge 2$) and each processor has the permitted maximum speed $(1 + \Delta/3m)\eta$, MBS is c-competitive for the objective of minimizing the IbFt+E, where* $c = \left(\frac{9}{8} + \frac{3\Delta}{8}\right) \cdot \left(1 + (1 + \Delta/3m)^{\alpha}\right) = O(1)$ *and* $0 < \Delta \le \frac{1}{3\alpha}$.

Algorithm 1: MBS (Multiprocessor with Bounded Speed)

Input: total m number of processors $\{u_1, \ldots, u_k, \ldots, u_m\}$, n_a NoAJ $\{j_1, \ldots, j_i, \ldots, j_{n_a}\}$ and the importance of all n_a active jobs $\{imp(j_1), \ldots, imp(j_i), \ldots, imp(j_{n_a})\}$.
Output: number of jobs allocated to every processor, the speed of all processors, at any time and execution speed share of each active job.
Repeat until all processors become idle:
1. **If** any job j_i arrives
2. **if** $m \geq n_a$
3. allocate job j_i to a idle processor u
4. otherwise, when $m < n_a$
5. allocate job j_i to a processor u with $min\ \sum_{f=1}^{n_{ua}} \left(\frac{imp_u(j_f)}{exs_u(j_f)}\right)$
6. $imp_{ua} = imp_{ua} + imp_u(j_i)$
7. $s_{ua} = (1+\Delta/3m) \cdot min\left(\left(\frac{imp_{ua}}{\dagger}\right)^{1/\alpha}, \eta\right)$, where $0 < \Delta \leq \left(\frac{1}{3\alpha}\right)$ and $\dagger > 1$ is a constant value
8. Otherwise, if any job j_i completes on any processor u and other active jobs are available for execution on that processor then
9. $imp_{ua} = imp_{ua} - imp_u(j_i)$
10. $s_{ua} = (1+\Delta/3m) \cdot min\left(\left(\frac{imp_{ua}}{\dagger}\right)^{1/\alpha}, \eta\right)$, where $0 < \Delta \leq \left(\frac{1}{3\alpha}\right)$ and $\dagger \geq 1$ is a constant value
11. the speed received by any job j_i, which is executing on a processor u, is $s_{ua} \cdot \left(\frac{imp_u(j_i)}{imp_{ua}}\right)$
12. otherwise, processors continue to execute remaining jobs

5.2. Necessary Conditions to be Fulfilled

A potential function is needed to calculate the c-competitiveness of an algorithm. An algorithm is called c-competitive if at any time t, the sum of augmentation in the objective cost of algorithm and the modification in the value of potential is at the most c times the augmentation in the objective cost of the optimal adversary algorithm. A potential function $\Phi(t)$ is required to demonstrate that MBS is c-competitive. A c-competitive algorithm should satisfy the conditions:

Boundary Condition: The value of potential function is zero before the release of any job and after the completion of all jobs.

Job Arrival and Completion Condition: The value of potential function remains same on arrival or completion of a job.

Running Condition: At time when the above condition do not exist, the sum of the (rate of change) RoC of G_a and the RoC of Φ is at the most c times the RoC of G_o.

$$\frac{dG_a(t)}{dt} + \gamma \cdot \frac{d\Phi}{dt} \leq c \cdot \frac{dG_o(t)}{dt}, \text{ where } \gamma > 0. \quad (1)$$

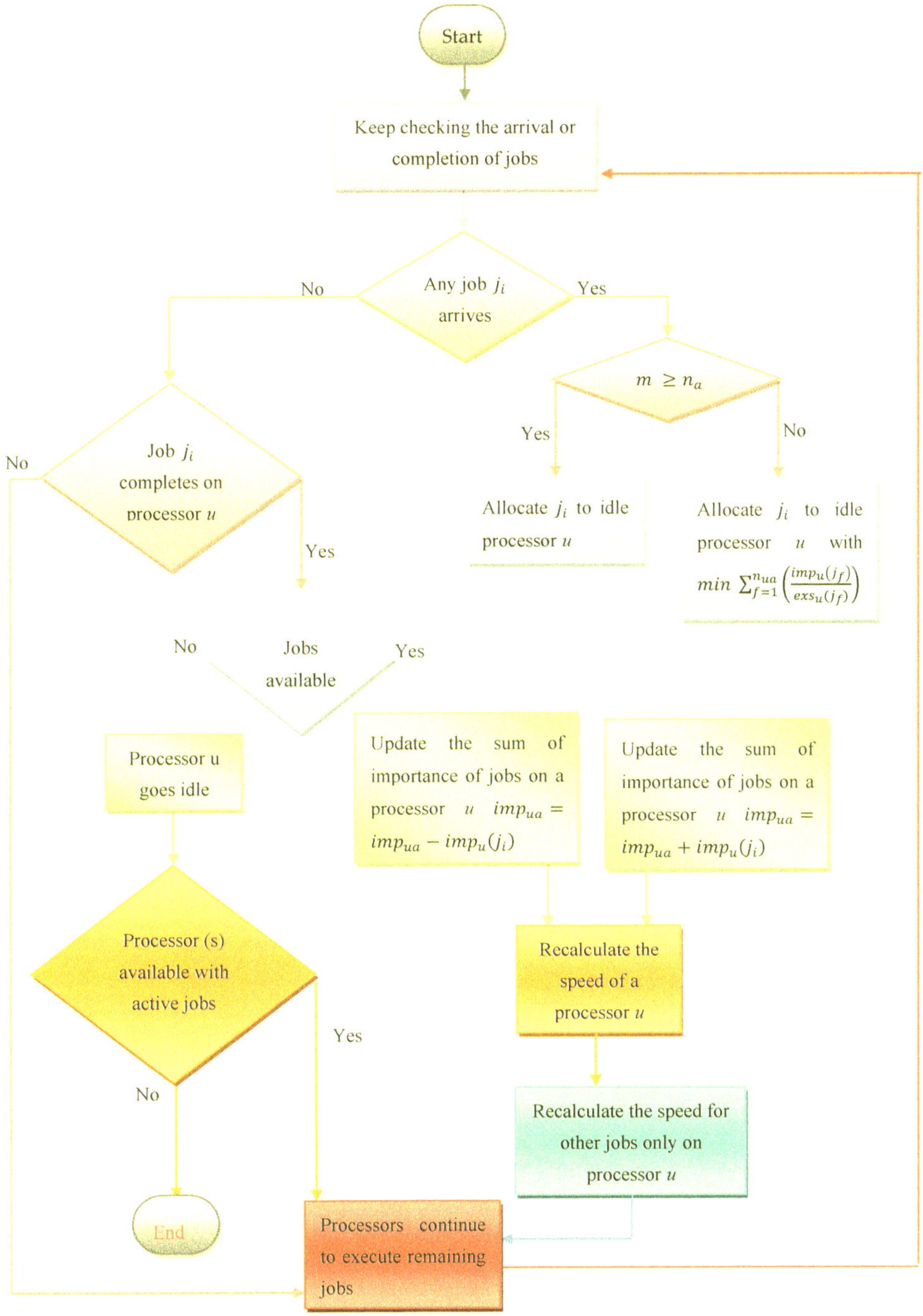

Figure 2. Flow chart of the MBS scheduling algorithm.

5.3. Potential Function $\Phi(t)$

An active job j is lagging, if $(pwk_a(j,\ t) - pwk_o(j,\ t)) > 0$. Since t is the instantaneous time, this factor is dropped from the rest of the analysis. For any processor u, let $LG_u = \{j_1,\ j_2,\ \ldots,\ j_{lg_u}\}$ be a group of lagging jobs using MBS and these jobs are managed in the ascending order of latest time (when any job gets changed into lagging job). $LG = \sum_{u=1}^{m} LG_u$ is a set of all lagging jobs on all m processors. Further, $imp_{lg_u} = \sum_{i=1}^{lg_u} imp_u(j_i)$ is the sum of the importance of lagging jobs on a processor u. Following this, $imp_{lg} = \sum_{u=1}^{m} imp_{lg_u}$ is the sum of the importance of lagging jobs on all m processors. Our potential function $\Phi(t)$ for IbFt+E is the addition of all potential values of m processors.

$$\Phi(t) = \sum_{u=1}^{m} \Phi_u(t) \tag{2}$$

$$\Phi_u(t) = \begin{cases} \sum_{i=1}^{lg_u}\left(\left(\sum_{k=1}^{i} imp_u(j_k)\right)^{1-2\delta}\right)\cdot\omega_i & if\ \sum_{k=1}^{i} imp_u(j_k) \le \eta^{1/2\delta} \\ \sum_{i=1}^{lg_u}\left(\frac{1}{1-\delta}\right)\cdot\left(\sum_{k=1}^{i} imp_u(j_k)\cdot\eta^{-1}\right)\cdot\omega_i & otherwise \end{cases} \tag{3}$$

$$\begin{gathered} Where\ \omega_i = max\{0,\ (pwk_a(j_i,\ t) - pwk_o(j_i,\ t))\} \\ \delta = \tfrac{1}{2\alpha}\ \text{and} \end{gathered} \tag{4}$$

$$\left(\sum_{k=1}^{i} imp_u(j_k)\right)^{1-2\delta} \text{and} \left(\frac{1}{1-\delta}\right)\left(\sum_{k=1}^{i} imp_u(j_k)\cdot\eta^{-1}\right) \tag{5}$$

are the coefficients c_i of j_i on processor u

MBS is analyzed per machine basis. Firstly, the verification of boundary condition: the value of Φ is zero after finishing of all jobs and prior to release of any job on any processor. There will be no active job on any processor in both situations. Therefore, the boundary condition is true. Secondly, the verification of arrival and completion condition: at time t, on release of a new job j_i in I, j_i without execution is appended at end of I. ω_i is zero as $pwk_a(j_i,\ t) - pwk_o(j_i,\ t) = 0$. The coefficient of all other jobs does not change and Φ remains unchanged. At the time of completion of a job j_i, ω_i becomes zero and other coefficients of lagging jobs either remains unchanged or decreases, so, Φ does not increase. Thus the arrival and completion criteria holds true. The third and last criterion to confirm is running condition, with no job arrival or completion.

According to previous discussion, for any processor u, let $\frac{dG_{ua}}{dt} = imp_{ua} + {s_{ua}}^{\alpha}$ and $\frac{dG_{uo}}{dt} = imp_{uo} + {s_{uo}}^{\alpha}$ be the alteration of IbFt+E in an infinitesimal period of time $[t,\ t+dt]$ by MBS and Opt, respectively. The alteration of Φ because of Opt and MBS in an infinitesimal period of time $[t,\ t+dt]$ by u is $\frac{d\Phi_{uo}}{dt}$ and $\frac{d\Phi_{ua}}{dt}$, respectively. The whole alteration in Φ because of Opt and MBS in infinitesimal period of time $[t,\ t+dt]$ by u is $\frac{d\Phi_u}{dt} = \frac{d\Phi_{uo}}{dt} + \frac{d\Phi_{ua}}{dt}$. As this is multiprocessor system therefore to bound the RoC of Φ by Opt and MBS, the analysis is divided in two cases based on n_a and m, and then every case is further divided in three sub cases depending on whether $imp_{ua} > \eta^{\alpha}$ and $imp_{l_u} > \eta^{\alpha}$, afterwards each sub case is further divided in two sub cases depending on $imp_{lg} > \left(imp_a - \left(\frac{3}{3+\Delta}\right)\cdot imp_a\right)$ and $imp_{lg} \le \left(imp_a - \left(\frac{3}{3+\Delta}\right)\cdot imp_a\right)$, where $0 < \Delta < 1$, $\mu = \frac{3}{3+\Delta}$. The potential analysis is done on individual processor basis, the reason behind it is that all the processors will not face the same case at the same time; rather different processors may face same or different cases.

Lemma 1. *For the positive real numbers x, y, A and B, if $x^{-1} + y^{-1} = 1$ holds then [2]:*

$$x^{-1}\cdot A^{x} + y^{-1}\cdot B^{y} \ge A\cdot B \tag{6}$$

Lemma 2. *If $n_a \le m$ and $imp_{lg_u} \le \eta^{\alpha}$*

(a) $\frac{d\Phi_{uo}}{dt} \le \frac{{s_{uo}}^{\alpha}}{\alpha} + (1-2\delta)\cdot imp_{lg_u}$; (b) $\frac{d\Phi_{ua}}{dt} \le -\left(s_{ua}\cdot {imp_{lg_u}}^{1-2\delta}\right)$

Proof. If $n_a \leq m$ then every processor executes not more than one job, i.e., every job is processed on individual processor.

(a) It is required to upper-bound $\frac{d\Phi_{uo}}{dt}$ for a processor u. To calculate the upper-bound, the worst-case is considered which occurs if Opt executes a job on u with the largest coefficient $c_{lg_u} = imp_{lg_u}{}^{1-2\delta}$. At this time, ω_i increases at the rate of s_{uo} (because of Opt on u). The count of lagging jobs on some u may be only one.

$$\frac{d\Phi_{uo}}{dt} \leq c_{lg_u} \cdot s_{uo} \leq imp_{lg_u}{}^{1-2\delta} \cdot s_{uo} \tag{7}$$

Using Young's inequality, Lemma 1 (Equation (6)) in (7) such that $A = s_{uo}$, $B = \left(imp_{lg_u}\right)^{1-2\delta}$, $x = \alpha$ and $y = \frac{1}{1-2\delta}$ we have:

$$\frac{d\Phi_{uo}}{dt} \leq \frac{s_{uo}{}^{\alpha}}{\alpha} + (1-2\delta) \cdot imp_{lg_u} \tag{8}$$

(b) Next, it is required to upper-bound $\frac{d\Phi_{ua}}{dt}$ for a processor u. To compute the upper-bound, consider that a lagging job j_i on u is executed at the rate of $\left(s_{ua} \cdot \frac{imp_u(j_i)}{\sum_{k=1}^{n_{ua}} imp_u(j_k)}\right)$ or $\left(s_{ua} \cdot \frac{imp_u(j_i)}{imp_{ua}}\right)$, therefore, the change in ω_i is at the rate of $\left(-s_{ua} \cdot \frac{imp_u(j_i)}{imp_{ua}}\right)$.

$$\frac{d\Phi_{ua}}{dt} = \sum_{i=1}^{lg_u} \left(\left(\sum_{k=1}^{i} imp_u(j_k) \right)^{1-2\delta} \right) \cdot \left(-s_{ua} \cdot \frac{imp_u(j_i)}{imp_{ua}} \right)$$

As only one job executes on a processor, therefore $\frac{imp_u(j_i)}{imp_{ua}} = 1$ and $lg_u = i = 1$,

$$\begin{aligned} \frac{d\Phi_{ua}}{dt} &= \left(imp_{lg_u}{}^{1-2\delta}\right) \cdot (-s_{ua}) \\ \frac{d\Phi_{ua}}{dt} &= -\left(s_{ua} \cdot imp_{lg_u}{}^{1-2\delta}\right) \end{aligned} \tag{9}$$

□

Lemma 3. *If $n_a \leq m$ and $imp_{lg_u} > \eta^{\alpha}$*

(a) $\frac{d\Phi_{uo}}{dt} \leq \left(\frac{1}{1-\delta}\right) \cdot imp_{lg_u}$; (b) $\frac{d\Phi_{ua}}{dt} = -\frac{(1+\Delta/3m)}{(1-\delta)} \cdot imp_{lg_u}$

Proof. If $n_a \leq m$ then every processor executes not more than one job, i.e., every job is processed on individual processor.

(a) It is required to upper-bound $\frac{d\Phi_{uo}}{dt}$ for a processor u. To calculate the upper-bound, the worst-case is considered which occurs if Opt executes a job on u with the largest coefficient $c_{lg_u} = \left(\frac{1}{1-\delta}\right) \cdot imp_{lg_u} \cdot \eta^{-1}$. At this time, ω_i increases at the rate of s_{uo} (because of Opt on u) where $s_{uo} \leq \eta$. The count of lagging jobs on any u may be only one.

$$\begin{aligned} \frac{d\Phi_{uo}}{dt} \leq c_{lg_u} \cdot s_{uo} \leq c_{lg_u} \cdot \eta &= \left(\frac{1}{1-\delta}\right) \cdot imp_{lg_u} \cdot \eta^{-1} \cdot \eta \\ \frac{d\Phi_{uo}}{dt} &\leq \left(\frac{1}{1-\delta}\right) \cdot imp_{lg_u} \end{aligned} \tag{10}$$

(b) Next, it is required to upper-bound $\frac{d\Phi_{ua}}{dt}$ for a processor u. To compute the upper-bound, consider that a lagging job j_i on u is executed at the rate of $\left(s_{ua} \cdot \frac{imp_u(j_i)}{\sum_{k=1}^{n_{ua}} imp_u(j_k)}\right)$ or $\left(s_{ua} \cdot \frac{imp_u(j_i)}{imp_{ua}}\right)$, therefore the change in ω_i is at the rate of $\left(-s_{ua} \cdot \frac{imp_u(j_i)}{imp_{ua}}\right)$. $imp_{ua} \geq imp_{gl_u} > \eta^{\alpha}$, $s_{ua} = (1 + \Delta/3m) \cdot \eta$

$$\frac{d\Phi_{ua}}{dt} = \sum_{i=1}^{lg_u} \left(\frac{1}{1-\delta}\right) \cdot \left(\sum_{k=1}^{i} imp_u(j_k) \cdot \eta^{-1} \right) \cdot \left(-s_{ua} \cdot \frac{imp_u(j_i)}{imp_{ua}} \right)$$

As only one job executes on a processor, therefore $\frac{imp_u(j_i)}{imp_{ua}} = 1$ and $lg_u = i = 1$,

$$\begin{aligned}\frac{d\Phi_{ua}}{dt} &= \left(\frac{1}{1-\delta}\right)\cdot\left(imp_{lg_u}\cdot\eta^{-1}\right)\cdot(-s_{ua}) \\ &= -\left(\frac{1}{1-\delta}\right)\cdot\left(s_{ua}\cdot imp_{lg_u}\cdot\eta^{-1}\right) \\ &= -\left(\frac{1}{1-\delta}\right)\cdot\left((1+\Delta/3m)\cdot\eta\cdot imp_{lg_u}\cdot\eta^{-1}\right)\end{aligned}$$

$$\frac{d\Phi_{ua}}{dt} = -\frac{(1+\Delta/3m)}{(1-\delta)}\cdot imp_{lg_u} \quad (11)$$

□

Lemma 4. *If $n_a > m$ and $imp_{lg_u} \leq \eta^{\alpha}$*

(a) $\frac{d\Phi_{uo}}{dt} \leq \frac{s_{uo}{}^{\alpha}}{\alpha} + (1-2\delta)\cdot imp_{lg_u}$; (b) $\frac{d\Phi_{ua}}{dt} \leq -\frac{s_{ua}}{(2-2\delta)}\cdot\left(\frac{imp_{lg_u}{}^{2-2\delta}}{imp_{ua}}\right)$

Proof. If $n_a > m$ then:

(a) It is required to upper-bound $\frac{d\Phi_{uo}}{dt}$ for a processor u. To calculate the upper-bound, the worst-case is considered which occurs if Opt is executing a job on u with the largest coefficient $c_{lg_u} = imp_{lg_u}{}^{1-2\delta}$. At this time, ω_i increases at the rate of s_{uo} (because of Opt on u).

$$\frac{d\Phi_{uo}}{dt} \leq c_{lg_u}\cdot s_{uo} = imp_{lg_u}{}^{1-2\delta}\cdot s_{uo} \quad (12)$$

Using Young's inequality, Lemma 1 (Equation (6)) in (12) such that $A = s_{uo}$, $B = imp_{lg_u}{}^{1-2\delta}$, $x = \alpha$ and $y = \frac{1}{1-2\delta}$ we have:

$$\frac{d\Phi_{uo}}{dt} \leq \frac{s_{uo}{}^{\alpha}}{\alpha} + (1-2\delta)\cdot imp_{lg_u} \quad (13)$$

(b) Next, it is required to upper-bound $\frac{d\Phi_{ua}}{dt}$ for a processor u, to compute the upper-bound consider that a lagging job j_i on u is executed at the rate of $\left(s_{ua}\cdot\frac{imp_u(j_i)}{\sum_{k=1}^{n_{ua}} imp_u(j_k)}\right)$ or $\left(s_{ua}\cdot\frac{imp_u(j_i)}{imp_{ua}}\right)$, therefore the change in ω_i is at the rate of $\left(-s_{ua}\cdot\frac{imp_u(j_i)}{imp_{ua}}\right)$. To make the discussion straightforward, let $h_{ui} = \sum_{k=1}^{i} imp_u(j_k)$, $h_{u0} = 0$, $h_{ulg_u} = imp_{lg_u}$ and $imp_u(j_i) = h_{ui} - h_{ui-1}$. (by using Equation (3):

$$\begin{aligned}\frac{d\Phi_{ua}}{dt} &= \sum_{i=1}^{lg_u}\left(\left(\sum_{k=1}^{i} imp_u(j_k)\right)^{1-2\delta}\right)\cdot\left(-s_{ua}\cdot\frac{imp_u(j_i)}{imp_{ua}}\right) \\ &= -\frac{s_{ua}}{imp_{ua}}\sum_{i=1}^{lg_u}\left((h_{ui})^{1-2\delta}\right)\cdot(h_{ui}-h_{ui-1}) \\ &\leq -\frac{s_{ua}}{imp_{ua}}\sum_{i=1}^{lg_u}\int_{h_{ui-1}}^{h_{ui}} f^{1-2\delta}\,df \\ &\leq -\frac{s_{ua}}{imp_{ua}}\int_{0}^{h_{ulg_u}} f^{1-2\delta}\,df \\ &= -\frac{s_{ua}}{imp_{ua}}\cdot\frac{h_{ulg_u}{}^{2-2\delta}}{(2-2\delta)} \\ &= -\frac{s_{ua}}{imp_{ua}}\cdot\frac{imp_{lg_u}{}^{2-2\delta}}{(2-2\delta)}\end{aligned}$$

$$\frac{d\Phi_{ua}}{dt} \leq -\frac{s_{ua}}{(2-2\delta)}\cdot\left(\frac{imp_{lg_u}{}^{2-2\delta}}{imp_{ua}}\right) \quad (14)$$

□

Lemma 5. *If $n_a > m$ and $imp_{lg_u} > \eta^{\alpha}$*

(a) $\frac{d\Phi_{uo}}{dt} \leq \left(\frac{1}{1-\delta}\right)\cdot imp_{lg_u}$; (b) $\frac{d\Phi_{ua}}{dt} \leq -\frac{(1+\Delta/3m)}{(2-2\delta)}\cdot\left(\frac{imp_{lg_u}{}^{2}}{imp_{ua}}\right)$

Proof. If $n_a > m$ then:

(a) It is required to upper-bound $\frac{d\Phi_{uo}}{dt}$ for a processor u. To calculate the upper-bound, the worst-case is considered which occurs if Opt executes a job on u with the largest coefficient $c_{lg_u} = \left(\frac{1}{1-\delta}\right) \cdot imp_{lg_u} \cdot \eta^{-1}$ (as $imp_{ua} \geq imp_{lg_u} > \eta^{\alpha}$). At this time, ω_i increases at the rate of s_{uo} (because of Opt on u).

$$\begin{aligned}\frac{d\Phi_{uo}}{dt} &\leq c_{lg_u} \cdot s_{uo} \\ &= \left(\tfrac{1}{1-\delta}\right) \cdot imp_{lg_u} \cdot \eta^{-1} \cdot s_{uo} \\ &\leq \left(\tfrac{1}{1-\delta}\right) \cdot imp_{lg_u} \cdot \eta^{-1} \cdot \eta \ \{\because s_{uo} \leq \eta\}\end{aligned}$$

$$\frac{d\Phi_{uo}}{dt} \leq \left(\frac{1}{1-\delta}\right) \cdot imp_{lg_u} \tag{15}$$

(b) Next, it is required to upper-bound $\frac{d\Phi_{ua}}{dt}$ for a processor u. To compute the upper-bound, consider that a lagging job j_i on u is executed at the rate of $\left(s_{ua} \cdot \frac{imp_u(j_i)}{\sum_{k=1}^{n_{ua}} imp_u(j_k)}\right)$ or $\left(s_{ua} \cdot \frac{imp_u(j_i)}{imp_{ua}}\right)$, therefore the change in ω_i is at the rate of $\left(-s_{ua} \cdot \frac{imp_u(j_i)}{imp_{ua}}\right)$. To make the discussion uncomplicated, let $h_{ui} = \sum_{k=1}^{i} imp_u(j_k)$, $h_{u0} = 0$, $h_{ulg_u} = imp_{lg_u} > \eta^{\alpha}$, $imp_{ua} \geq imp_{lg_u} > \eta^{\alpha}$ and $imp_u(j_i) = h_{ui} - h_{ui-1}$. Let $z < l_u$ be the largest integer such that $h_{u_z} \leq \eta^{\alpha}$. (using Equation (3)):

$$\begin{aligned}\frac{d\Phi_{ua}}{dt} &= \sum_{i=1}^{lg_u} c_i \cdot \left(-s_{ua} \cdot \frac{imp_u(j_i)}{imp_{ua}}\right) \\ &= -\left(\tfrac{s_{ua}}{imp_{ua}}\right) \cdot \left(\sum_{i=1}^{z} \left(imp_u(j_i) \cdot (h_{ui})^{1-2\delta}\right) + \sum_{i=z+1}^{lg_u} \left(\left(\tfrac{1}{1-\delta}\right) \cdot \left(imp_u(j_i) \cdot h_{ui} \cdot \eta^{-1}\right)\right)\right) \\ &\leq -\left(\tfrac{s_{ua}}{imp_{ua}}\right) \cdot \left(\left(\int_0^{h_{u_z}} f^{1-2\delta} df\right) + \left(\left(\tfrac{1}{1-\delta}\right) \cdot \eta^{-1} \cdot \left(\int_{h_{u_z}}^{h_{ulg_u}} f df\right)\right)\right) \\ &= -\left(\tfrac{s_{ua}}{imp_{ua}}\right) \cdot \left(\frac{h_{u_z}^{2-2\delta}}{(2-2\delta)} + \frac{h_{ulg_u}^{2} - h_{u_z}^{2}}{(2-2\delta) \cdot \eta}\right) \\ &= -\left(\tfrac{s_{ua}}{imp_{ua}}\right) \cdot \left(\frac{h_{u_z}^{2}}{(2-2\delta) h_{u_z}^{1/\alpha}} + \frac{h_{ulg_u}^{2} - h_{u_z}^{2}}{(2-2\delta) \cdot \eta}\right) \\ &\leq -\frac{(1+\Delta/3m) \cdot \eta}{imp_{ua}} \cdot \left(\frac{h_{u_z}^{2}}{(2-2\delta)\eta} + \frac{h_{ulg_u}^{2} - h_{u_z}^{2}}{(2-2\delta) \cdot \eta}\right) \{\because h_{u_z} \leq \eta^{\alpha}\} \\ &= -\frac{(1+\Delta/3m)}{imp_{ua}} \cdot \frac{h_{ulg_u}^{2}}{(2-2\delta)} \\ &= -\frac{(1+\Delta/3m)}{(2-2\delta)} \cdot \left(\frac{imp_{lg_u}^{2}}{imp_{ua}}\right)\end{aligned}$$

$$\frac{d\Phi_{ua}}{dt} \leq -\frac{(1+\Delta/3m)}{(2-2\delta)} \cdot \left(\frac{imp_{lg_u}^{2}}{imp_{ua}}\right) \tag{16}$$

□

Lemma 6. *At all time t, when Φ does not comprise discrete alteration $\frac{dG_{ua}}{dt} + \gamma \cdot \frac{d\Phi_u}{dt} \leq c \cdot \frac{dG_{uo}}{dt}$, where $c = \left(\frac{9}{8} + \frac{3\Delta}{8}\right) \cdot \left(1 + (1 + \Delta/3m)^{\alpha}\right)$. Assume that $\gamma = \frac{1}{16} \cdot \left(1 + (1 + \Delta/3m)^{\alpha}\right)$.*

Proof. The analysis is divided in two cases based on $n_a > m$ or $n_a \leq m$, and then each case is again alienated in three sub-cases depending on whether $imp_{ua} > \eta^{\alpha}$ or $imp_{ua} \leq \eta^{\alpha}$ and $imp_{lg_u} > \eta^{\alpha}$ or $imp_{lg_u} \leq \eta^{\alpha}$, afterwards each sub-case is again alienated in two sub-cases depending on whether $imp_{lg_u} > \left(imp_{ua} - \left(\frac{3}{3+\Delta}\right) \cdot imp_{ua}\right)$ or $imp_{lg_u} \leq \left(imp_{ua} - \left(\frac{3}{3+\Delta}\right) \cdot imp_{ua}\right)$, where $0 < \mu = \left(\frac{3}{3+\Delta}\right) < 1$ and $\Delta = (1/3\alpha)$. As a job in MBS which is not lagging must be an active job in Opt,

$$imp_{uo} \geq imp_{ua} - imp_{lg_u} \geq imp_{ua} - (imp_{ua} - \mu \cdot imp_{ua}) \geq \mu \ imp_{ua} \Rightarrow imp_{ua} \leq \frac{imp_{uo}}{\mu} \tag{17}$$

$$\mu = \left(\frac{3}{3+\Delta}\right) \tag{18}$$

$$\gamma = \frac{1}{16} \cdot \left(1 + (1 + \Delta/3m)^{\alpha}\right) \tag{19}$$

$$c = \left(\frac{9}{8} + \frac{3\Delta}{8}\right) \cdot \left(1 + (1 + \Delta/3m)^{\alpha}\right) \tag{20}$$

□

Case I: When $n_a \leq m$ and $imp_{ua} \leq \eta^{\alpha}$, since $imp_{lg_u} \leq imp_{ua}$ we have $imp_{lg_u} \leq \eta^{\alpha}$, and $s_{ua}(t) = (1 + \Delta/3m) \cdot min\left(imp_{ua}{}^{1/\alpha}, \eta\right) = (1 + \Delta/3m) \cdot imp_{ua}{}^{1/\alpha}$.

(a) If $imp_{lg_u} > \left(imp_{ua} - \left(\frac{3}{3+\Delta}\right) \cdot imp_{ua}\right)$ then the total RoC of Φ because of Opt and MBS is $\frac{d\Phi_u}{dt} = \frac{d\Phi_{uo}}{dt} + \frac{d\Phi_{ua}}{dt}$.

(using Equations (8) and (9))

$$\frac{d\Phi_u}{dt} \leq \left(\frac{s_{uo}{}^{\alpha}}{\alpha} + (1 - 2\delta) \cdot imp_{lg_u}\right) - \left(s_{ua} \cdot imp_{lg_u}{}^{1-2\delta}\right) \tag{21}$$

(by using Equations (1) and (21))

$$\begin{aligned}
&\frac{dG_{ua}}{dt} + \gamma \cdot \frac{d\Phi_u}{dt} \\
&\leq \left(imp_{ua} + s_{ua}{}^{\alpha} + \gamma \cdot \left(\left(\frac{s_{uo}{}^{\alpha}}{\alpha} + (1-2\delta) \cdot imp_{lg_u}\right) - \left(s_{ua} \cdot imp_{lg_u}{}^{1-2\delta}\right)\right)\right) \\
&= \left(imp_{ua} + (1 + \Delta/3m)^{\alpha} \cdot imp_{ua} + \frac{\gamma}{\alpha} \cdot s_{uo}{}^{\alpha} + \gamma \cdot (1-2\delta) \cdot imp_{lg_u} - \gamma \cdot (1 + \Delta/3m) \cdot imp_{ua}{}^{1/\alpha} \cdot imp_{lg_u}{}^{1-2\delta}\right) \\
&\leq \left(\frac{\gamma}{\alpha} \cdot s_{uo}{}^{\alpha} + \left(1 + (1 + \Delta/3m)^{\alpha}\right) \cdot imp_{ua} + \gamma \cdot (1-2\delta) \cdot imp_{ua} - \gamma \cdot (1 + \Delta/3m) \cdot imp_{ua}{}^{1/\alpha} \cdot \left(\left(1 - \left(\frac{3}{3+\Delta}\right)\right) \cdot imp_{ua}\right)^{1-2\delta}\right) \\
&= \left(\frac{\gamma}{\alpha} \cdot s_{uo}{}^{\alpha} + imp_{ua} \cdot \left(1 + (1 + \Delta/3m)^{\alpha} + \gamma \cdot (1-2\delta) - \gamma \cdot (1 + \Delta/3m) \cdot \left(\frac{\Delta}{3+\Delta}\right)^{1-2\delta}\right)\right) \\
&\leq \frac{\gamma}{\alpha} \cdot s_{uo}{}^{\alpha} + imp_{ua} \cdot \left(1 + (1 + \Delta/3m)^{\alpha} + \gamma - \gamma \cdot \left(\frac{\Delta}{3+\Delta}\right)^{1-2\delta}\right) \\
&\leq \frac{\gamma}{\alpha} \cdot s_{uo}{}^{\alpha} + imp_{ua} \cdot \left(1 + (1 + \Delta/3m)^{\alpha} + \gamma - \gamma \cdot \left(\frac{\Delta}{3+\Delta}\right)\right) \\
&= \frac{\gamma}{\alpha} \cdot s_{uo}{}^{\alpha} + imp_{ua} \cdot \left(1 + (1 + \Delta/3m)^{\alpha} + \gamma \cdot \left(\frac{3}{3+\Delta}\right)\right) \\
&\leq \frac{\gamma}{\alpha} \cdot s_{uo}{}^{\alpha} + imp_{ua} \cdot \left(1 + (1 + \Delta/3m)^{\alpha} + \gamma\right) \\
&= \frac{\gamma}{\alpha} \cdot s_{uo}{}^{\alpha} + imp_{ua} \cdot \left(1 + (1 + \Delta/3m)^{\alpha} + \frac{1}{16} \cdot \left(1 + (1 + \Delta/3m)^{\alpha}\right)\right) && \text{(by using Equation (19))} \\
&\leq \frac{\gamma}{\alpha} \cdot s_{uo}{}^{\alpha} + \frac{imp_{uo}}{\mu} \cdot \left(\frac{17}{16} \cdot \left(1 + (1 + \Delta/3m)^{\alpha}\right)\right) && \text{(by using Equation (17))} \\
&= \frac{\gamma}{\alpha} \cdot s_{uo}{}^{\alpha} + imp_{uo} \cdot \left(\frac{17}{16} \cdot \left(1 + \frac{\Delta}{3}\right) \cdot \left(1 + (1 + \Delta/3m)^{\alpha}\right)\right) && \text{(by using Equation (18))} \\
&\leq \frac{\gamma}{\alpha} \cdot s_{uo}{}^{\alpha} + imp_{uo} \cdot \left(\left(\frac{9}{8} + \frac{3\Delta}{8}\right) \cdot \left(1 + (1 + \Delta/3m)^{\alpha}\right)\right) && \text{(by using Equation (20))}
\end{aligned}$$

$$\frac{dG_{ua}}{dt} + \gamma \cdot \frac{d\Phi_u}{dt} \leq \frac{\gamma}{\alpha} \cdot s_{uo}{}^{\alpha} + imp_{uo} \cdot c \tag{22}$$

Since $c = \left(\frac{9}{8} + \frac{3\Delta}{8}\right) \cdot \left(1 + (1 + \Delta/3m)^{\alpha}\right)$ and $\gamma = \frac{1}{16} \cdot \left(1 + (1 + \Delta/3m)^{\alpha}\right)$, we have

$$\Rightarrow c = \left(\frac{9}{8} + \frac{3\Delta}{8}\right) \cdot 16\gamma \Rightarrow \frac{\gamma}{c} = \frac{1}{18 + 6\Delta} < 1 \Rightarrow \frac{\gamma}{c} < 1 \Rightarrow \gamma < c$$

$$\text{Since } \gamma < c \text{ and } \alpha > 1 \Rightarrow 1 > \frac{1}{\alpha} \Rightarrow \frac{\gamma}{\alpha} < c \tag{23}$$

(by using Equation (23) in Equation (22))

$$\frac{dG_{ua}}{dt} + \gamma \cdot \frac{d\Phi_u}{dt} \leq c \cdot s_{uo}{}^{\alpha} + c \cdot imp_{uo} = c \cdot \left(s_{uo}{}^{\alpha} + imp_{uo}\right) = c \cdot \frac{dG_{uo}}{dt}$$

Hence the running condition is fulfilled for $n_a \leq m$, $imp_{ua} \leq \eta^{\alpha}$, $imp_{lg_u} \leq \eta^{\alpha}$, $imp_{lg_u} > \left(imp_{ua} - \left(\frac{3}{3+\Delta}\right) \cdot imp_{ua}\right)$, $c = \left(\frac{9}{8} + \frac{3\Delta}{8}\right) \cdot \left(1 + (1 + \Delta/3m)^{\alpha}\right)$.

(b) If $imp_{lg_u} \leq \left(imp_{ua} - \left(\frac{3}{3+\Delta}\right) \cdot imp_{ua}\right)$ then the total RoC of Φ because of Opt and MBS depends on $\frac{d\Phi_{uo}}{dt}$ since $\frac{d\Phi_{ua}}{dt} \leq 0$.

(by using Equation (8))

$$\frac{d\Phi_u}{dt} \le \left(\frac{{s_{uo}}^{\alpha}}{\alpha} + (1-2\delta)\cdot imp_{l_{g_u}}\right) \tag{24}$$

(by using Equations (1) and (24))

$\frac{dG_{ua}}{dt} + \gamma\cdot\frac{d\Phi_u}{dt} \le imp_{ua} + {s_{ua}}^{\alpha} + \gamma\cdot\left(\frac{{s_{uo}}^{\alpha}}{\alpha} + (1-2\delta)\cdot imp_{l_{g_u}}\right)$
$= imp_{ua} + (1+\Delta/3m)^{\alpha}\cdot imp_{ua} + \frac{\gamma}{\alpha}\cdot {s_{uo}}^{\alpha} + \gamma\cdot(1-2\delta)\cdot imp_{l_{g_u}}$
$\le \frac{\gamma}{\alpha}\cdot {s_{uo}}^{\alpha} + imp_{ua} + (1+\Delta/3m)^{\alpha}\cdot imp_{ua} + \gamma\cdot(1-2\delta)\cdot imp_{ua}$
$= \frac{\gamma}{\alpha}\cdot {s_{uo}}^{\alpha} + imp_{ua}\cdot\left(1 + (1+\Delta/3m)^{\alpha} + \gamma\cdot(1-2\delta)\right)$
$\le \frac{\gamma}{\alpha}\cdot {s_{uo}}^{\alpha} + \frac{imp_{uo}}{\mu}\cdot\left(1 + (1+\Delta/3m)^{\alpha} + \gamma\cdot(1-2\delta)\right)$ (by using Equation (17))
$\le \frac{\gamma}{\alpha}\cdot {s_{uo}}^{\alpha} + \left(\frac{1}{\mu}\cdot\left(\left(1 + (1+\Delta/3m)^{\alpha}\right) + \gamma\right)\right)\cdot imp_{uo}$
$= \frac{\gamma}{\alpha}\cdot {s_{uo}}^{\alpha} + \left(\frac{1}{\mu}\cdot\left(\left(1 + (1+\Delta/3m)^{\alpha}\right) + \frac{1}{16}\cdot\left(1 + (1+\Delta/3m)^{\alpha}\right)\right)\right)\cdot imp_{uo}$ (by using Equation (19))
$= \frac{\gamma}{\alpha}\cdot {s_{uo}}^{\alpha} + \left(\frac{17}{16}\cdot\left(1+\frac{\Delta}{3}\right)\cdot\left(1 + (1+\Delta/3m)^{\alpha}\right)\right)\cdot imp_{uo}$ (by using Equation (18))
$\le \frac{\gamma}{\alpha}\cdot {s_{uo}}^{\alpha} + \left(\left(\frac{9}{8}+\frac{3\Delta}{8}\right)\cdot\left(1 + (1+\Delta/3m)^{\alpha}\right)\right)\cdot imp_{uo}$
$\le c\cdot {s_{uo}}^{\alpha} + c\cdot imp_{uo}$ (by using Equations (18) and (23))
$\frac{dG_{ua}}{dt} + \gamma\cdot\frac{d\Phi_u}{dt} \le c\cdot({s_{uo}}^{\alpha} + w_{uo})$
$\frac{dG_{ua}}{dt} + \gamma\cdot\frac{d\Phi_u}{dt} \le c\cdot\frac{dG_{uo}}{dt}$

Hence the running condition is satisfied for $n_a \le m$, $imp_{ua} \le \eta^{\alpha}$, $imp_{l_{g_u}} \le \eta^{\alpha}$, $imp_{l_{g_u}} \le \left(imp_{ua} - \left(\frac{3}{3+\Delta}\right)\cdot imp_{ua}\right)$, $c = \left(\frac{9}{8}+\frac{3\Delta}{8}\right)\cdot\left(1 + (1+\Delta/3m)^{\alpha}\right)$.

Case II: When $n_a \le m$, $imp_{ua} > \eta^{\alpha}$, $imp_{l_{g_u}} \le \eta^{\alpha}$, and $s_{ua}(t) = (1+\Delta/3m)\cdot min\left({imp_{ua}}^{1/\alpha}, \eta\right) = (1+\Delta/3m)\eta$.

(a) If $imp_{l_{g_u}} > \left(imp_{ua} - \left(\frac{3}{3+\Delta}\right)\cdot imp_{ua}\right)$ then the total RoC of Φ because of Opt and MBS is $\frac{d\Phi_u}{dt} = \frac{d\Phi_{uo}}{dt} + \frac{d\Phi_{ua}}{dt}$.

(by using Equations (8) and (9))

$$\frac{d\Phi_u}{dt} \le \left(\frac{{s_{uo}}^{\alpha}}{\alpha} + (1-2\delta)\cdot imp_{l_{g_u}}\right) - \left(s_{ua}\cdot {imp_{l_{g_u}}}^{1-2\delta}\right) \tag{25}$$

(by using Equations (1) and (25))

$\frac{dG_{ua}}{dt} + \gamma\cdot\frac{d\Phi_u}{dt}$
$\le imp_{ua} + {s_{ua}}^{\alpha} + \gamma\cdot\left(\left(\frac{{s_{uo}}^{\alpha}}{\alpha} + (1-2\delta)\cdot imp_{l_{g_u}}\right) - \left(s_{ua}\cdot {imp_{l_{g_u}}}^{1-2\delta}\right)\right)$
$= imp_{ua} + (1+\Delta/3m)^{\alpha}\cdot\eta^{\alpha} + \frac{\gamma}{\alpha}\cdot {s_{uo}}^{\alpha} + \gamma\cdot(1-2\delta)\cdot imp_{l_{g_u}} - \gamma\cdot(1+\Delta/3m)\cdot\eta\cdot {imp_{l_{g_u}}}^{1-2\delta}$
$\le imp_{ua} + (1+\Delta/3m)^{\alpha}\cdot imp_{ua} + \frac{\gamma}{\alpha}\cdot {s_{uo}}^{\alpha} + \gamma\cdot(1-2\delta)\cdot imp_{ua} - \gamma\cdot(1+\Delta/3m)\cdot {imp_{l_{g_u}}}^{2\delta}\cdot {imp_{l_{g_u}}}^{1-2\delta}$
$= \frac{\gamma}{\alpha}\cdot {s_{uo}}^{\alpha} + \left(1 + (1+\Delta/3m)^{\alpha} + \gamma\cdot(1-2\delta)\right)\cdot imp_{ua} - \gamma\cdot(1+\Delta/3m)\cdot imp_{l_{g_u}}$
$\le \frac{\gamma}{\alpha}\cdot {s_{uo}}^{\alpha} + \left(1 + (1+\Delta/3m)^{\alpha} + \gamma\right)\cdot imp_{ua} - \gamma\cdot(1+\Delta/3m)\cdot\left(1 - \left(\frac{3}{3+\Delta}\right)\right)\cdot imp_{ua}$
$\le \frac{\gamma}{\alpha}\cdot {s_{uo}}^{\alpha} + \left(1 + (1+\Delta/3m)^{\alpha} + \gamma - \gamma\cdot\left(\frac{\Delta}{3+\Delta}\right)\right)\cdot imp_{ua}$
$= \frac{\gamma}{\alpha}\cdot {s_{uo}}^{\alpha} + \left(1 + (1+\Delta/3m)^{\alpha} + \gamma\cdot\left(\frac{3}{3+\Delta}\right)\right)\cdot imp_{ua}$
$\le \frac{\gamma}{\alpha}\cdot {s_{uo}}^{\alpha} + \left(1 + (1+\Delta/3m)^{\alpha} + \gamma\right)\cdot imp_{ua}$
$\le \frac{\gamma}{\alpha}\cdot {s_{uo}}^{\alpha} + \left(1 + (1+\Delta/3m)^{\alpha} + \frac{1}{16}\cdot\left(1 + (1+\Delta/3m)^{\alpha}\right)\right)\cdot\frac{imp_{uo}}{\mu}$ (by using Equations (17) and (19))
$= \frac{\gamma}{\alpha}\cdot {s_{uo}}^{\alpha} + \left(\frac{17}{16}\cdot\left(1+\frac{\Delta}{3}\right)\cdot\left(1 + (1+\Delta/3m)^{\alpha}\right)\right)\cdot imp_{uo}$ (by using Equation (18))
$\le \frac{\gamma}{\alpha}\cdot {s_{uo}}^{\alpha} + \left(\left(\frac{9}{8}+\frac{3\Delta}{8}\right)\cdot\left(1 + (1+\Delta/3m)^{\alpha}\right)\right)\cdot imp_{uo}$ (by using Equations (20) and (23))
$\le c\cdot {s_{uo}}^{\alpha} + c\cdot imp_{uo}$
$= c\cdot({s_{uo}}^{\alpha} + imp_{uo})$
$\frac{dG_{ua}}{dt} + \gamma\cdot\frac{d\Phi_u}{dt} \le c\cdot\frac{dG_{uo}}{dt}$

Hence the running condition is fulfilled for $n_a \le m$, $imp_{ua} > \eta^{\alpha}$, $imp_{l_{g_u}} \le \eta^{\alpha}$, $imp_{l_{g_u}} > \left(imp_{ua} - \left(\frac{3}{3+\Delta}\right)\cdot imp_{ua}\right)$, $c = \left(\frac{9}{8}+\frac{3\Delta}{8}\right)\cdot\left(1 + (1+\Delta/3m)^{\alpha}\right)$.

(b) If $imp_{lg_u} \leq \left(imp_{ua} - \left(\frac{3}{3+\Delta}\right)\cdot imp_{ua}\right)$ then the total RoC of Φ because of Opt and MBS depends on $\frac{d\Phi_{uo}}{dt}$ since $\frac{d\Phi_{ua}}{dt} \leq 0$. (by using Equation (7))

$$\frac{d\Phi_u}{dt} \leq \left(\frac{s_{uo}{}^{\alpha}}{\alpha} + (1-2\delta)\cdot imp_{lg_u}\right) \tag{26}$$

(by using Equations (1) and (26))

$\frac{dG_{ua}}{dt} + \gamma\cdot\frac{d\Phi_u}{dt} \leq imp_{ua} + s_{ua}{}^{\alpha} + \gamma\cdot\left(\frac{s_{uo}{}^{\alpha}}{\alpha} + (1-2\delta)\cdot imp_{lg_u}\right)$
$= imp_{ua} + (1+\Delta/3m)^{\alpha}\cdot\eta^{\alpha} + \frac{\gamma}{\alpha}\cdot s_{uo}{}^{\alpha} + \gamma\cdot(1-2\delta)\cdot imp_{lg_u}$
$\leq imp_{ua} + (1+\Delta/3m)^{\alpha}\cdot imp_{ua} + \frac{\gamma}{\alpha}\cdot s_{uo}{}^{\alpha} + \gamma\cdot(1-2\delta)\cdot imp_{lg_u}$
$\leq \frac{\gamma}{\alpha}\cdot s_{uo}{}^{\alpha} + imp_{ua} + (1+\Delta/3m)^{\alpha}\cdot imp_{ua} + \gamma\cdot(1-2\delta)\cdot imp_{ua}$
$= \frac{\gamma}{\alpha}\cdot s_{uo}{}^{\alpha} + imp_{ua}\cdot\left(1 + (1+\Delta/3m)^{\alpha} + \gamma\cdot(1-2\delta)\right)$
$\leq \frac{\gamma}{\alpha}\cdot s_{uo}{}^{\alpha} + \frac{imp_{uo}}{\mu}\cdot\left(1 + (1+\Delta/3m)^{\alpha} + \gamma\right)$ (by using Equation (17))
$= \frac{\gamma}{\alpha}\cdot s_{uo}{}^{\alpha} + \left(1 + (1+\Delta/3m)^{\alpha} + \frac{1}{16}\cdot\left(1 + (1+\Delta/3m)^{\alpha}\right)\right)\cdot\frac{imp_{uo}}{\mu}$ (by using Equation (19))
$= \frac{\gamma}{\alpha}\cdot s_{uo}{}^{\alpha} + \left(\frac{17}{16}\cdot\left(1+\frac{\Delta}{3}\right)\cdot\left(1 + (1+\Delta/3m)^{\alpha}\right)\right)\cdot imp_{uo}$ (by using Equation (18))
$\leq \frac{\gamma}{\alpha}\cdot s_{uo}{}^{\alpha} + \left(\left(\frac{9}{8} + \frac{3\Delta}{8}\right)\cdot\left(1 + (1+\Delta/3m)^{\alpha}\right)\right)\cdot imp_{uo}$
$\leq c\cdot s_{uo}{}^{\alpha} + c\cdot imp_{uo}$ (by using Equations (20) and (23))
$= c\cdot(s_{uo}{}^{\alpha} + imp_{uo})$
$\frac{dG_{ua}}{dt} + \gamma\cdot\frac{d\Phi_u}{dt} \leq c\cdot\frac{dG_{uo}}{dt}$

Hence the running condition is satisfied for $n_a \leq m$, $imp_{ua} > \eta^{\alpha}$, $imp_{lg_u} \leq \eta^{\alpha}$, $imp_{lg_u} \leq \left(imp_{ua} - \left(\frac{3}{3+\Delta}\right)\cdot imp_{ua}\right)$, $c = \left(\frac{9}{8} + \frac{3\Delta}{8}\right)\cdot\left(1 + (1+\Delta/3m)^{\alpha}\right)$.

Case III: When $n_a \leq m$, $imp_{ua} > \eta^{\alpha}$, $imp_{lg_u} > \eta^{\alpha}$, and $s_{ua}(t) = (1+\Delta/3m)\cdot min\left(imp_{ua}{}^{1/\alpha}, \eta\right) = (1+\Delta/3m)\eta$.

(a) If $imp_{lg_u} > \left(imp_{ua} - \left(\frac{3}{3+\Delta}\right)\cdot imp_{ua}\right)$ then the total RoC of Φ because of Opt and MBS is $\frac{d\Phi_u}{dt} = \frac{d\Phi_{uo}}{dt} + \frac{d\Phi_{ua}}{dt}$.

(by using Equations (10) and (11))

$$\frac{d\Phi_u}{dt} \leq \left(\frac{1}{1-\delta}\right)\cdot imp_{lg_u} - \frac{(1+\Delta/3m)}{(1-\delta)}\cdot imp_{lg_u} \tag{27}$$

(by using Equations (1) and (27))

$\frac{dG_{ua}}{dt} + \gamma\cdot\frac{d\Phi_u}{dt}$
$\leq imp_{ua} + s_{ua}{}^{\alpha} + \gamma\cdot\left(\left(\frac{1}{1-\delta}\right)\cdot imp_{lg_u} - \frac{(1+\Delta/3m)}{(1-\delta)}\cdot imp_{lg_u}\right)$
$= imp_{ua} + (1+\Delta/3m)^{\alpha}\cdot\eta^{\alpha} + \gamma\cdot\left(\frac{1}{1-\delta}\right)\cdot imp_{lg_u} - \gamma\cdot\frac{(1+\Delta/3m)}{(1-\delta)}\cdot imp_{lg_u}$
$\leq imp_{ua} + (1+\Delta/3m)^{\alpha}\cdot imp_{ua} + \gamma\cdot\left(\frac{1}{1-\delta}\right)\cdot imp_{ua} - \gamma\cdot\frac{(1+\Delta/3m)}{(1-\delta)}\cdot\left(imp_{ua} - \left(\frac{3}{3+\Delta}\right)\cdot imp_{ua}\right)$
$= imp_{ua}\cdot\left(1 + (1+\Delta/3m)^{\alpha} + \gamma\cdot\left(\frac{1}{1-\delta}\right) - \gamma\cdot\left(\frac{1}{1-\delta}\right)\cdot(1+\Delta/3m)\cdot\left(\frac{\Delta}{3+\Delta}\right)\right)$
$\leq imp_{ua}\cdot\left(1 + (1+\Delta/3m)^{\alpha} + \gamma\cdot\left(\frac{1}{1-\delta}\right) - \gamma\cdot\left(\frac{1}{1-\delta}\right)\cdot\left(\frac{\Delta}{3+\Delta}\right)\right)$
$= imp_{ua}\cdot\left(1 + (1+\Delta/3m)^{\alpha} + \gamma\cdot\left(\frac{1}{1-\delta}\right)\cdot\left(\frac{3}{3+\Delta}\right)\right)$
$\leq \frac{imp_{uo}}{\mu}\cdot\left(1 + (1+\Delta/3m)^{\alpha} + \frac{1}{16}\cdot\left(1 + (1+\Delta/3m)^{\alpha}\right)\cdot\left(\frac{1}{1-\delta}\right)\right)$ (by using Equations (17) and (19))

$$\frac{dG_{ua}}{dt} + \gamma\cdot\frac{d\Phi_u}{dt} \leq \frac{imp_{uo}}{\mu}\cdot\left(\left(1 + (1+\Delta/3m)^{\alpha}\right) + \frac{1}{16}\cdot\left(1 + (1+\Delta/3m)^{\alpha}\right)\cdot\left(\frac{2\alpha}{2\alpha-1}\right)\right) \tag{28}$$

$$\text{Since } \alpha > 1 \Rightarrow \frac{2\alpha}{2\alpha-1} = \frac{2\alpha-1+1}{2\alpha-1} = 1 + \frac{1}{2\alpha-1} < 2 \Rightarrow 1 < \left(\frac{2\alpha}{2\alpha-1}\right) < 2 \tag{29}$$

(by using Equations (29) and (28))

$$\begin{aligned}\frac{dG_{ua}}{dt} + \gamma\cdot\frac{d\Phi_u}{dt} &\le \frac{imp_{uo}}{\mu}\cdot\left(\left(1 + (1 + \Delta/3m)^{\alpha}\right) + \frac{2}{16}\cdot\left(1 + (1 + \Delta/3m)^{\alpha}\right)\right)\\ &= \left(\frac{9}{8} + \frac{3\Delta}{8}\right)\cdot\left(1 + (1 + \Delta/3m)^{\alpha}\right)\cdot imp_{uo} \text{ (by using Equation (18))}\\ &= c\cdot imp_{uo} \text{ (by using Equation (20))}\\ &\le c\cdot({s_{uo}}^{\alpha} + imp_{uo})\\ \frac{dG_{ua}}{dt} + \gamma\cdot\frac{d\Phi_u}{dt} &\le c\cdot\frac{dG_{uo}}{dt}\end{aligned}$$

Hence the running condition is fulfilled for $n_a \le m$, $imp_{ua} > \eta^{\alpha}$, $imp_{lg_u} > \eta^{\alpha}$, $imp_{lg_u} > \left(imp_{ua} - \left(\frac{3}{3+\Delta}\right)\cdot imp_{ua}\right)$, $c = \left(\frac{9}{8} + \frac{3\Delta}{8}\right)\cdot\left(1 + (1 + \Delta/3m)^{\alpha}\right)$.

(b) If $imp_{lg_u} \le \left(imp_{ua} - \left(\frac{3}{3+\Delta}\right)\cdot imp_{ua}\right)$ then the total RoC of Φ because of Opt and MBS depends on $\frac{d\Phi_{uo}}{dt}$ since $\frac{d\Phi_{ua}}{dt} \le 0$.

(by using Equation (10))

$$\frac{d\Phi_u}{dt} \le \left(\frac{1}{1-\delta}\right)\cdot imp_{lg_u} \tag{30}$$

(by using Equations (1) and (30))

$$\begin{aligned}\frac{dG_{ua}}{dt} + \gamma\cdot\frac{d\Phi_u}{dt} &\le imp_{ua} + {s_{ua}}^{\alpha} + \gamma\cdot\left(\frac{1}{1-\delta}\right)\cdot imp_{lg_u}\\ &= imp_{ua} + (1 + \Delta/3m)^{\alpha}\cdot\eta^{\alpha} + \gamma\cdot\left(\frac{1}{1-\delta}\right)\cdot imp_{lg_u}\\ &\le imp_{ua} + (1 + \Delta/3m)^{\alpha}\cdot imp_{ua} + \gamma\cdot\left(\frac{1}{1-\delta}\right)\cdot imp_{ua}\\ &= imp_{ua}\cdot\left(1 + (1 + \Delta/3m)^{\alpha} + \gamma\cdot\left(\frac{2\alpha}{2\alpha-1}\right)\right)\\ &\le \frac{imp_{uo}}{\mu}\cdot\left(1 + (1 + \Delta/3m)^{\alpha} + 2\gamma\right) \text{ (by using Equations (17) and (29))}\\ &= \frac{imp_{uo}}{\mu}\cdot\left(\left(1 + (1 + \Delta/3m)^{\alpha}\right) + \frac{2}{16}\cdot\left(1 + (1 + \Delta/3m)^{\alpha}\right)\right) \text{ (by using Equation (19))}\\ &= \left(\frac{9}{8} + \frac{3\Delta}{8}\right)\cdot\left(1 + (1 + \Delta/3m)^{\alpha}\right)\cdot imp_{uo} \text{ (by using Equation (18))}\\ &= c\cdot imp_{uo} \text{ (by using Equation (20))}\\ &\le c\cdot({s_{uo}}^{\alpha} + imp_{uo})\\ \frac{dG_{ua}}{dt} + \gamma\cdot\frac{d\Phi_u}{dt} &\le c\cdot\frac{dG_{uo}}{dt}\end{aligned}$$

Hence the running condition is satisfied if $n_a \le m$, $imp_{ua} > \eta^{\alpha}$, $imp_{lg_u} > \eta^{\alpha}$, $imp_{lg_u} \le \left(imp_{ua} - \left(\frac{3}{3+\Delta}\right)\cdot imp_{ua}\right)$, for $c = \left(\frac{9}{8} + \frac{3\Delta}{8}\right)\cdot\left(1 + (1 + \Delta/3m)^{\alpha}\right)$.

Case IV: When $n_a > m$ and $imp_{ua} \le \eta^{\alpha}$, since $imp_{lg_u} \le imp_{ua}$ we have $imp_{lg_u} \le \eta^{\alpha}$ and $s_{ua}(t) = (1 + \Delta/3m)\cdot min\left({imp_{ua}}^{1/\alpha}, \eta\right) = (1 + \Delta/3m){imp_{ua}}^{1/\alpha}$.

If $imp_{lg_u} > \left(imp_{ua} - \left(\frac{3}{3+\Delta}\right)\cdot imp_{ua}\right)$ then total RoC of Φ because of Opt and MBS is $\frac{d\Phi_u}{dt} = \frac{d\Phi_{uo}}{dt} + \frac{d\Phi_{ua}}{dt}$.

(by using Equations (13) and (14))

$$\frac{d\Phi_u}{dt} \le \left(\frac{{s_{uo}}^{\alpha}}{\alpha} + (1-2\delta)\cdot imp_{lg_u}\right) - \left(\frac{s_{ua}}{(2-2\delta)}\cdot\left(\frac{{imp_{lg_u}}^{2-2\delta}}{imp_{ua}}\right)\right) \tag{31}$$

(by using Equations (1) and (31))

$$\begin{aligned}&\frac{dG_{ua}}{dt} + \gamma\cdot\frac{d\Phi_u}{dt}\\ &\le imp_{ua} + {s_{ua}}^{\alpha} + \gamma\cdot\left(\left(\frac{{s_{uo}}^{\alpha}}{\alpha} + (1-2\delta)\cdot imp_{lg_u}\right) - \left(\frac{s_{ua}}{(2-2\delta)}\cdot\left(\frac{{imp_{lg_u}}^{2-2\delta}}{imp_{ua}}\right)\right)\right)\\ &= imp_{ua} + (1 + \Delta/3m)^{\alpha}\cdot imp_{ua} + \frac{\gamma}{\alpha}\cdot{s_{uo}}^{\alpha} + \gamma\cdot(1-2\delta)\cdot imp_{lg_u} - \gamma\cdot\frac{(1+\Delta/3m){imp_{ua}}^{1/\alpha}}{(2-2\delta)}\cdot\left(\frac{{imp_{lg_u}}^{2-2\delta}}{imp_{ua}}\right)\\ &\le \frac{\gamma}{\alpha}\cdot{s_{uo}}^{\alpha} + \left(1 + (1 + \Delta/3m)^{\alpha}\right)\cdot imp_{ua} + \gamma\cdot imp_{ua} - \gamma\cdot\frac{(1+\Delta/3m){imp_{ua}}^{1/\alpha}}{(2-2\delta)}\cdot\left(\frac{\left(imp_{ua} - \left(\frac{3}{3+\Delta}\right)\cdot imp_{ua}\right)^{2-2\delta}}{imp_{ua}}\right)\\ &= \frac{\gamma}{\alpha}\cdot{s_{uo}}^{\alpha} + imp_{ua}\cdot\left(1 + (1 + \Delta/3m)^{\alpha} + \gamma - \gamma\cdot\frac{(1+\Delta/3m)}{(2-2\delta)}\cdot\left(\frac{\Delta}{3+\Delta}\right)^{2-2\delta}\right)\end{aligned}$$

$$\leq \tfrac{\gamma}{\alpha}\cdot s_{uo}{}^{\alpha} + imp_{ua}\cdot\left(1 + (1+\Delta/3m)^{\alpha} + \gamma - \gamma\cdot\frac{\left(\frac{\Delta}{3+\Delta}\right)^{2-2\delta}}{(2-2\delta)}\right)$$

$$\frac{dG_{ua}}{dt} + \gamma\cdot\frac{d\Phi_u}{dt} \leq \frac{\gamma}{\alpha}\cdot s_{uo}{}^{\alpha} + imp_{ua}\cdot\left(1 + (1+\Delta/3m)^{\alpha} + \gamma - \gamma\cdot\frac{\left(\frac{\Delta}{3+\Delta}\right)^{2}}{(2-2\delta)}\right) \quad (32)$$

$$\text{Since } \alpha > 1 \Rightarrow \tfrac{2\alpha}{2\alpha-1} = \tfrac{2\alpha-1+1}{2\alpha-1} = 1 + \tfrac{1}{2\alpha-1} > 1 \Rightarrow \left(\tfrac{1}{2-2\delta}\right) = \tfrac{\alpha}{2\alpha-1} = \tfrac{1}{2}\cdot\left(\tfrac{2\alpha}{2\alpha-1}\right) > \tfrac{1}{2} \quad (33)$$

(by using Equations (32) and (33))

$$\begin{aligned}
\tfrac{dG_{ua}}{dt} + \gamma\cdot\tfrac{d\Phi_u}{dt} &\leq \tfrac{\gamma}{\alpha}\cdot s_{uo}{}^{\alpha} + imp_{ua}\cdot\left(1 + (1+\Delta/3m)^{\alpha} + \gamma - \tfrac{\gamma}{2}\cdot\left(\tfrac{\Delta}{3+\Delta}\right)^{2}\right)\\
&= \tfrac{\gamma}{\alpha}\cdot s_{uo}{}^{\alpha} + imp_{ua}\cdot\left(1 + (1+\Delta/3m)^{\alpha} + \gamma\left(1 - \tfrac{1}{2}\cdot\left(\tfrac{\Delta}{3+\Delta}\right)^{2}\right)\right)\\
&= \tfrac{\gamma}{\alpha}\cdot s_{uo}{}^{\alpha} + imp_{ua}\cdot\left(1 + (1+\Delta/3m)^{\alpha} + \gamma\left(\tfrac{2\Delta^2+11\Delta+18}{2\Delta^2+12\Delta+18}\right)\right)\\
&\leq \tfrac{\gamma}{\alpha}\cdot s_{uo}{}^{\alpha} + imp_{ua}\cdot\left(1 + (1+\Delta/3m)^{\alpha} + \gamma\right)
\end{aligned}$$

(by using Equations (17) and (19))

$$\begin{aligned}
&\leq \tfrac{\gamma}{\alpha}\cdot s_{uo}{}^{\alpha} + \tfrac{imp_{uo}}{\mu}\cdot\left(1 + (1+\Delta/3m)^{\alpha} + \tfrac{1}{16}\cdot\left(1 + (1+\Delta/3m)^{\alpha}\right)\right)\\
&= \tfrac{\gamma}{\alpha}\cdot s_{uo}{}^{\alpha} + \tfrac{imp_{uo}}{\mu}\cdot\left(\tfrac{17}{16}\cdot\left(1 + (1+\Delta/3m)^{\alpha}\right)\right)\\
&= \tfrac{\gamma}{\alpha}\cdot s_{uo}{}^{\alpha} + \left(\tfrac{17}{16}\cdot\left(1+\tfrac{\Delta}{3}\right)\cdot\left(1 + (1+\Delta/3m)^{\alpha}\right)\right)\cdot imp_{uo} \quad \text{(by using Equation (18))}\\
&\leq \tfrac{\gamma}{\alpha}\cdot s_{uo}{}^{\alpha} + \left(\left(\tfrac{9}{8}+\tfrac{3\Delta}{8}\right)\cdot\left(1 + (1+\Delta/3m)^{\alpha}\right)\right)\cdot imp_{uo}\\
&\leq c\cdot s_{uo}{}^{\alpha} + c\cdot imp_{uo} \quad \text{(by using Equations (20) and (23))}\\
&= c\cdot\left(s_{uo}{}^{\alpha} + imp_{uo}\right)
\end{aligned}$$

$$\tfrac{dG_{ua}}{dt} + \gamma\cdot\tfrac{d\Phi_u}{dt} \leq c\cdot\tfrac{dG_{uo}}{dt}$$

Hence the running condition is fulfilled for $n_a > m$, $imp_{ua} \leq \eta^{\alpha}$, $imp_{lg_u} \leq \eta^{\alpha}$, $imp_{lg_u} > \left(imp_{ua} - \left(\frac{3}{3+\Delta}\right)\cdot imp_{ua}\right)$, $c = \left(\frac{9}{8}+\frac{3\Delta}{8}\right)\cdot\left(1+(1+\Delta/3m)^{\alpha}\right)$.

(a) If $imp_{lg_u} \leq \left(imp_{ua} - \left(\frac{3}{3+\Delta}\right)\cdot imp_{ua}\right)$ then total RoC of Φ because of Opt and MBS depends on $\frac{d\Phi_{uo}}{dt}$ since $\frac{d\Phi_{ua}}{dt} \leq 0$.

(by using Equation (13))

$$\frac{d\Phi_u}{dt} \leq \left(\frac{s_{uo}{}^{\alpha}}{\alpha} + (1-2\delta)\cdot imp_{lg_u}\right) \quad (34)$$

(by using Equations (1) and (34))

$$\begin{aligned}
\tfrac{dG_{ua}}{dt} + \gamma\cdot\tfrac{d\Phi_u}{dt} &\leq imp_{ua} + s_{ua}{}^{\alpha} + \gamma\cdot\left(\tfrac{s_{uo}{}^{\alpha}}{\alpha} + (1-2\delta)\cdot imp_{lg_u}\right)\\
&= imp_{ua} + (1+\Delta/3m)^{\alpha}\cdot imp_{ua} + \tfrac{\gamma}{\alpha}\cdot s_{uo}{}^{\alpha} + \gamma\cdot(1-2\delta)\cdot imp_{lg_u}\\
&\leq \tfrac{\gamma}{\alpha}\cdot s_{uo}{}^{\alpha} + imp_{ua} + (1+\Delta/3m)^{\alpha}\cdot imp_{ua} + \gamma\cdot(1-2\delta)\cdot imp_{ua}\\
&= \tfrac{\gamma}{\alpha}\cdot s_{uo}{}^{\alpha} + imp_{ua}\cdot\left(1 + (1+\Delta/3m)^{\alpha} + \gamma\cdot(1-2\delta)\right)\\
&\leq \tfrac{\gamma}{\alpha}\cdot s_{uo}{}^{\alpha} + imp_{ua}\cdot\left(1 + (1+\Delta/3m)^{\alpha} + \gamma\right)
\end{aligned}$$

(by using Equations (17) and (19))

$$\begin{aligned}
&\leq \tfrac{\gamma}{\alpha}\cdot s_{uo}{}^{\alpha} + \tfrac{imp_{uo}}{\mu}\cdot\left(1 + (1+\Delta/3m)^{\alpha} + \tfrac{1}{16}\cdot\left(1 + (1+\Delta/3m)^{\alpha}\right)\right)\\
&= \tfrac{\gamma}{\alpha}\cdot s_{uo}{}^{\alpha} + \tfrac{imp_{uo}}{\mu}\cdot\left(\tfrac{17}{16}\cdot\left(1 + (1+\Delta/3m)^{\alpha}\right)\right)\\
&= \tfrac{\gamma}{\alpha}\cdot s_{uo}{}^{\alpha} + \left(\tfrac{17}{16}\cdot\left(1+\tfrac{\Delta}{3}\right)\cdot\left(1 + (1+\Delta/3m)^{\alpha}\right)\right)\cdot imp_{uo} \quad \text{(by using Equation (18))}\\
&\leq \tfrac{\gamma}{\alpha}\cdot s_{uo}{}^{\alpha} + \left(\left(\tfrac{9}{8}+\tfrac{3\Delta}{8}\right)\cdot\left(1 + (1+\Delta/3m)^{\alpha}\right)\right)\cdot imp_{uo}\\
&\leq c\cdot s_{uo}{}^{\alpha} + c\cdot imp_{uo} \quad \text{(by using Equations (20) and (23))}\\
&= c\cdot\left(s_{uo}{}^{\alpha} + imp_{uo}\right)
\end{aligned}$$

$$\tfrac{dG_{ua}}{dt} + \gamma\cdot\tfrac{d\Phi_u}{dt} \leq c\cdot\tfrac{dG_{uo}}{dt}$$

Hence the running condition is satisfied for $n_a > m$, $imp_{ua} \leq \eta^{\alpha}$, $imp_{lg_u} \leq \eta^{\alpha}$, $imp_{lg_u} \leq \left(imp_{ua} - \left(\frac{3}{3+\Delta}\right)\cdot imp_{ua}\right)$, $c = \left(\frac{9}{8}+\frac{3\Delta}{8}\right)\cdot\left(1+(1+\Delta/3m)^{\alpha}\right)$.

Case V: When $n_a > m$ and $imp_{ua} > \eta^{\alpha}$, $imp_{l_{g_u}} \leq \eta^{\alpha}$, and $s_{ua}(t) = (1+\Delta/3m)\cdot min\left(imp_{ua}{}^{1/\alpha},\ \eta\right) = (1+\Delta/3m)\eta$.

(a) If $imp_{l_{g_u}} > \left(imp_{ua} - \left(\frac{3}{3+\Delta}\right)\cdot imp_{ua}\right)$ then the total RoC of Φ because of Opt and MBS is $\frac{d\Phi_u}{dt} = \frac{d\Phi_{uo}}{dt} + \frac{d\Phi_{ua}}{dt}$.

(by using Equations (13) and (14))

$$\frac{d\Phi_u}{dt} \leq \left(\frac{s_{uo}{}^{\alpha}}{\alpha} + (1-2\delta)\cdot imp_{l_{g_u}}\right) - \left(\frac{s_{ua}}{(2-2\delta)}\cdot\left(\frac{imp_{l_{g_u}}{}^{2-2\delta}}{imp_{ua}}\right)\right) \tag{35}$$

(by using Equations (1) and (35))

$$\begin{aligned}
&\frac{dG_{ua}}{dt} + \gamma\cdot\frac{d\Phi_u}{dt} \\
&\leq imp_{ua} + s_{ua}{}^{\alpha} + \gamma\cdot\left(\left(\frac{s_{uo}{}^{\alpha}}{\alpha} + (1-2\delta)\cdot imp_{l_{g_u}}\right) - \left(\frac{s_{ua}}{(2-2\delta)}\cdot\left(\frac{imp_{l_{g_u}}{}^{2-2\delta}}{imp_{ua}}\right)\right)\right) \\
&= imp_{ua} + (1+\Delta/3m)^{\alpha}\cdot\eta^{\alpha} + \frac{\gamma}{\alpha}\cdot s_{uo}{}^{\alpha} + \gamma\cdot(1-2\delta)\cdot imp_{l_{g_u}} - \gamma\cdot\frac{(1+\Delta/3m)\eta}{(2-2\delta)}\cdot\left(\frac{imp_{l_{g_u}}{}^{2-2\delta}}{imp_{ua}}\right) \\
&\leq imp_{ua} + (1+\Delta/3m)^{\alpha}\cdot imp_{ua} + \frac{\gamma}{\alpha}\cdot s_{uo}{}^{\alpha} + \gamma\cdot(1-2\delta)\cdot imp_{ua} - \gamma\cdot\frac{(1+\Delta/3m)\eta}{(2-2\delta)}\cdot\left(\frac{imp_{l_{g_u}}{}^{2-2\delta}}{imp_{ua}}\right) \\
&\leq \frac{\gamma}{\alpha}\cdot s_{uo}{}^{\alpha} + imp_{ua}\cdot\left(1 + (1+\Delta/3m)^{\alpha} + \gamma\right) - \gamma\cdot\frac{(1+\Delta/3m)imp_{l_{g_u}}{}^{1/\alpha}}{(2-2\delta)}\cdot\left(\frac{imp_{l_{g_u}}{}^{2-2\delta}}{imp_{ua}}\right) \\
&\leq \frac{\gamma}{\alpha}\cdot s_{uo}{}^{\alpha} + imp_{ua}\cdot\left(1 + (1+\Delta/3m)^{\alpha} + \gamma\right) - \gamma\cdot\frac{1}{(2-2\delta)}\cdot\frac{\left(1-\left(\frac{3}{3+\Delta}\right)\right)^2\cdot imp_{ua}{}^2}{imp_{ua}}
\end{aligned}$$

$$\frac{dG_{ua}}{dt} + \gamma\cdot\frac{d\Phi_u}{dt} \leq \frac{\gamma}{\alpha}\cdot s_{uo}{}^{\alpha} + imp_{ua}\cdot\left(1 + (1+\Delta/3m)^{\alpha} + \gamma - \gamma\cdot\frac{1}{(2-2\delta)}\cdot\left(\frac{\Delta}{3+\Delta}\right)^2\right) \tag{36}$$

(by using Equations (36) and (33))

$$\begin{aligned}
\frac{dG_{ua}}{dt} + \gamma\cdot\frac{d\Phi_u}{dt} &\leq \frac{\gamma}{\alpha}\cdot s_{uo}{}^{\alpha} + imp_{ua}\cdot\left(1 + (1+\Delta/3m)^{\alpha} + \gamma - \frac{\gamma}{2}\cdot\left(\frac{\Delta}{3+\Delta}\right)^2\right) \\
&= \frac{\gamma}{\alpha}\cdot s_{uo}{}^{\alpha} + imp_{ua}\cdot\left(1 + (1+\Delta/3m)^{\alpha} + \gamma\left(1 - \frac{1}{2}\cdot\left(\frac{\Delta}{3+\Delta}\right)^2\right)\right) \\
&= \frac{\gamma}{\alpha}\cdot s_{uo}{}^{\alpha} + imp_{ua}\cdot\left(1 + (1+\Delta/3m)^{\alpha} + \gamma\left(\frac{2\Delta^2+11\Delta+18}{2\Delta^2+12\Delta+18}\right)\right) \\
&\leq \frac{\gamma}{\alpha}\cdot s_{uo}{}^{\alpha} + imp_{ua}\cdot\left(1 + (1+\Delta/3m)^{\alpha} + \gamma\right)
\end{aligned}$$

(by using Equations (17) and (19))

$$\begin{aligned}
&\leq \frac{\gamma}{\alpha}\cdot s_{uo}{}^{\alpha} + \frac{imp_{uo}}{\mu}\cdot\left(1 + (1+\Delta/3m)^{\alpha} + \frac{1}{16}\cdot\left(1 + (1+\Delta/3m)^{\alpha}\right)\right) \\
&= \frac{\gamma}{\alpha}\cdot s_{uo}{}^{\alpha} + \frac{imp_{uo}}{\mu}\cdot\left(\frac{17}{16}\cdot\left(1 + (1+\Delta/3m)^{\alpha}\right)\right) \\
&= \frac{\gamma}{\alpha}\cdot s_{uo}{}^{\alpha} + \left(\frac{17}{16}\cdot\left(1+\frac{\Delta}{3}\right)\cdot\left(1 + (1+\Delta/3m)^{\alpha}\right)\right)\cdot imp_{uo} \text{ (by using Equation (18))} \\
&\leq \frac{\gamma}{\alpha}\cdot s_{uo}{}^{\alpha} + \left(\left(\frac{9}{8}+\frac{3\Delta}{8}\right)\cdot\left(1 + (1+\Delta/3m)^{\alpha}\right)\right)\cdot imp_{uo} \\
&\leq c\cdot s_{uo}{}^{\alpha} + c\cdot imp_{uo} \text{ (by using Equations (20) and (23))} \\
&= c\cdot\left(s_{uo}{}^{\alpha} + imp_{uo}\right)
\end{aligned}$$

$$\frac{dG_{ua}}{dt} + \gamma\cdot\frac{d\Phi_u}{dt} \leq c\cdot\frac{dG_{uo}}{dt}$$

Hence the running condition is fulfilled for $n_a > m$, $imp_{ua} > \eta^{\alpha}$, $imp_{l_{g_u}} \leq \eta^{\alpha}$, $imp_{l_{g_u}} > \left(imp_{ua} - \left(\frac{3}{3+\Delta}\right)\cdot imp_{ua}\right)$, $c = \left(\frac{9}{8}+\frac{3\Delta}{8}\right)\cdot\left(1 + (1+\Delta/3m)^{\alpha}\right)$.

(a) If $imp_{l_{g_u}} \leq \left(imp_{ua} - \left(\frac{3}{3+\Delta}\right)\cdot imp_{ua}\right)$ then total RoC of Φ due to Opt and MBS depends on $\frac{d\Phi_{uo}}{dt}$ since $\frac{d\Phi_{ua}}{dt} \leq 0$.

(by using Equation (13))

$$\frac{d\Phi_u}{dt} \leq \left(\frac{s_{uo}{}^{\alpha}}{\alpha} + (1-2\delta)\cdot imp_{l_{g_u}}\right) \tag{37}$$

(by using Equations (1) and (37))

$$\begin{aligned}
\frac{dG_{ua}}{dt} + \gamma\cdot\frac{d\Phi_u}{dt} &\leq imp_{ua} + s_{ua}{}^{\alpha} + \gamma\cdot\left(\frac{s_{uo}{}^{\alpha}}{\alpha} + (1-2\delta)\cdot imp_{l_{g_u}}\right) \\
&= imp_{ua} + (1+\Delta/3m)^{\alpha}\cdot\eta^{\alpha} + \frac{\gamma}{\alpha}\cdot s_{uo}{}^{\alpha} + \gamma\cdot(1-2\delta)\cdot imp_{l_{g_u}} \\
&\leq \frac{\gamma}{\alpha}\cdot s_{uo}{}^{\alpha} + imp_{ua} + (1+\Delta/3m)^{\alpha}\cdot imp_{ua} + \gamma\cdot imp_{ua}
\end{aligned}$$

$= \frac{\gamma}{\alpha} \cdot s_{uo}{}^{\alpha} + imp_{ua} \cdot \left(1 + (1 + \Delta/3m)^{\alpha} + \gamma\right)$
(by using Equations (17) and (19))
$\leq \frac{\gamma}{\alpha} \cdot s_{uo}{}^{\alpha} + \frac{imp_{uo}}{\mu} \cdot \left(1 + (1 + \Delta/3m)^{\alpha} + \frac{1}{16} \cdot \left(1 + (1 + \Delta/3m)^{\alpha}\right)\right)$
$= \frac{\gamma}{\alpha} \cdot s_{uo}{}^{\alpha} + \frac{imp_{uo}}{\mu} \cdot \left(\frac{17}{16} \cdot \left(1 + (1 + \Delta/3m)^{\alpha}\right)\right)$
$= \frac{\gamma}{\alpha} \cdot s_{uo}{}^{\alpha} + \left(\frac{17}{16} \cdot \left(1 + \frac{\Delta}{3}\right) \cdot \left(1 + (1 + \Delta/3m)^{\alpha}\right)\right) \cdot imp_{uo}$ (by using Equation (18))
$\leq \frac{\gamma}{\alpha} \cdot s_{uo}{}^{\alpha} + \left(\left(\frac{9}{8} + \frac{3\Delta}{8}\right) \cdot \left(1 + (1 + \Delta/3m)^{\alpha}\right)\right) \cdot imp_{uo}$
$\leq c \cdot s_{uo}{}^{\alpha} + c \cdot imp_{uo}$ (by using Equations (20) and (23))
$= c \cdot \left(s_{uo}{}^{\alpha} + imp_{uo}\right)$
$\frac{dG_{ua}}{dt} + \gamma \cdot \frac{d\Phi_u}{dt} \leq c \cdot \frac{dG_{uo}}{dt}$

Hence the running condition is satisfied for $n_a > m$, $imp_{ua} \leq \eta^{\alpha}$, $imp_{lg_u} \leq \eta^{\alpha}$, $imp_{lg_u} \leq \left(imp_{ua} - \left(\frac{3}{3+\Delta}\right) \cdot imp_{ua}\right)$, $c = \left(\frac{9}{8} + \frac{3\Delta}{8}\right) \cdot \left(1 + (1 + \Delta/3m)^{\alpha}\right)$.

Case VI: When $n_a > m$ and $imp_{ua} > \eta^{\alpha}$, $imp_{lg_u} > \eta^{\alpha}$, and $s_{ua}(t) = (1 + \Delta/3m) \cdot min\left(imp_{ua}{}^{1/\alpha}, \eta\right) = (1 + \Delta/3m)\eta$.

(a) If $imp_{lg_u} > \left(imp_{ua} - \left(\frac{3}{3+\Delta}\right) \cdot imp_{ua}\right)$ then total RoC of Φ because of Opt and MBS is $\frac{d\Phi_u}{dt} = \frac{d\Phi_{uo}}{dt} + \frac{d\Phi_{ua}}{dt}$.

(by using Equations (15) and (16))

$$\frac{d\Phi_u}{dt} \leq \left(\left(\frac{1}{1-\delta}\right) \cdot imp_{lg_u}\right) - \left(\frac{(1 + \Delta/3m)}{(2 - 2\delta)} \cdot \left(\frac{imp_{lg_u}{}^2}{imp_{ua}}\right)\right) \quad (38)$$

(by using Equations (1) and (38))
$\frac{dG_{ua}}{dt} + \gamma \cdot \frac{d\Phi_u}{dt}$
$\leq imp_{ua} + s_{ua}{}^{\alpha} + \gamma \cdot \left(\left(\left(\frac{1}{1-\delta}\right) \cdot imp_{lg_u}\right) - \left(\frac{(1+\Delta/3m)}{(2-2\delta)} \cdot \left(\frac{imp_{lg_u}{}^2}{imp_{ua}}\right)\right)\right)$
$\leq imp_{ua} + (1 + \Delta/3m)^{\alpha} \cdot \eta^{\alpha} + \gamma \cdot \left(\frac{1}{1-\delta}\right) \cdot imp_{ua} - \gamma \cdot \frac{(1+\Delta/3m)}{(2-2\delta)} \cdot \left(\frac{imp_{lg_u}{}^2}{imp_{ua}}\right)$
$\leq imp_{ua} + (1 + \Delta/3m)^{\alpha} \cdot imp_{ua} + \gamma \cdot \left(\frac{1}{1-\delta}\right) \cdot imp_{ua} - \gamma \cdot \frac{1}{(2-2\delta)} \cdot \left(\frac{\left(imp_{ua} - \left(\frac{3}{3+\Delta}\right) \cdot imp_{ua}\right)^2}{imp_{ua}}\right)$
$= imp_{ua} + (1 + \Delta/3m)^{\alpha} \cdot imp_{ua} + \gamma \cdot \left(\frac{1}{1-\delta}\right) \cdot imp_{ua} - \gamma \cdot \frac{\left(\frac{\Delta}{3+\Delta}\right)^2}{(2-2\delta)} \cdot imp_{ua}$
$= \left(1 + (1 + \Delta/3m)^{\alpha} + \gamma \cdot \left(\frac{2\alpha}{2\alpha-1}\right) - \gamma \cdot \frac{\left(\frac{\Delta}{3+\Delta}\right)^2}{(2-2\delta)}\right) \cdot imp_{ua}$
$\leq \left(1 + (1 + \Delta/3m)^{\alpha} + 2\gamma - \frac{\gamma}{2} \cdot \left(\frac{\Delta}{3+\Delta}\right)^2\right) \cdot imp_{ua}$ (by using Equations (29) and (33))
$= \left(1 + (1 + \Delta/3m)^{\alpha} + \gamma \cdot \left(2 - \frac{1}{2} \cdot \left(\frac{\Delta}{3+\Delta}\right)^2\right)\right) \cdot imp_{ua}$
$= \left(1 + (1 + \Delta/3m)^{\alpha} + \gamma \cdot \left(1 + \frac{2\Delta^2+11\Delta+18}{2\Delta^2+12\Delta+18}\right)\right) \cdot imp_{ua}$
$\leq \left(1 + (1 + \Delta/3m)^{\alpha} + 2\gamma\right) \cdot imp_{ua}$
$\leq \frac{imp_{uo}}{\mu} \cdot \left(1 + (1 + \Delta/3m)^{\alpha} + \frac{2}{16} \cdot \left(1 + (1 + \Delta/3m)^{\alpha}\right)\right)$ (by using Equations (17) and (19))
$= \frac{imp_{uo}}{\mu} \cdot \left(\frac{18}{16} \cdot \left(1 + (1 + \Delta/3m)^{\alpha}\right)\right)$
$= \left(\frac{18}{16} \cdot \left(1 + \frac{\Delta}{3}\right) \cdot \left(1 + (1 + \Delta/3m)^{\alpha}\right)\right) \cdot imp_{uo}$ (by using Equation (18))
$= \left(\left(\frac{9}{8} + \frac{3\Delta}{8}\right) \cdot \left(1 + (1 + \Delta/3m)^{\alpha}\right)\right) \cdot imp_{uo}$
$= c \cdot imp_{uo}$ (by using Equation (20))
$\leq c \cdot \left(s_{uo}{}^{\alpha} + imp_{uo}\right)$
$\frac{dG_{ua}}{dt} + \gamma \cdot \frac{d\Phi_u}{dt} \leq c \cdot \frac{dG_{uo}}{dt}$

Hence the running condition is fulfilled for $n_a > m$, $imp_{ua} > \eta^{\alpha}$, $imp_{lg_u} > \eta^{\alpha}$, $imp_{lg_u} > \left(imp_{ua} - \left(\frac{3}{3+\Delta}\right) \cdot imp_{ua}\right)$, $c = \left(\frac{9}{8} + \frac{3\Delta}{8}\right) \cdot \left(1 + (1 + \Delta/3m)^{\alpha}\right)$.

(a) If $imp_{lg_u} \leq \left(imp_{ua} - \left(\frac{3}{3+\Delta}\right)\cdot imp_{ua}\right)$ then total RoC of Φ due to Opt and MBS depends on $\frac{d\Phi_{uo}}{dt}$ since $\frac{d\Phi_{ua}}{dt} \leq 0$.

(by using Equations (15))

$$\frac{d\Phi_u}{dt} \leq \left(\left(\frac{1}{1-\delta}\right)\cdot imp_{lg_u}\right) \tag{39}$$

(by using Equations (1) and (39))

$$\begin{aligned}
\frac{dG_{ua}}{dt} + \gamma\cdot\frac{d\Phi_u}{dt} &\leq imp_{ua} + s_{ua}{}^{\alpha} + \gamma\cdot\left(\left(\frac{1}{1-\delta}\right)\cdot imp_{lg_u}\right)\\
&= imp_{ua} + (1 + \Delta/3m)^{\alpha}\cdot\eta^{\alpha} + \gamma\cdot\left(\frac{1}{1-\delta}\right)\cdot imp_{lg_u}\\
&\leq imp_{ua} + (1 + \Delta/3m)^{\alpha}\cdot imp_{ua} + \gamma\cdot\left(\frac{1}{1-\delta}\right)\cdot imp_{ua}\\
&\leq imp_{ua}\left(1 + (1 + \Delta/3m)^{\alpha} + \gamma\cdot\left(\frac{1}{1-\delta}\right)\right)\\
&= imp_{ua}\cdot\left(1 + (1 + \Delta/3m)^{\alpha} + \gamma\cdot\left(\frac{2\alpha}{2\alpha-1}\right)\right)\\
&\leq \frac{imp_{uo}}{\mu}\cdot\left(1 + (1 + \Delta/3m)^{\alpha} + 2\gamma\right) \text{ (by using Equations (17) and (29))}\\
&= \frac{imp_{uo}}{\mu}\cdot\left(\left(1 + (1 + \Delta/3m)^{\alpha}\right) + \frac{2}{16}\cdot\left(1 + (1 + \Delta/3m)^{\alpha}\right)\right) \text{ (by using Equation (19))}\\
&= \left(\frac{9}{8} + \frac{3\Delta}{8}\right)\cdot\left(1 + (1 + \Delta/3m)^{\alpha}\right)\cdot imp_{uo} \text{ (by using Equation (18))}\\
&= c\cdot imp_{uo} \text{ (by using Equation (20))}\\
&\leq c\cdot(s_{uo}{}^{\alpha} + imp_{uo})
\end{aligned}$$

$$\frac{dG_{ua}}{dt} + \gamma\cdot\frac{d\Phi_u}{dt} \leq c\cdot\frac{dG_{uo}}{dt}$$

Hence the running condition is satisfied for $n_a > m$, $imp_{ua} > \eta^{\alpha}$, $imp_{lg_u} > \eta^{\alpha}$, $imp_{lg_u} \leq \left(imp_{ua} - \left(\frac{3}{3+\Delta}\right)\cdot imp_{ua}\right)$, $c = \left(\frac{9}{8} + \frac{3\Delta}{8}\right)\cdot\left(1 + (1 + \Delta/3m)^{\alpha}\right)$.

The analysis of all cases and sub cases in Lemma 6 prove that the first condition, running condition is fulfilled. Aggregating the discourse about all conditions job arrival and completion condition, boundary condition and Lemma 6, it is concluded that Theorem 1 follows. The competitive values of related algorithms and MBS on $\alpha = 2$ and 3 are shown in Table 3. Among all online clairvoyant and ON-C scheduling algorithms, the competitiveness of MBS is least, which reflects that the MBS outperforms other algorithms.

6. Illustrative Example

To observe the performance of MBS, a group of four processors and a set of seven jobs are considered. The best known result in the online non-clairvoyant scheduling algorithms is provided by the Azar et al. [40] in NC-PAR. NC-PAR is a super-constant lower bound on the competitive ratio of any deterministic algorithm even for fractional flow-time in the case of uniform densities. The processing of jobs using algorithms MBS and NC-PAR [40] is simulated and the results are stated in Table 4 as well as in Figures 3–11. The jobs arrived along with their importance but the size of jobs was computed on the completion of jobs. The response time (Rt) is the time interval between the starting time of execution and arrival time of a job. The turnaround time is the time duration between completion time and arrival time of a job. Most of the jobs using MBS have lesser turnaround time than using NC-PAR. The Rt of the jobs using MBS is better than NC-PAR. In Figures 3 and 4, the allocation and execution sequence of jobs on four processors is depicted with the help of triangles and rectangles using NC-PAR and MBS, respectively. As per the Figures 3 and 4, the importance of the jobs in NC-PAR increased with time where as in MBS the importance remains constant during the life time of the jobs. It is clearly evident from the Figures 3 and 4 that on any processor using NC-PAR at a time only one job has been executed, whereas using MBS the processor has been shared by more than one job. The hardware specifications are mentioned in the Table 5.

Table 4. Job details and execution data using MBS and NC-PAR.

Job Details				MBS [This Paper]			NC-PAR [40]		
Job	**Size**	**Importance**	**Arrival Time**	**Completion Time**	**Response Time**	**Turnaround Time**	**Completion Time**	**Response Time**	**Turnaround Time**
J1	35	8	1	14	0	13	14	0	13
J2	64	10	2	24	0	22	23	0	21
J3	15	5	4	10	0	6	12	0	8
J4	83	11	6	30	0	24	33	0	27
J5	45	5	7	29	0	22	29	6	22
J6	17	4	8	23	0	15	23	7	15
J7	56	6	10	32	0	22	43	14	33
Average Values				23.143	0	17.714	25.286	3.857	19.857

Table 5. Hardware specifications.

Simulation Parameters	Values
CPU	Intel(R) Core(TM) i5-4210U CPU @ 1.70 GHz
RAM	4.00 GB RAM
Hard Drive	1.0 TB
Operating System	Red Hat Linux 6.1
Kernel	Linux kernel version 2.2.12

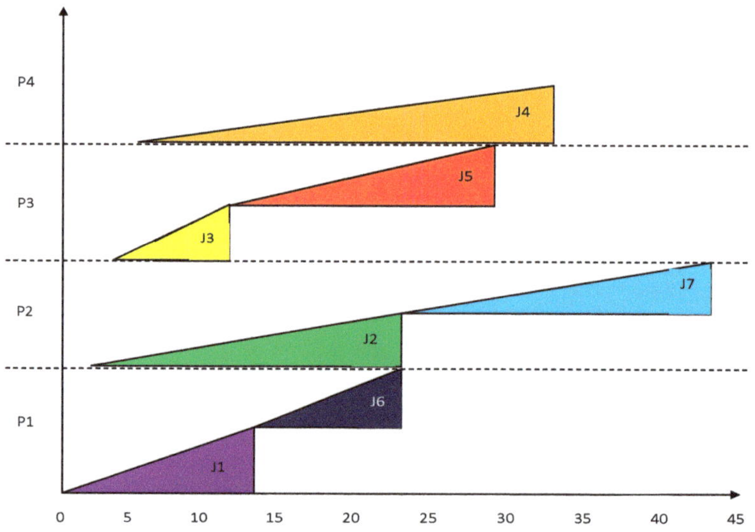

Figure 3. Scheduling of jobs using NC-PAR.

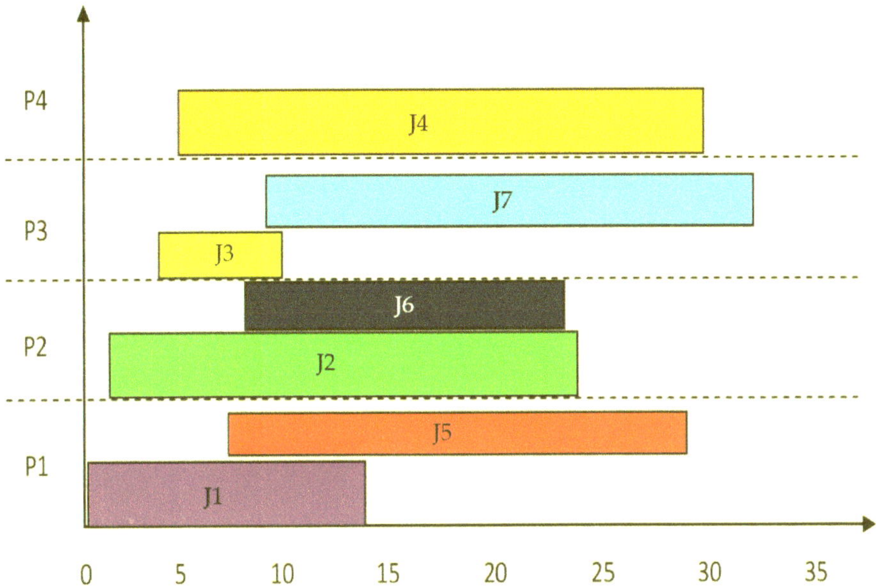

Figure 4. Scheduling of jobs using MBS.

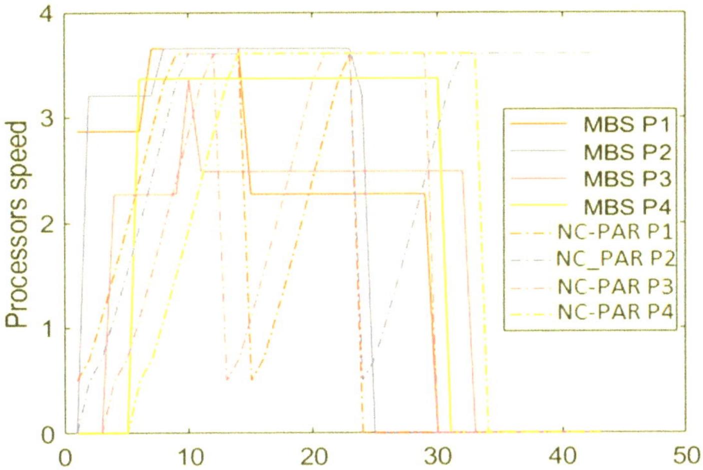

Figure 5. Speed of processors using MBS and NC-PAR.

Figures 5 and 6 present the speed of different processors and combined speed of all processors with respect to time using MBS and NC-PAR, respectively. As per the graphs of Figure 5, the speed of a processor using MBS goes high initially but later it reduces and most of the time the speed of processors using MBS is constant, but when processors executes jobs using NC-PAR the speed of processors have heavy fluctuations, which shows that some extra energy may be needed for such frequent fluctuation in NC-PAR. The graphs of the Figure 6 shows that the combined speed of processors using NC-PAR increased and decreased linearly whereas using MBS it increased and decreased stepwise. The count of local maxima and minima in the speed growth graphs (Figure 7) of NC-PAR is more than MBS. Therefore, not only individual processor's speed but also the combined speed of all the processors is reflecting the heavy fluctuation in NC-PAR and varying-constant mixed behaviour of MBS.

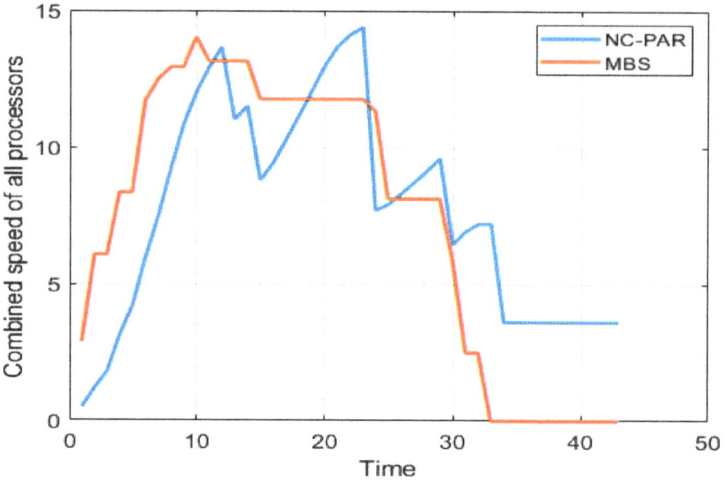

Figure 6. Combined speed of all processors using MBS and NC-PAR.

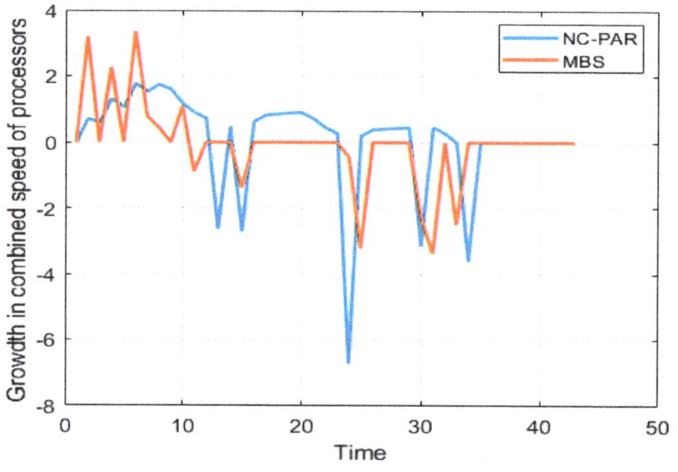

Figure 7. Growth of combined speed of all processors using MBS and NC-PAR.

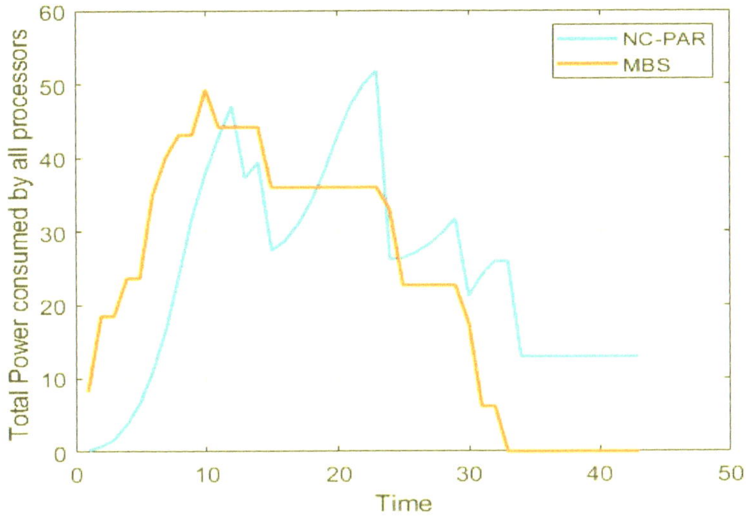

Figure 8. Total power consumed by all processors using MBS and NC-PAR.

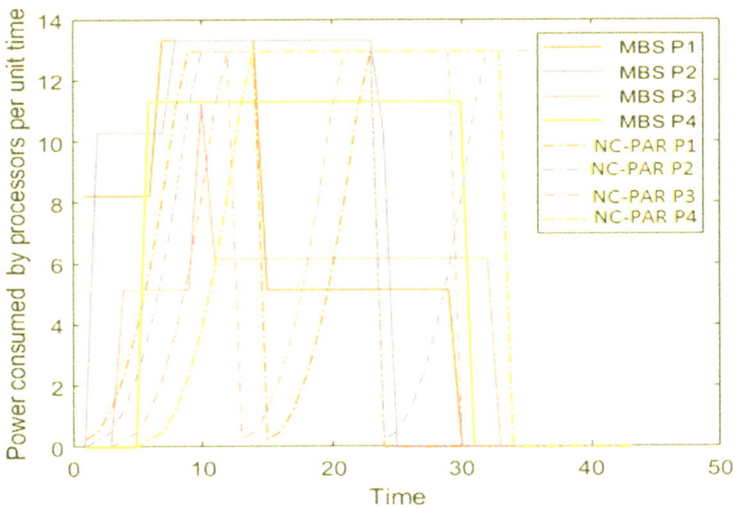

Figure 9. Power consumed by processors using MBS and NC-PAR.

In this simulation analysis the traditional power function is used and the value of α is 2. The processors are having the maximum limit of speed which is considered 3.6. The value of $\Delta = (3\alpha)^{-1}$ is considered for the analysis. The power consumed is square of the speed, i.e., proportional to the speed this fact can be viewed by comparing the graphs of Figures 5 and 9. Figure 8, shows that initially MBS consumed more power but power consumption decreased with respect to increase in time, whereas in case of NC-PAR there is no fix pattern, but power consumption is higher most of the time than in MBS.

The graphs of Figure 10 demonstrate the objective of the algorithm (important based flow time plus energy). It reveals that except one processor P1, all other processor have lesser objective value, when these processors executed jobs by using MBS than NC-PAR. The combined objective of all processor is given in the Figure 11, which strengthen the previous observation of Figure 10 (the objective values using MBS is lesser than using NC-PAR). It can be concluded from the different observations and the Figure 11, that the algorithm MBS performs better than NC-PAR.

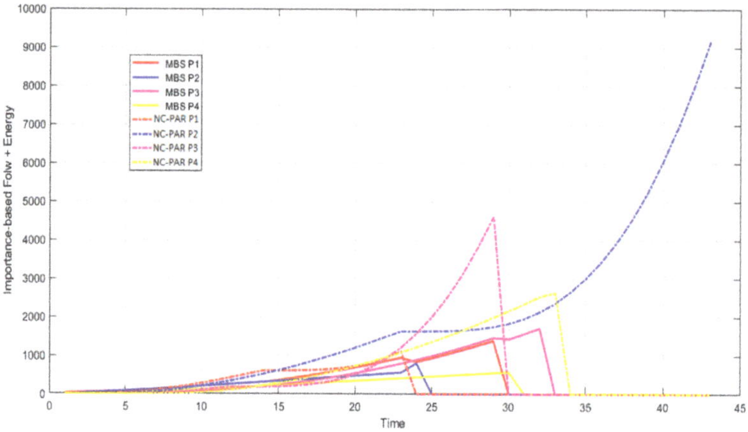

Figure 10. Importance-based flow time + energy consumed using MBS and NC-PAR.

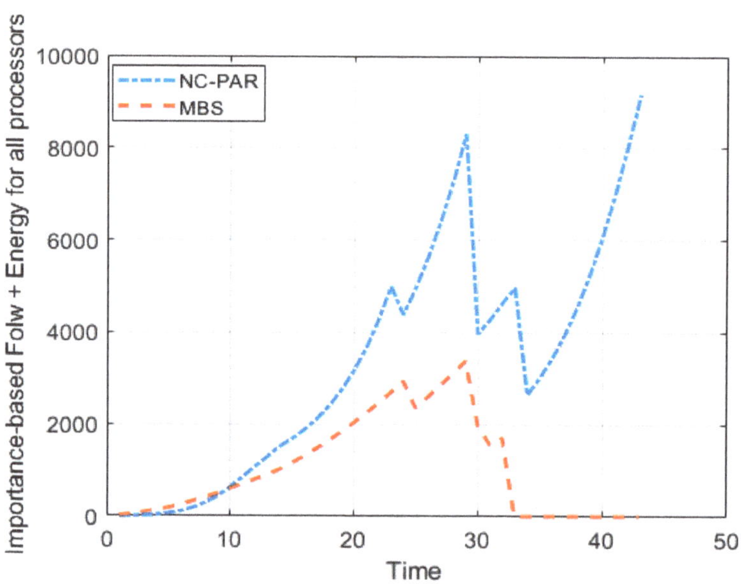

Figure 11. Total importance-based flow time + energy consumed using MBS and NC-PAR.

7. Conclusions and Future Work

To date, the problem of ON-C scheduling algorithms with an objective to minimize the IbFt+E for multiprocessor setting is studied less extensively. A scheduling algorithm multiprocessor with bounded speed (MBS) is proposed, which uses importance-based/weighted round robin (WRR) for job selection. MBS extends the theoretical study of an ON-C multiprocessor DSS scheduling problem with an objective to minimize the IbFt+E using the bounded speed model, where every processor's maximum speed using MBS is $(1 + \Delta/3m)\eta$ and using offline adversary Opt is η. The speed of any processor changes if there is a variation in the total importance of jobs on that processor. The competitiveness of MBS is $\left(\frac{9}{8} + \frac{3\Delta}{8}\right) \cdot \left(1 + \left(1 + \Delta/3m\Delta/3m\Delta/3m\Delta/3m\Delta/3m\Delta/3m\Delta/3m\Delta/3m\right)^{\alpha}\right) = O(1)$ against an offline adversary, using the potential function analysis and traditional power function. The performance of MBS is compared with best known algorithm NC-PAR [40]. A set of jobs and processors are used to simulate the working of MBS and NC-PAR. The average turnaround and response time of jobs, when they are executed by using MBS is lesser than NC-PAR. The speed scaling strategy and power consumption in MBS is better than NC-PAR. For all processors at any time, MBS provides the lesser value of the sum of important-based flow time and energy consumed than NC-PAR. Competitiveness of NC-PAR is 3 for $\alpha = 2$ and 3.5 for $\alpha = 3$, whereas the value of competitive ratio c of MBS for $\Delta = (3\alpha)^{-1}$, $m = 2$ and $\alpha = 2$ is 2.442; for $\Delta = (3\alpha)^{-1}$, $m = 2$ and $\alpha = 3$ is 2.399; for $\Delta = (3\alpha)^{-1}$, $m > 2$, $\alpha = 2$ is $2.375 < c < 2.442$; for $\Delta = (3\alpha)^{-1}$, $m > 2$, $\alpha = 3$ is $2.333 < c < 2.399$. These results demonstrate that the scheduling algorithm MBS outperforms other algorithms. The competitive value of MBS is least to date. Before these outcomes, there were no results acknowledged for the multi-processor machines in the ON-C model with identified importance, even for unit importance jobs [40]. The further enhancement of this study will be to implement the MBS in real environment. One open problem is to achieve a reasonably less competitive algorithm than MBS. In this study, author considers non-migratory and sequential jobs and this work may be extended to find a scheduling for migratory and non-sequential jobs. Other factors (such as memory requirement) may also be considered for analysis in future extension.

Author Contributions: All authors have worked on this manuscript together. Writing—original draft, P.S.; writing—review and editing, B.K., O.P.M., H.H.A. and G.H. All authors have read and agreed to the published version of the manuscript.

Funding: This research received no external funding.

Conflicts of Interest: The authors declare no conflicts of interest.

References

1. Belady, C. In the Data Center, Power and Cooling Costs More Than the IT Equipment It Supports, Electronics Cooling Magazine. Available online: http://www.electronics-cooling.com/2007/02/in-the-data-center-power-and-cooling-costs-more-than-the-it-equipment-it-supports/ (accessed on 10 January 2020).
2. Chan, H.L.; Edmonds, J.; Lam, T.W.; Lee, L.K.; Marchetti-Spaccamela, A.; Prush, K. Non-clairvoyant speed scaling for flow and energy. *Algorithmica* **2011**, *61*, 507–517. [CrossRef]
3. Merritt, R. CPU Designers' Debate Multi-Core Future, EE Times. 2 June 2008. Available online: http://www.eetimes.com/document.asp?doc_id=1167932 (accessed on 15 January 2020).
4. U.S. Environmental Protection Agency, EPA Report on Server and Data Center Energy Efficiency. Available online: https://www.energystar.gov/ia/partners/prod_development/downloads/EPA_Report_Exec_Summary_Final.pdf (accessed on 14 January 2020).
5. Singh, P.; Wolde-Gabriel, B. Executed-time Round Robin: EtRR an online non-clairvoyant scheduling on speed bounded processor with energy management. *J. King Saud Univ. Comput. Inf. Sci.* **2016**, *29*, 74–84. [CrossRef]
6. Bansal, N.; Chan, H.L.; Pruhs, K. Speed scaling with an arbitrary power function. In Proceedings of the Annual ACM-SIAM Symposium on Discrete Algorithms, New York, NY, USA, 4–6 January 2009; pp. 693–701.
7. Bansal, N.; Kimbrel, T.; Pruhs, K. Dynamic speed scaling to manage energy and temperature. *J. ACM* **2007**, *54*, 1–39. [CrossRef]
8. Singh, P.; Khan, B.; Vidyarthi, A.; Haes Alhelou, H.; Siano, P. Energy-aware online non-clairvoyant scheduling using speed scaling with arbitrary power function. *Appl. Sci.* **2019**, *9*, 1467. [CrossRef]
9. Lam, T.W.; Lee, L.K.; To, I.K.K.; Wong, P.W.H. Nonmigratory multiprocessor scheduling for response time and energy. *IEEE Trans. Parallel Distrib. Syst.* **2008**, *19*, 1–13.
10. Motwani, R.; Phillips, S.; Torng, E. Nonclairvoyant scheduling. *Theor. Comput. Sci.* **1994**, *30*, 17–47. [CrossRef]
11. Yao, F.; Demers, A.; Shenker, S. A scheduling model for reduced CPU energy. In Proceedings of the Annual Symposium on Foundations of Computer Science, Berkeley, CA, USA, 23–25 October 1995; pp. 374–382.

12. Koren, G.; Shasha, D. Dover: An optimal on-line scheduling algorithm for overloaded uniprocessor real-time systems. *SIAM J. Comput.* **1995**, *24*, 318–339. [CrossRef]
13. Leonardi, S.; Raz, D. Approximating total flow time on parallel machines. In Proceedings of the ACM Symposium on Theory of Computing, El Paso, TX, USA, 4–6 May 1997; pp. 110–119.
14. Kalyanasundaram, B.; Pruhs, K. Speed is as powerful as clairvoyant. *J. ACM* **2000**, *47*, 617–643. [CrossRef]
15. Edmonds, J. Scheduling in the dark. *Theor. Comput. Sci.* **2000**, *235*, 109–141. [CrossRef]
16. Kalyanasundaram, B.; Pruhs, K. Minimizing flow time nonclairvoyantly. *J. ACM* **2003**, *50*, 551–567. [CrossRef]
17. Becchetti, L.; Leonardi, S. Nonclairvoyant scheduling to minimize the total flow time on single and parallel machines. *J. ACM* **2004**, *51*, 517–539. [CrossRef]
18. Chekuri, C.; Goel, A.; Khanna, S.; Kumar, A. Multiprocessor scheduling to minimize flow time with epsilon resource augmentation. In Proceedings of the 36th Annual ACM Symposium on Theory of Computing, Chicago, IL, USA, 13–15 June 2004; pp. 363–372.
19. Chen, J.J.; Hsu, H.R.; Chuang, K.H.; Yang, C.L.; Pang, A.C.; Kuo, T.W. Multiprocessor energy efficient scheduling with task migration considerations. In Proceedings of the 16th Euromicro Conference on Real-Time Systems, Catania, Italy, 2 July 2004; pp. 101–108.
20. Albers, S.; Fujiwara, H. Energy-efficient algorithms for flow time minimization. *ACM Trans. Algorithms* **2007**, *3*, 49. [CrossRef]
21. Bansal, N.; Pruhs, K.; Stein, C. Speed scaling for weighted flow time. In Proceedings of the 18th Annual ACM-SIAM Symposium on Discrete Algorithms, New Orleans, LA, USA, 7–9 January 2007; pp. 805–813.
22. Lam, T.W.; Lee, L.K.; To, I.K.K.; Wong, P.W.H. Competitive non-migratory scheduling for flow time and energy. In Proceedings of the 20th ACM Symposium on Parallelism in Algorithms and Architectures, Munich, Germany, 14–16 June 2008; pp. 256–264.
23. Chadha, J.; Garg, N.; Kumar, A.; Muralidhara, V. A competitive algorithm for minimizing weighted flow time on unrelated processors with speed augmentation. In Proceedings of the Annual ACM Symposium on Theory of Computing, Bethesda, MD, USA, 31 May–2 June 2009; pp. 679–684.
24. Chan, S.H.; Lam, T.W.; Lee, L.K.; Liu, C.M.; Ting, H.F. Sleep management on multiple processors for energy and flow time. In Proceedings of the 38th International Colloquium on Automata, Languages and Programming, Zurich, Switzerland, 4–8 July 2011; pp. 219–231.
25. Albers, S.; Antoniadis, A.; Greiner, G. On multi-processor speed scaling with migration. In Proceedings of the 23rd Annual ACM Symposium on Parallelism in Algorithms and Architectures, San Jose, CA, USA, 4–6 June 2011; pp. 279–288.
26. Gupta, A.; Im, S.; Krishnaswamy, R.; Moseley, B.; Pruhs, K. Scheduling heterogeneous processors isn't as easy as you think. In Proceedings of the 23rd Annual ACM-SIAM Symposium on Discrete Algorithms, Kyoto, Japan, 17–19 January 2012; pp. 1242–1253.
27. Chan, S.H.; Lam, T.W.; Lee, L.K.; Zhu, J. Nonclairvoyant sleep management and flow-time scheduling on multiple processors. In Proceedings of the 25th Annual ACM Symposium on Parallelism in Algorithms and Architectures, Montreal, QC, Canada, 23–25 July 2013; pp. 261–270.
28. Lawler, E.L.; Lenstra, J.K.; Kan, A.R.; Shmoys, D.B. Sequencing and Scheduling: Algorithms and Complexity. In *Handbooks in Operations Research and Management Science*; Elsevier: Amsterdam, The Netherlands, 1993; Volume 4, pp. 445–522.
29. Hall, H. Approximation algorithm for scheduling. In *Approximation Algorithm for NP-Hard Problems*; Hochbaum, D.S., Ed.; PWS Publishing Company: Boston, MA, USA, 1997; pp. 1–45.
30. Sgall, J. On-line scheduling. In *Online Algorithms, The State of the Art*; Fiat, A., Woeginger, G.J., Eds.; Springer: Berlin/Heidelberg, Germany, 1998; pp. 196–231.
31. Karger, D.; Stein, C.; Wein, J. *Scheduling Algorithms, CRC Handbook of Theoretical Computer Science*; CRC Press: Boca Raton, FL, USA, 1999.
32. Irani, S.; Pruhs, K. Algorithmic problems in power management. *ACM SIGACT News* **2005**, *36*, 63–76. [CrossRef]
33. Albers, S. Energy efficient algorithms. *Commun. ACM* **2010**, *53*, 86–96. [CrossRef]
34. Albers, S. Algorithms for dynamic speed scaling. In Proceedings of the 28th International Symposium of Theoretical Aspects of Computer Science, Dortmund, Germany, 10–12 March 2011; pp. 1–11.

35. Gupta, A.; Krishnaswamy, R.; Pruhs, K. Scalably scheduling power-heterogeneous processors. In Proceedings of the 37th International Colloquium on Automata, Languages and Programming, Bordeaux, France, 5–10 July 2010; pp. 312–323.
36. Fox, K.; Im, S.; Moseley, B. Energy efficient scheduling of parallelizable jobs. In Proceedings of the 24th Annual Symposium on Discrete Algorithms, New Orleans, LA, USA, 6–8 January 2013; pp. 948–957.
37. Thang, N.K. Lagrangian duality in online scheduling with resource augmentation and speed scaling. In Proceedings of the 21st European Symposium on Algorithms, Sophia Antipolis, France, 2–4 September 2013; pp. 755–766.
38. Im, S.; Kulkarni, J.; Munagala, K.; Pruhs, K. SelfishMigrate: A scalable algorithm for non-clairvoyantly scheduling heterogeneous processors. In Proceedings of the 55th IEEE Annual Symposium on Foundations of Computer Science, Philadelphia, PA, USA, 18–21 October 2014; pp. 531–540.
39. Bell, P.C.; Wong, P.W.H. Multiprocessor speed scaling for jobs with arbitrary sizes and deadlines. *J. Comb. Optim.* **2015**, *29*, 739–749. [CrossRef]
40. Azar, Y.; Devanue, N.R.; Huang, Z.; Panighari, D. Speed scaling in the non-clairvoyant model. In Proceedings of the 27th Annual ACM Symposium on Parallelism in Algorithms and Architectures, Portland, OR, USA, 13–15 June 2015; pp. 133–142.

applied sciences

MDPI

Review

Microgrid Cyber-Security: Review and Challenges toward Resilience

Bushra Canaan, Bruno Colicchio and Djaffar Ould Abdeslam *

IRIMAS Laboratory, University of Haute Alsace, 61 rue Albert Camus, 68093 Mulhouse, France; bushra.canaan@uha.fr (B.C.); bruno.colicchio@uha.fr (B.C.)

* Correspondence: djaffar.ould-abdeslam@uha.fr; Tel.: +33-3-8933-6020

Received: 25 July 2020; Accepted: 12 August 2020; Published: 14 August 2020

Abstract: The importance of looking into microgrid security is getting more crucial due to the cyber vulnerabilities introduced by digitalization and the increasing dependency on information and communication technology (ICT) systems. Especially with a current academic unanimity on the incremental significance of the microgrid's role in building the future smart grid, this article addresses the existing approaches attending to cyber-physical security in power systems from a microgrid-oriented perspective. First, we start with a brief descriptive review of the most commonly used terms in the latest relevant literature, followed by a comprehensive presentation of the recent efforts explored in a manner that helps the reader to choose the appropriate future research direction among several fields.

Keywords: cyber-physical security; microgrid; cyber-attacks

1. Introduction

The sustainable flow of energy, or in other words, energy security, in the field of electrical supply systems, does not always rely on the physical availability of resources. Today's technical challenges are extended to include the constant equilibration of demand–supply systems in terms of electricity quality and cyber security.

Upgrading the electrical network has not been as dynamic as it should be. With technology being implemented in almost every area of our modern life and smart applications growing in scope and complexity, the power sector makes its steps towards the smart grid at a pace of an extreme cautiousness.

The bidirectional flow of power and information generated and monitored by highly advanced types of equipment and mechanisms signifies the new generation of the energy networks. Smart grids are expected to deliver tangible progress to our conventional power systems on both aspects of efficiency and reliability, all together with integrating the maximum share of renewable resources reinforced by distributed intelligent and demand-side management techniques [1,2].

Changes will give the consumers and prosumers a wider range of choices and accord them with the possibility to actively participate in the optimizing operation of the system, by means of providing them with detailed instructions on how to better use their supply and act as authorized partners.

Smart grid benefits can also be extended to enrich the coupled economic sector through reducing operational costs and losses, generating new job opportunities, and reformulating the face of the energy market with time-based pricing and a more accurate speculation of demand and response profiles [3], in a time where electricity price forecasts have become a fundamental input and an important tool for decision-making mechanisms of the energy service provider companies.

But yet, the complexity level of the actual power networks and the critical role that it plays in every domain form a double-edged challenge—especially when the introduced technologies might itself be the source of threat.

New types of communication and data-management systems must handle not just the different emerging media trends and smart equipment (e.g., computer-based or microprocessor-based), it also needs to cope with existing legacy systems [4] in a manner that is adjustable to scalability and above all, resistant to cyber intrusion [5]. To this end, smart grids have to come as a complementary solution and not an eliminating or excluding one. These technical uncertainties, plus the additional investment costs, have evoked the political reluctance practiced by energy operators against this shift.

Europe has been working on energy transition and smart grids since 2005, starting by creating the smart grid technology platform which has set the year 2020 as a horizon to complete the process [2]. There were also several initiatives that carried out the development of experimental testbeds for smart grids solutions which aimed to highlight the most critical challenges and potentials accompanied by this evolution and their influence on the European power systems. Nevertheless, a further and more holistic analysis that is based on a profound technical understanding of each individual system architecture and basically includes the impact of both social and economic aspects on such heterogeneous systems, is yet to be accomplished in order to be able to trade-off between the existing approaches and pilot experiences, choosing a unique and valid experience that is suitable to be scaled up and replicated [6].

On the other hand, a very promising approach to overcome the majority of previous issues appears through energy communities, in which current grid problems are managed in a coordinated way such that avoiding costly network reinforcement along with maintaining aspired values of the smart grid. That is why we might be able to envisage the future smart grid as a sort of aggregation of multiple integrated entities or microgrids supervised, monitored, and controlled via a reliable communication-based layer. Accordingly, the increasing interest in microgrid development as the core of the smart grid systems is completely justified [7], although this increasing interdependency between physical and nonphysical power system components, which forms the so-called cyber-physical systems, raises a whole new level of complications.

In this work we closely examine the existing approaches to address the cyber-physical security in power systems with focusing on microgrids.

The structure of the paper is organized as follows; the second section describes the gradient evolution of the concept of the cyber threat, starting from the attacks targeting industrial control down to the electrical grid. Later, the third section elaborates on standardized definitions and terminology choices for the contemporary problematic challenges. In section four, we move on to the actual issues and case studies that occupy the researchers' attention from different viewpoints. Finally, we conclude by providing some insights about the unsettled challenges in addition to realistic recommendations in the light of the presented argument.

2. Industrial Cybersecurity Incidents Emergence

The 21st century witnessed the initiation of various cyber incidents affecting sensitive infrastructures. The discovered complexity of cyber-attacks on Industrial Control Systems (ICS) revealed the dexterity level of the attackers in Industrial Con [8].

The smart grid internet interconnection subjects the grid to different forms of hazards, particularly with regard to Advanced Persistent Threats (APT), Distributed-Denial-of-Service (DDoS), botnets, and zero-days. Stuxnet, Duqu, Red October, or Black Energy are only a few examples of the advent mayhems touching industrial security since 2010 [3].

Stuxnet, the worm that caused the first reported cyber-physical incident, was discovered by a senior researcher at Kaspersky Lab, Roel Schouwenberg, in June 2010. With a purpose that was beyond stealing, erasing or modifying data, Stuxnet endeavored to cause material sabotage in the supervisory control and data acquisition (SCADA) system as a physical industrial control system. It was regarded as the first cyber-warfare weapon to encompass a complex piece of malware that has infected an estimated 50,000 to 100,000 computers mostly found in Iran, Indonesia, India, and Azerbaijan [9].

Duqu and Flame, another two worms intended towards industrial control systems, were observed more than a year after Stuxnet. Despite the similarities in code source with Stuxnet, they had different objectives. Duqu was designed to track and gather useful information that would help to compromise the opted industrial control set. Flame or Flamer was a more sophisticated malware, especially developed for cyber espionage on these networks. Spotted cases were mainly located in Iran and other countries of the Middle East [10].

In December 2015, a cyber attack on Ukraine's power system has procured a wide-area outage, affecting around 225,000 customers. The attack was associated with a new variant of Black Energy Trojan named Disakil [3]. According to reports issued by power companies, the SANS institute and Electricity Information Sharing and the Analysis Center (E-ISAC), the problem started several months before the actual attack by installing the malware through phishing emails. At this period, the hackers only monitored and collected valuable information about the system operation during what is usually called the reconnaissance phase. On the day of the incident, the attackers took control over the Human–Machine Interface (HMI) and cut the power by opening a certain number of breakers. In order to intercept the service restoration, a denial of service (DoS) attack on the communication network, additionally to the classic telephone lines, was employed to prevent the clients from reporting the problem. Even applications that determined the outage extent were blocked by the malware that was able to recognize the system softwares [11,12].

One year earlier, the same threat agents were identified by the Industrial Control Systems Computer Emergency Response Team (ICS-CERT) during an attempt to penetrate the U.S. electric sector. Despite the fact that the attack, in this case, never happened, it definitely attracted attention on the future potentials of the cyber threats on a sector of utmost vitality [9].

3. Definitions and Overview

3.1. Cyber-Physical Security

The IEA (International Energy Agency) defines energy security as "the uninterrupted availability of energy sources at an affordable price". Traditionally, security used to be achieved on two fundamental levels; short-term security that deals with the stability of the demand–supply procurers, and the dynamism that enables the energy system to adapt as quickly as possible to sudden changes in the grid loads. Moreover, long-term security focuses on investments that support economic and sustainable development requirements.

Recently, with the arrival of smart grids which are essentially defined according to IEEE 2030-2011 standard, as a composition of three interoperability infrastructures, as set forth in Figure 1. This suggested interdependency has led the security problem to grow in complex imposing supplementary challenges threatening of introducing easier ways of causing damage to the fundamental security concerns, all along with creating new ones.

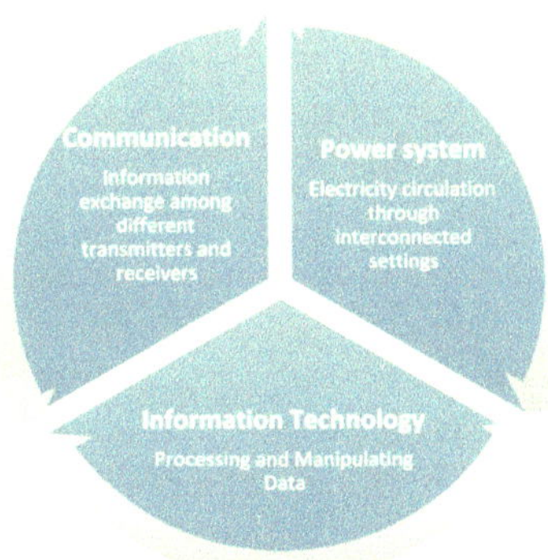

Figure 1. Smart grid architecture in compliance with IEEE 2030.

Consequently, recent security assessment has focused on identifying the potential vulnerabilities introduced by the cyber layer and analyzes the possible impacts on energy systems, which has given birth to a brand new research area called cyber-physical security. A cyber-physical system is co-engineered collaborating domains of physical and computational counterparts, in which the crucial system tasks are basically handled with its physical part, while informatically enhanced processes-normally referred to as cyber- are responsible for maximizing the exploration of intelligent devices and application [13].

The reason why academia recently chosen to add the term "physical" to the equation is to shed light over the emerging threats imposed by connecting these two fundamentally different infrastructures together, which practically may lead to problems that do not particularly belong to a failure of either systems [14]. In light of these assumptions, further investigation is still needed to either confirm or deny the putative relationships [15].

The most indispensable objectives of security requirements considerations of any data transferring communication in the IT network security are known as CIA-triad, which stands for Confidentiality, Integrity, and Availability, respectively. According to The National Institute of Standards and Technology (NIST)'s guide on cybersecurity strategy, architecture, and high-level requirements, Confidentiality refers to "Preserving authorized restrictions on information access and disclosure, including means for protecting personal privacy and proprietary information ... " [44, U.S.C., Sec. 3542], and a loss of confidentiality results in unauthorized disclosure of information. Whereas Integrity is "Guarding against improper information modification or destruction, and includes ensuring information non-repudiation and authenticity ... " [44, U.S.C., Sec. 3542] in other words, integrity is the unauthorized modification or destruction of transferred information. Availability, on the other hand, means "Ensuring timely and reliable access to and use of information ... " [44, U.S.C., Sec. 3542] as if altering availability will lead to the disruption of the access to or use of information or an information system.

Smart grid security is also built upon the previous trestles, but with a difference in priority order, where availability comes on top of the requirements, followed by integrity, accountability, and finally confidentiality. Other referencing emphasizes the accountability as additional security criteria [16]. This sequence of importance goes back to the severity of impacts resulting from tampering with these criteria.

Attackers can penetrate the smart grid communication systems using vulnerable entry points in the logical border surrounding a network, known as the Electronic Security Perimeter (ESP). Interventions may occur with the help of numerous mediums, such as the Universal Serial Bus (USB) thumb drive, viruses, and even software patches and updates [17].

Despite the fact that cyber intrusions on cyber-physical systems (CPSs) can be found under different terms, such as bias injection attack, zero dynamics attack, denial of service (DoS) attacks, eavesdropping attack, replay attack, stealthy attack, covert attack, and dynamic false data injection attacks [18]. These attacks can still be classified according to the one or multiple security criteria they are jeopardizing, as set forth in Figure 2.

Intentionally introduced faults or malicious attacks triggered by the cyber layer leaving serious impacts on not only the technical aspects, but also on economic and social correlations in power network operations, are the focus of this research.

Effects range from tampering smart meters data or manipulating the forecasted load profiles up to reaching equipment damage or even complete blackouts [19].

However, achieving such results is never an easy business. Indeed, physical prerequisites and the current state of the power system architecture with contemporary defense mechanisms, such as controllers prepared to re-examine each input parameter against a selection of acceptable values preventing possible physical damages [20], burden the attacker with the mandatory acquisition of a customized knowledge about the physical nature of the system added to the already required computer-related competencies. But then again, this does not mean that the conventional ways of protection, such as the ones adopted to restrain the spread of fault effects by isolating of a malfunctioning entity, are enough to prevent an attacker from achieving an unacceptable condition in the grid [20].

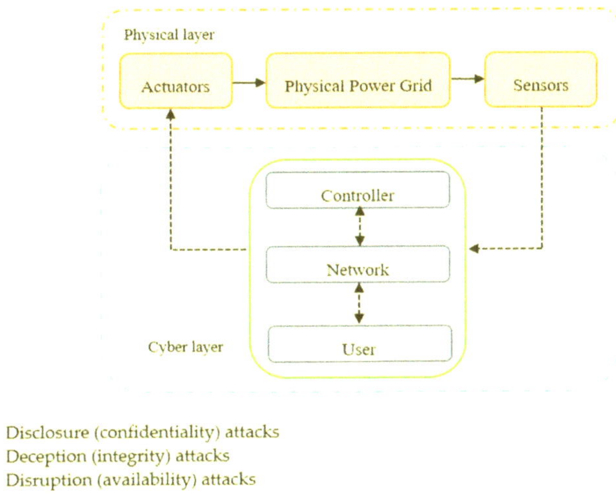

Figure 2. Three types of cyber-attacks.

In that vein, reasonable strategies to fend off such incidence fall into two complementary categories. The first one is about developing measures that tend to detect malicious attacks and tackle down the cause of infection in the system in order to deal with either the compromised unit or entity through isolation or the direct cause from wherein the adversary could have accessed the network. The second important aspect is cyber resiliency, in which we anticipate the behavior of our system under attack and elaborate on what could be done to expeditiously recover from these attacks in a passive protection fashion.

At any rate, we must keep in mind that keeping the system utterly safe, over and above maintaining a level of simplicity allowing the intuitive understanding of the entangled operation, is a paradox that preoccupies the power system researchers and engineers.

3.2. Modern Distribution Network Vulnerabilities

Distribution systems play a major role in the electricity sector value chain linking transmission to consumption and providing direct contact with consumers [6]. Knowing that their systems were originally designed for passive energy delivery (in one direction), Distribution System Operators (DSOs) find themselves nowadays forced to cope up with the tremendous changes pertaining to the electrical networks, on especially on the medium to low voltage scale.

Unlike in transmission systems that have adopted the Energy Management Strategy (EMS) early in the 1970s, the application of proper EMS at the distribution level was not put into action until recently, since it did not have much of added avail [21].

Following the foregoing tendency, measures continue to offer incentives that consolidate the integration of all the flexible distributed resources into the market, side by side, with new demand–response technologies on the demand side [5].

Dispatchable generation units owned by the DSO, which could be turned on and off by the energy operator to match a scheduled output that meets the network requirements, are a very useful avenue that has been widely exploited over the years in peak shaving and declining stress over the network components at times of high demand. Nevertheless, the surplus of the distributed generation (DG), especially the non-dispatchable (renewable) type, can adversely affect the performance of the distribution systems causing power quality issues, augmented fault levels, voltage violation, protection issues, in addition to line overloading or congestion [21].

Certain DSOs have set rules of thumb that determine the adequate segment of DG that should enter the distribution networks depending on the hosting capacity of each of them. In general, an estimated 15% of the network's peak demand could be connected to the distribution network without causing significant problems [22].

The needed elements for DG metering and monitoring change from country to country or even between regions. Hence, more or less data might participate in the decision that determines whether a DG participates in the energy markets or not, in respect to its impact on the local network, keeping in mind that larger DG installations could also have an extended disconcerting impact on the regional or national transmission system [21].

3.3. Microgrids as a Cyber-Physical System (CPS)

Despite the tendency to associate the term microgrid with the power sector, we find that the concept represents itself in a larger context related to the energy community with different means of energy production, transition, and storage, all along with achieving the mutual goals of boosting technical and economic resilience [23].

Through the years, different definitions have been placed in the technical literature to describe the concept of a microgrid. The first one was proposed in [24,25] imagining the microgrid as the ultimate solution for the reliable integration and control of the ensemble of Distributed Energy Resources (DERs), including Energy Storage Systems (ESSs) and controllable loads [26].

Similarly, in [27,28], microgrid paradigm is foreseen as a very appealing strategy to overcome challenges in integrating the massive renewable resources resulting from summing up all community-scale capacities, which is still being kept on hold due to the inflexibility of the current networks. Furthermore, these individual DERs are often too small to enter the electricity market, which is another problem that has been solved thanks to this new topology.

This goes in line perfectly with what is stated by the US Department of Energy, with only one difference stressing the clear barriers with respect to the distribution network, in the way that it permits the microgrid to have the ability to operate not only within grid-connected mode but also in autonomous island mode [29], which in turn was found, in numerous studies, to be considered as a sine qua non to denote a microgrid [9].

With microgrid pushing the power system over the edges of decentralization, a geographically localized distributed power model makes more sense regarding risk-management in terms of regional resilience and preventing cascading failure in the event of weather events, cyber-attacks, etc. [28]. Knowing that the electricity supply for small urban or industrial communities (isolated microgrid) where the main grid connection is inaccessible was never a novel trend in the world of electrical alimentation [27].

There were numerous attempts to create a standardized configuration of the smart grid's building block, namely microgrid. However, its structure is yet considered to be arbitrary and any technically well functioning connection is valid [13,27]. It is important to notice that the microgrid's ability to fit in different configurations and to be customized as a function of the present requirements and constraints is the exact same reason why it is so hard to classify it in a fixed frame. Figure 3 illustrates a generalized structure for modern microgrids.

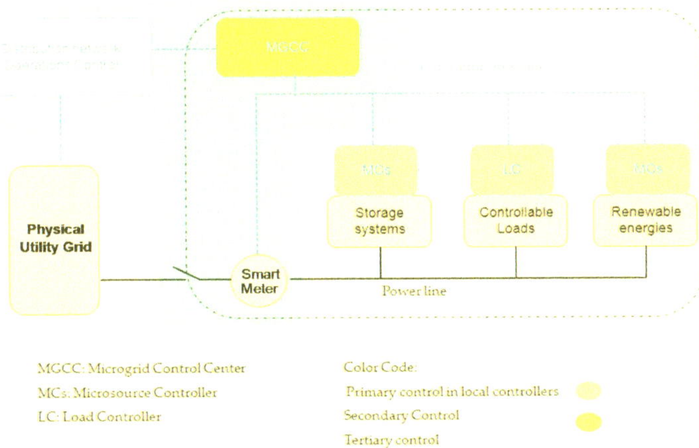

Figure 3. Modern basic structure of a microgrid.

From an operational perspective, there are only two types of microgrids (MGs):

A grid-connected MG, which is built to operate in either islanded or connected mode, might have one or several connection points with the grid. A single point of connection is very common though [8].

An isolated or stand-alone microgrid does not even have a point of common coupling (PCC) with the main grid [19]. Microgrids' implementation into utility does not always have to follow the classical case where a single MG is connected directly. Other alternatives can still exist, such as multiple tie line-based interconnected microgrids and small MGs within larger ones, so as the biggest takes the role of the governor large areas electrical power system [28].

The operational efficiency of microgrids necessitates sophisticated but most importantly secure measurement, communication, and control realized by various controlling methods, sensors, actuators, and field devices [18]. Moreover, microgrids are a highly sensitive cyber-physical system [13], in which the physical part is strongly influenced by the integrity of the cyber part, due to more entry point, very low required latency and the absence of multi-stage security detection. Consequently, attackers have more of a chance to cause serious problems in microgrids, leading to overall catastrophic consequences [30].

Recent papers have gone through securing the cyber-physical structure of the microgrid from different standpoints. Preliminary efforts probing cyber-attacks against the power systems would usually treat these attacks as a sort of noise or disturbance. So they tried their best to eliminate these disturbances using filtration techniques [31,32]. However, these techniques are based on pre-defined statistics which lose their effectiveness when facing slightly more fine tuned attacks [11].

In the following sections, we reviewed the recently proposed approaches from different domains.

4. Perspective-Based Interventions Addressing Cyber Attacks

4.1. Microgrid Communication

Being a cyber-physical system, microgrids inherent equally advantages and disadvantages of the combined systems. Communication network is essential to effectively incorporate many of desired features of the smart grid, such as the distributed automated system, distributed energy resource protection, islanding, and the display of network state and performance.

Standard communication problems also appear in microgrids, as it suffers from incompatibility between different types of heterogeneous communication technologies [13], besides the increasing reliance on Wi-Fi and internet-based communications, which are more susceptible to cyber interference but still essential for ancillary services related to microgrids, such as weather forecast data, fuel prices, peak hours, etc. [13]. Taking into consideration the expanding amount of data transferred between microgrid's components, different connected microgrids, or with an external centralized control and monitoring point, upon the design of the control structure. Satellite data (GPS) might also be a sort of communicated signals under danger in synchronous microgrid with phasor measurement units (PMUs).

On the other side, intrusion detection, firewall, and other selected solutions from the traditional security measures against rudimentary attacks targeting conventional data networks can also be included in smart microgrid applications [33].

While some prefer to leave the power generation control network isolated from the public network as a countermeasure against cyber contingencies, the leverage of open transmission protocols and computers with common operating systems that performed as intelligent electronic devices (IED) cannot be neglected nor eliminated today. Especially with essential improvements on automation efficiency and control system costs [16].

As an attempt to study and simulate the influence of an attacked communication network on electric power systems, earlier efforts went to model the attacks as a time delay to be accounted within the control loop, a subject that has been widely explored even outside the scope of cyber-attacks. For example, in an islanded microgrid, the authors in [34] have examined the communication delay limits beyond which we might risk having instability issues. They proposed an impact mitigation approach that revolves around gain scheduling for PI (proportional–integral) used in the secondary frequency controller that can be adopted in other microgrids as long as they can be modeled in the same small-signal model.

However, these assumptions on the nature of attack impacts are oversimplified and do not fully cover the new debouching aspect of joint cyber-physical models [35,36]. Others argue on the matter of communication latency's impact on microgrid control on the first place, building on an example that puts out shreds of evidence on having an inconspicuous and highly nonlinear relationship of delay rates between the source causing the delay and the resulting delays in the networks [37]. They also state the fact that, except for the simplest of cases, deriving tight bounds between delays, or other relevant metrics such as loss rates, is nearly impossible, especially when the models' analytical accuracy declines as the network size grows from single-hop settings to relay networks.

Taking the communication problem to a larger extent, a Cyber-Physical Power System (CPPS), ref. [38] digs into what might be a better communication configuration in terms of preventing a cascading failure, and in a comparison, based on transmission efficiency threshold values, they find that double-star communication networks perform better than the mesh communication networks.

Preventing cascading failure in cyber-physical power systems (CPPS) through a comprehensive analysis of the mechanisms and dynamical characteristics of interdependent networks was also the focus of the research presented in [2]. The writers have reviewed the different existing approaches and methods of power and communication systems coupling and interconnection and then proposed a novel interdependent model with the "degree–electrical degree" assortative link pattern that has proved its effectiveness in reducing the probability of large-scale blackouts caused by random attacks. Whereas, in the case of malicious attacks, simulation outputs have demonstrated the superiority of the random link model. Results also highlight in a more general manner the importance of coupling strength between the two layers over the choice of the interdependent model. The more dependent the power system is on the communication system, the more fragile it becomes.

Among a very diverse variety of problems discussed in the Information and Communication Technology (ICT) field, the particularity of synchrophasor systems vulnerabilities against cyber-attacks was highlighted due to the growing interest in synchrophasor technology applications [39]. Notably, the absence of built-in security structures in the widely adopted IEEE C37.118 communication framework that sets up the standards for PMUs and Phasor Data Concentrators (PDCs) is making it highly exposed to cyber threats [40].

Experiments involved resiliency examination of the communication system structure based on IEEE C37.118 under different attack scenarios, accompanied by estimation of possible impacts on synchrophasor application that uses this standard [40].

In [41], vulnerability analysis went deeper into the IEEE C37.118 framework structure to its weakest components, which the transport protocol layer, as they discuss the susceptibility of two commonly used protocols in transport layers (i.e., Transmission control Protocol (TCP) and User DatagramProtocol UDP) against DoS and FDI attacks summarizing the requirements to be used for creating a successful cyber intrusion as well as to prevent it.

A comprehensive comparison with ICE 61,850 that took into consideration the security implication of both standards stressing the advantages and disadvantages that encounter the synchrophasor application developers was also performed in [42].

Figure 4 summarizes the proposed approaches explaining the main mechanisms and pathways considered in acting against cyber intervention in the communication domain.

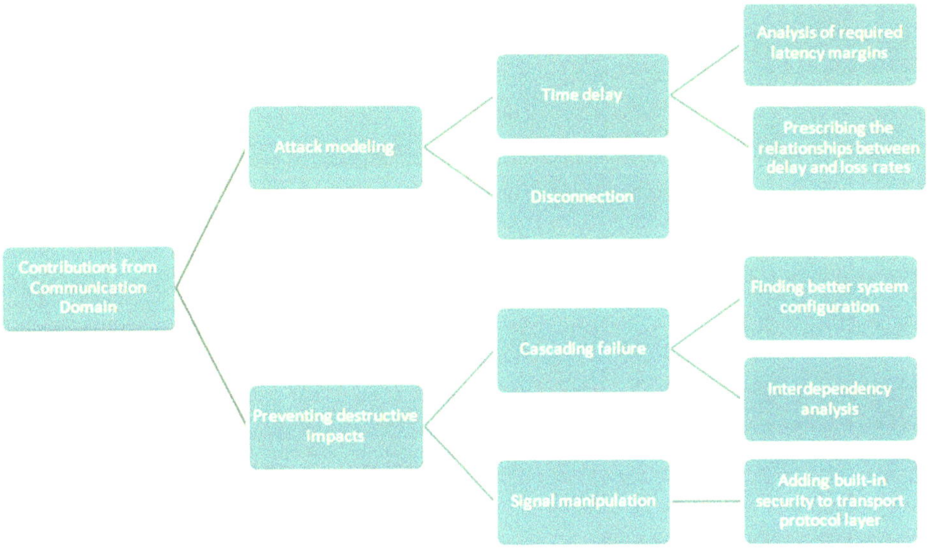

Figure 4. Contributions from the communication domain.

4.2. Impact Analysis

Another way of proving that cyber-physical solutions for the power sector are not just a solution waiting for a problem, remarkable efforts in the field of impact estimation and threat modeling were made to dispel the doubts on the capability of cyber-attacks to cause actual physical damage.

The research presented in [20] demonstrates the possibility of authentic corruption caused by two types of cyber-attacks (availability and integrity attacks) jeopardizing the ICT and the GPS systems required for the sane functioning of a microgrid in three different operating modes (connected, islanded and sync-islanded). The severity of the physical impact which may vary between local blackout, the main instability violation of power quality equipment damage and human danger witnessed in the example microgrid is strongly related to its own architecture.

Other researchers were more interested in exploring the effects on a specific area in the microgrid systems, such as secondary frequency control function in [43] and distributed load sharing [44]. But since these studies are limited by the chosen system, more efforts had to go further into developing threat modeling methodology that fits into the purpose of risk characterization in different systems architectures.

As an attempt to fill the gap, ref. [39] explores the possible arising genres of threat in the components of different systems predicted on missing security properties, and how powerful this could be on the system security entirely.

4.3. Microgrid Control

4.3.1. Control Structure

The sound operation of power management strategy (PMS) is more critical in microgrids. Reasons narrow down to the imperative adjustment of multiple interconnected DG units with significant differences in power capacities and generation system characteristics that become more and more Electronically Interfaced (EI-DG), requiring a faster response to keep dynamically changing characteristics (voltage/angle) within the appropriate margins [26].

A microgrid's control systems are a typical target for attackers. Based on the purpose of which they are implemented, microgrid's control structure can be profoundly different and the control features are limited or customized to a desired subset of functions from a larger group of options [45]. In grid-connected mode, for example, frequency and voltage values are regulated by the host grid at the point of common coupling (PCC) whereas tasks like the DER's active and reactive power accommodation, energy management and load sharing, in addition to a safe transition between connected–islanded modes are still elaborated by the microgrid's control systems. During islanded mode or in remote conditions that are completely isolated from the grid, the microgrid's local controllers take full responsibility for all stability measures which also vary depending on the microgrid type (Islanded AC, synchronized islanded AC, naurally islanded DC) [27]. Typically, having one of the DERs operating as the isochronous generator forming the microgrid voltage and frequency is quite common in islanded microgrids. In this case, the rest of the DERs could participate in supporting voltage and frequency if needed [9], whilst the complete absence of a dominant source of energy generation during the autonomous mode of operation adversely amplifies the complexity of the assigned task list.

The majority of prior researches pictures the microgrid's control paradigm in a hierarchical manner, following the successfully adopted structure in the legacy grid [23].

Hierarchical control levels of microgrids are usually anticipated in three layers: primary, secondary, and tertiary. There are no definite technical boundaries between strategies of each level rather than a sort of indication based on relevant considerations, such as response rapidity interval, purpose-oriented control, and central-distributed control.

However, without losing generality, we can say that primary and secondary control strategies are practically associated with operational stability and accordance between microgrid's components, while harmonization with the host grid is applied by tertiary control [27].

Primary control features the fastest response with the smallest decision time step, voltage, and frequency regulation as well as protection executed on this level, which are entirely based on local measurements and droop mechanism with no communication needed [27]. That is why operations on this level are conserved away from cyber incidents.

On the contrary, secondary control operates on a slower time scale, often with a reduced communication bandwidth by using sampled measurements. It collaborates consistently with the other two levels to satisfy the requirements set by the tertiary control. The secondary control measures values across the microgrid, and accordingly, updates the desired setpoints for the primary controllers [46].

Tertiary control on the highest hierarchy collects state information of the energy system through the communication infrastructure and makes decisions to optimize the overall performances of microgrid with the longer decision time step. It may also be responsible on the economic dispatch of controllable resources and coordination with the distribution system operator with Energy Management System (EMS) ensuring power balance constraint, security constraint, and operational constraints [23].

Previous control functions can be achieved through either centralized or distributed implementation of the control architecture. The discussion on privileging one control method over the other is still questioned by several papers focusing on different aspects [7,23,43].

Distributed control was originally proposed as a solution to boost scalability in modern networks by means of facilitating the introduction of supplementary DERs. It splits control tasks between units instead of the substantive upgrading of single excessive computational capacities. Moreover, the sparsity of communication networks utilized in distributed control schemes reduces the infrastructure cost [43]. Not to mention, that is also considered to be more resilient as single-point failure does not lead to cascading failure, unlike centralized or what might be called hierarchical control, in which messages that carry out measurements and instructions from and to all system components should pass by a dedicated central controller. Correspondingly, centralized control schemes have a better understanding of the microgrid functions since it has an embedded version of the system model in the central controller which in turn will trigger an optimal application of EMS objectives including the economic performance simultaneously with satisfying real-time operational constraints.

In [7], authors review basic branches of distributed control optimization and their application with a brief reflection on the cybersecurity consideration, promoting distributed control on the bases of mitigation obstacles relevant to communication risks and stakeholders' resistance to sharing critical data.

Among distributed optimization methods, consensus control has gained more attention in microgrid's control community recently. The initial notion was inspired by biological phenomena that revolve around providing each unit of vision on the overall objective to a limit where different DERs converge to a single value. Here, decisions are built upon local measurements and peer-to-peer communication, offering this model extra flexibility, adding to the already well established feature in the distributed structure. Cooperative control is also a very feasible solution for stability control in terms of voltage and frequency equilibration and economic control with cost consensus for generation units across the network [23].

4.3.2. Automation Control against Cyber-Attacks

Since control systems were conventionally developed to detect, process, and mitigate the systematic and unpredicted errors, there is no wonder it has been the focus of numerous research cases in the field of attack predictability, detection and protection.

Microgrids are prone to the same types of attacks found in the utility grid. DoS events provoke multiple issues without a doubt, but at the same time, they are easily detected by the system operator which will probably recognize in an adequate rapidity that it is under attack. Similarly, the superior severity of the FDI attacks is largely attributed to the detection method's complexity and variability upon the adopted control structure [47].

Broadly, detection and mitigation of conventional attacks are already well explored in the literature. FDI that succeeds in penetrating the network while maintaining discretion without altering the system observability disturbance alarms, also known as stealth attacks [48], are able to cause unpredictable stability issues and the worse is that they are practically impossible to detect [49].

From a defender perspective, recent research attainable choices are perceived into either addressing the fault detection and isolation in control loops (detection based) [50] or working on precaution measures based on threat modeling and security analysis (protection based) [51].

The popular method used to detect bad measurement data in power transmission systems is the Static State Estimator (SSE). It is generally based on a weighted least squared (WLS) solution and it is not immune against attacks itself [52,53].

State estimation is also important to microgrid control functionality and it is usually found in traditional energy management systems derived from steady-state models [18]. However, static state models were no longer able to capture the systems' dynamics accurately with the exacerbated numbers of DERs on the generation side and the debuting retrofits on the demand side.

The research presented in [18] emphasizes the importance of deploying a secure dynamic state estimation on the side of AC-connected microgrids as a portion of the distribution network. Similar to [54] they proposed an estimator algorithm for a standard structure-preserving model that incorporates system dynamics. Method validation illustrates the estimator's ability to give a secure dynamic state estimation when supplied with inaccurate measurements caused by either an attack that manipulates communication between transceivers and the microgrid operator, or attacks that manipulate measurement units themselves, even without considering an attack scenario.

Traces left on the operation of observers turn into an efficient key to be used in attack detection. Distributed state estimation method is used as a way of detecting cyber-attacks of the FDI type. In [55], a consensus-based controlled DC microgrid was investigated where each distributed generation unit had employed the Unknown Input Observer (UIO) to estimate the state of its neighboring units and isolating the fault source consequently.

Another control approach using UIO was proposed in [50]. A fully decentralized load frequency controller was developed and tested with a perspective to be applied to multi-AC and DC microgrids.

Given that the relative simplicity of the cyber-attack detection of the FDI type in distributed control schemes, authors in [49] have decided to raise the bar by firstly introducing a stealth attack that is able to deceive the conventional distributed voltage observer without triggering the detection mechanism. After that, they proposed a general algorithmic-based detection framework for DC microgrids where they added a cooperative vulnerability factor (CVF) to the voltage PI controller. Finally, and under worst-case scenarios, artificial disturbances were added to by coupling the CVF with the secondary current sublayer in order to enhance the chance of capturing the attacks.

Later, the same DC microgrid model was used in another experiment using artificial intelligence in [56]. A Nonlinear Auto-Regressive Exogenous (NARX) neural network was trained over the previously mentioned control method during offline operation, capturing and storing its behavior, only to be used then as an online estimator for DC voltages and output currents of each unit. The FDI attacks detectability of this method was built on the estimation errors making it suitable for a larger spectrum of DC microgrid, in contradiction to the cases presented in [49,57] that only suits those functioning with cooperative consensus-based algorithms.

The FDI problem shaping in terms of determining the aspects that could be altered by such an attack was the subject of [58], in which a detection method was built on the assumption of the attack capability to modify the invariant values required in the secondary distributed control layer.

A new technique for optimal dynamic state estimation, based on a distributed algorithm for multiple connected DC microgrids under FDI attack, was proposed and tested over malicious and normal load disturbance in [47], proving its capability of distinguishing between the two cases. Unlike previous literature that dealt with DC microgrids as quasi-static models, this work employed a dynamic microgrid paradigm where the three DC connected microgrids employed in the study collaborated under a control configuration, that enabled each of them to verify the security status of the other two, making it possible to isolate the potentially infected entity.

4.3.3. Protective Control

Without continuing further into the investigation on the nature of the imperiling data or the way that the attackers may use in order to achieve instability in the system, other research simply focused on adding redundancy security to the existing used control methods. Authors in [30] have proposed increasing the security by coding the signals that carry the information about the state's measurements with an error-correcting code Recursive Systematic Convolutional (RSC) code and then decoding it to enhance the performance of the proposed semidefinite programming based on optimal feedback controller, coupled with Kalman filter estimator by elimination of a portion of noise on an IEEE 4-bus distribution feeder considered as four grid-connected microgrids.

Reachability analyses were frequently employed to determine if an unstable state could be reached due to certain changes in the monitored variable. This was elaborated in [28] by designing a stability monitoring and control comprehensive framework, that guarantees resiliency against attacks through isolating the problematic bus, while covering critical loads compensated from neighboring microgrids.

4.4. Co-Simulation Testbeds

Experimental studies are still economically unfeasible for microgrids, as it is for large-scale smart grids. With only a few reliable platforms around the world capable of performing highly complicated tests, real-time simulation is a powerful alternative solution in this area of research [37].

Models running on this mode of simulation adhere to a very small step size in order to achieve an accurate result. Bypassing the right step size to a larger value produces an erroneous simulation while smaller values do not fit into the simulator constraints. This, in turn, creates a challenge, especially with models with high-speed switching devices [37]. The large number of switching devices is not the only source of trouble in simulating power systems. Broadly, the integrated power electronics devices in the smart grid simulation also tolerate high-frequency pulse width modulation signals, and that what explains the shift from traditional offline simulators that are time-consuming and not adjusted for slow phenomena [59].

Additional complications come to the surface concerning the representation of the cyber-physical components in the same computational environment. The inhomogeneous nature of both power and communication systems plus the obvious differences in components, transmission content, and working mechanisms make it very challenging to accommodate the desired realistic features in one frame. For instance, the time-varying or continuous solvers suit the power systems while communication networks' simulation necessitate a discrete or event-dependent simulation [60]. In other words, the behavior of the grid control is mainly well defined using mathematical formulations, which is not exactly the case in the nearly stochastic, unpredictable data transmission protocol layer that belongs to the accompanying ICT system [61].

Even when ending up finding the perfect tool to simulate each structure separately, interfacing the two simulators in a way that guarantees the integration of the distinct characteristics for both, without restraining the core to either of each is not always evident and often stipulates grappling with synchronization and data exchange.

With all these obstacles, it seemed just right that some works had chosen to probe the different types of taxonomies of the existing testbeds since that discovering the various tools and techniques implemented in the current testbeds, counting also the strong and weak points for each one, is crucial to build and develop new experimental platforms.

In this context, the term "co-simulation" refers to the practical and realistic co-existence of the examined subsystems whereupon they operate hand by hand to reflect smart grid interactions [62].

The co-simulation development can take two directions; the first one primarily focuses on a specific tool that has been familiarly dealt with, in the course of studying individual subsystems. Researchers who seek this approach are usually more concerned in deepening their understanding of the interaction control between the intended subsystems. The second approach is a platform-based one which implies fixing the attention on the development of a comprehensive framework with a standardized interface capable of embracing different tools. This attracts researchers who deal with utterly complicated simulation environments, especially when flexibility in connecting the subsystem is inevitably expected, without intervening in neither layouts [61].

4.5. Smart Meters and Data Security

Energy sector's pathway is clearly heading towards more distributed resources and control. Consumers are becoming more and more ready to invest in distribution-edge devices comprised of residential PV panels, storage devices, electrical vehicles, and their recharging points in addition to smart control tools. Smart Meters (SM) at the endpoint of distribution networks liaise with consumers and lends them an open window to interact with the utility. In an ideal scenario, smart segments must effectively communicate via Advanced Metering Infrastructure (AMI) to reach the perfect balance, as set forth in Figure 5. Since its first appearance in 1872 [63], the concept of electricity meter has remarkably evolved. Conventionally, electricity meters used to provide information only about electricity consumption in terms of total current amplitude, while intelligent meters are supposed to support a wide range of applications rather than just metering [64].

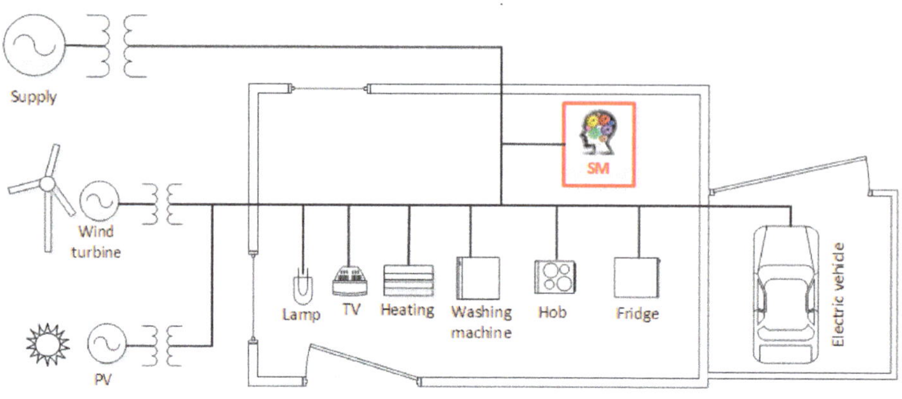

Figure 5. Smart meter role.

As specified in 2012/27/EC directive, Smart meters are "an electronic system that can measure energy consumption, providing more information than a conventional meter, and can transmit and receive data using a form of electronic communication" [65].

Once again, smart application functionalities are not clearly framed in official norms that usually define and impose quantifying criteria in terms of technical specifications. This is why working and elaborating on the standardization enclosure, especially for the most affiliated pieces of the smart grid, becomes an urgent need.

However, a good few guidelines were established to help the industry involved in developing the new generation of meters.

The European Smart Meters Industry Group (ESMIG) has fixed characteristics such as remote reading, bidirectional communication, support of advanced tariff systems and ability to run billing applications, and remote energy supply control to be the minimum features required in smart meters [64].

The need to address metering issues arises practically at the same time as the development of distribution electricity grids. Providing measurement, control, communication, power, display, and synchronization capabilities shall be no easy task for smart meters.

Smart meters at the moment are installed and deployed by the utility inside of consumer's facilities. They communicate real-time measurements with data concentrators and control centers that monitor and partially control the meters.

Machine-to-Machine (M2M) communications among appliances in accordance with information provided by service providers enhance demand response functionality, leading to a win–win situation [66]. Besides the aforementioned control and management advantages, collected data can also help the grid operators in an application such load forecasting.

The absence of human interventions is a key feature of advanced metering plug and play mode, is very desirable but unfortunately, at its earliest phases, comes with relatively high expenses.

The exposure to a different kinds of communication systems, including internet, in addition to the needed adaptability to work with different billing applications, that are probably open sourced not to mention the double ownership making smart meters the most vulnerable component of the smart grid.

Pursuing autonomy, future meters are being tested to enlarge their authority margin so the amount of transferred data to and from control centers can be reduced.

Impact on the electrical systems depends on the select functionality assigned to the smart meter. Of course, the availability of the entire service of a smart meter is still considered to leave the worst impact on systems, but data communicated via smart meters which provide considerably detailed information about consumers' consumption behavior or habits are the biggest new concerns. Confidentiality data can be exported in many grievous ways such as optimizing the attacker's understanding of the compromised target so that he can make a more severe attack, extort the service providers, or even sell to unauthorized parties [67].

Energy providers, on the other hand, had their own share of concerns: the manipulation of data at the user end either due to the intentional act by the consumers themselves or cyber-attacks could be used to steal power and billing manipulation, resulting in revenue loss [68]. For this reason, authenticated software should be implemented, not only on inside the meters, but also on the access side for a granted sound operation [66].

The consumer's trust is critical for SMs and AMI expansion: one of the most modern ways to resolve security-related issues is through blockchain or distributed ledgers technology.

Blockchain technology is a very promising solution with great potential to radically change the energy sector from the way we know it. It was firstly introduced in a financial context with cryptocurrency, widely known as "Bitcoin". Blockchain provides a trustworthy platform for peer-to-peer transaction using distributed storage for keeping track of the exchanged data. Smart contracts on top of the blockchain define individually the rules upon which contractors exchange resources (quantity, quality, price), eliminating third-party intermediaries and cutting down extra expenses and accelerating the operations rhythm [69].

Blockchain can contribute to maximizing social welfare for energy delivery through managing tamper-proof energy supply transactions in absolute transparency, providing the metering fundamentals as well as billing and clearing processes. It is also suitable for extended applications such as ownership certificates, asset management, proof of origins, copyrights and emission limits in addition to renewable energy quality standards [69]. All and more are features that also help to empower the role of small renewable generations that belong to prosumers and monetize their assets. Thus, it supports the two essential desired features in the smart grid security and distribution.

The study in [70] provides a systematic classification of the latest blockchain research projects and startups' experiences in the energy sector applications. It also analyzes the opportunities, potential challenges, and limitations of using a number of examples on peer-to-peer (P2P) energy trading in decentralized marketplaces with the latest technological inventions, notably the Internet of Things (IoT) and e-mobility.

Going back to the microgrids, authors in [71] put forward a blockchain-based framework for a microgrid as an aggregated prosumer to optimization-decentralized transactive energy management, and support secure the interactions among different energy sectors.

In practice, existing energy market mechanisms are still a bit far from getting replaced by blockchain models since they do not completely cope with current legal regulatory frames. Furthermore, the technology itself has not reached the desired certain maturity [69].

5. Discussion and Conclusions

In this article, we examined the existing approaches to address cyber-physical security from a microgrid perspective. As explained above, the work on the smart grid application, in general, lacks approach intersections, and is still being dealt with from separate domains in the research world. Although using the microgrid model to carry out experiments on the cyber-physical security has plenty of practical justifications attributed to the important role it plays in paving the way towards smart grids, the microgrid's context was mainly consulted owing to the relative simplicity in capturing and recording interventions, either as an injected attack or control modification. For example, the islanded microgrids broached by a fair number of papers, especially the DC type, have unarguable merits in terms of autonomy. However, this will only leave us with specially tailored methods and solutions that do not necessarily fit all cases.

Cybersecurity measures for energy systems still come as accessories and not as a built-in function. In particular, for most of part, the electricity-related equipment that gets evolved at an exponential rate makes it extremely difficult for cyber defenses' mechanisms to keep pace with this development in the absence of up-to-date standards and common market trends. At least, securing the smart grid requires a multidisciplinary approach, and economic and social development are usually forgotten or neglected aspects in this process. Even the most remarkable technology inventions are useless without being approved by clients.

Author Contributions: Cyber-Physical Security concept, Conceptualization and Writing, B.C. (Bushra Canaan); Supervision and Micro-Grid Concept, B.C. (Bruno Colicchio); Supervision and Validation, D.O.A. All authors have read and agreed to the published version of the manuscript.

Funding: This research was funded by the INTERREG-V program RES_TMO (Project No. 6.3).

Acknowledgments: This work is supported by the European INTERREG-V program RES_TMO (Project No. 6.3). We thank Barbara Koch (Freiburg University, Germany) project lead of RES_TMO and Upper Rhine Cluster for Sustainability Research.

Conflicts of Interest: The authors declare no conflict of interest.

References

1. Wang, W.; Lu, Z. Cyber security in the Smart Grid: Survey and challenges. *Comput. Netw.* **2013**, *57*, 1344–1371. [CrossRef]
2. Rashid, M.H. Energy Systems in Electrical Engineering. In *Smart Grids and Their Communication Systems*; Kabalci, E., Kabalci, Y., Eds.; Springer: Singapore, 2019; pp. 1–644.
3. Leszczyna, R. Standards on cyber security assessment of smart grid. *Int. J. Crit. Infrastruct. Prot.* **2018**, *22*, 70–89. [CrossRef]
4. Hossain, E. Communication Architectures and Models for Smart Grid. In *Smart Grid Communications and Networking*; Hossain, E., Han, Z., Poor, H.V., Eds.; Cambridge University Press: Cambridge, UK, 2012; pp. 1–103.
5. Prettico, M.; Flammini, G.; Andreadou, M.G.; Vitiello, N.; Fulli, S.; Masera, G. *Distribution System Operators Observatory 2018: Overview of the Electricity Distribution System in Europe*; Publications Office of the European Union: Ispra, Italy, 2019; pp. 1–77.
6. Prettico, G.; Gangale, F.; Mengolini, A.; Lucas, A.; Fulli, G. Distribution system operators from european electricity distribution systems to representative distribution networks. *JRC Tech. Rep. Luxemb.* **2018**, *99*, 273–280.
7. Yazdanian, M.; Mehrizi-Sani, A. Distributed control techniques in microgrids. *IEEE Trans. Smart Grid* **2014**. [CrossRef]
8. Sridhar, S.; Govindarasu, M. Model-based attack detection and mitigation for automatic generation control. *IEEE Trans. Smart Grid* **2014**. [CrossRef]
9. Chlela, M. *Cyber Security Enhancement Against Cyber-Attacks on Microgrid Controllers*; McGill University Montréal: Montréal, QC, Canada, 2017; pp. 1–177.
10. Knapp, E.D.; Samani, R. *Applied Cyber Security and the Smart Grid: Implementing Security Controls into the Modern Power Infrastructure*, 1st ed.; Syngress: Rockland, MA, USA, 2013.
11. Sun, C.C.; Hahn, A.; Liu, C.C. Cyber security of a power grid: State-of-the-art. *Int. J. Electr. Power Energy Syst.* **2017**, *99*, 45–56. [CrossRef]
12. Lee, R.M.; Assante, M.J.; Conway, T. *Analysis of the Cyber Attack on the Ukrainian Power Grid Defense Use Case.*; E-ISAC: Washington, DC, USA, 2016; pp. 1–23.
13. Rekik, M.; Chtourou, Z.; Gransart, C.; Atieh, A. A cyber-physical threat analysis for microgrids. In Proceedings of the 2018 15th International Multi-Conference on Systems, Signals & Devices (SSD), Hammamet, Tunisia, 19–22 March 2018. Available online: https://ieeexplore.ieee.org/document/8570411 (accessed on 6 May 2020).

14. Cai, Y.; Huang, T.C. Cascading failure analysis considering interaction between power grids and communication networks. *IEEE Trans. Smart Grid* **2016**, *7*, 530–538. [CrossRef]
15. Zhang, H.; Peng, M.; Guerrero, J.M.; Gao, X.; Liu, Y. Modelling and Vulnerability Analysis of Cyber-Physical Power Systems Based on Interdependent Networks. *Energies* **2019**, *12*, 3439. [CrossRef]
16. Liu, J.; Xiao, Y.; Gao, J. Achieving accountability in smart grid. *IEEE Syst. J.* **2014**, *8*, 493–508. [CrossRef]
17. Srivastava, A.; Morris, T.; Ernster, T.; Vellaithurai, C.; Pan, S.; Adhikari, U. Modeling cyber-physical vulnerability of the smart grid with incomplete information. *IEEE Trans. Smart Grid* **2013**, *4*, 235–244. [CrossRef]
18. Fooladivanda, D.; Hu, Q.; Chang, Y.H.; Sauer, P. Secure State Estimation and Control for Cyber Security of AC Microgrids. *arXiv* **2019**, arXiv:1908.05843.
19. Esmalifalak, M.; Shi, G.; Han, Z.; Song, L. Bad data injection attack and defense in electricity market using game theory study. *IEEE Trans. Smart Grid* **2013**, *4*, 160–169. [CrossRef]
20. Friedberg, I.; Laverty, D.; McLaughlin, F.; Smith, P. A Cyber-Physical Security Analysis of Synchronous-Islanded Microgrid Operation. In Proceedings of the 3rd International Symposium for ICS & SCADA Cyber Security Research 2015, Belfast, Swindon, UK, 17–18 September 2015.
21. Hayes, B.P. Distribution Generation Optimization and Energy Management. In *Distributed Generation Systems*; Gharehpetian, G.B., Agah, S.M.M., Eds.; Elsevier Inc.: Oxford, UK, 2017; pp. 415–451. [CrossRef]
22. Stavros, A.P.; Nikos, D.H.; Pierre, A.; Luiz, M.A.; Bernhard, B.; Clinton, G.C.-B.; Drossos, N.; Bayez, E.; Mingtian, F.; Vincent, G.; et al. Capacity of Distribution Feeders for Hosting Distributed Energy Resources. Papathanassiou 2014 Capacity ODCIGRE 2014. June 2014. Available online: http://cigreaustralia.org.au/assets/TB-SEPT-2014-3.1-Capacity-of-Distribution-Feeders-for-hosting-Distributed-Energy-Resources-DER-abstract.pdf (accessed on 5 June 2020).
23. Feng, X.; Shekhar, A.; Yang, F.; Hebner, R.E.; Bauer, P. Comparison of hierarchical control and distributed control for microgrid. *Electr. Power Compon. Syst.* **2017**. [CrossRef]
24. Lasseter, B. Microgrids distributed power generation. *Power Eng. Soc. Winter Meet.* **2001**, *1*, 146–149.
25. Lasseter, R. Microgrids. *IEEE Power Eng. Soc. Winter Meet.* **2002**, *1*, 305–308.
26. Katiraei, F.; Iravani, M.R. Power management strategies for a microgrid with multiple distributed generation units. *IEEE Trans. Power Syst.* **2006**, *21*, 1821–1831. [CrossRef]
27. Olivares, D.E. Trends in microgrid control. *IEEE Trans. Smart Grid* **2014**, *5*, 1905–1919. [CrossRef]
28. Buason, P.; Choi, H.; Valdes, A.; Liu, H.J. Cyber-physical systems of microgrids for electrical grid resiliency. *ICPS* **2019**, 492–497. [CrossRef]
29. Ton, D.; Bryan, E.; Marnay, C. Microgrids Program Overview, Power Systems Engineering Research and Development. *Aalb. 2015 Symp. Microgrids.* **2015**, 1–22.
30. Rana, M.M.; Li, L.; Su, S.W. Cyber attack protection and control of microgrids. *IEEE/CAA J. Autom. Sin.* **2018**, *5*, 602–609. [CrossRef]
31. Peach, N.; Basseville, M.; Nikiforov, I.V. Detection of Abrupt Changes: Theory and Applications. *J. R. Statal Soc. Ser. A (Stats in Soc.)* **1993**, *1*, 185. [CrossRef]
32. Jiao, Q.; Modares, H.; Lewis, F.L.; Xu, S.; Xie, L. Distributed $\mathcal{L}_2$-gain output-feedback control of homogeneous and heterogeneous systems. *Automatica* **2016**, *71*, 361–368. [CrossRef]
33. Manandhar, K.; Cao, X.; Hu, F.; Liu, Y. Detection of faults and attacks including false data injection attack in smart grid using Kalman filter. *IEEE Trans. Control Netw. Syst.* **2014**, *1*, 370–379. [CrossRef]
34. Liu, S.; Wang, X.; Liu, P.X. Impact of communication delays on secondary frequency control in an islanded microgrid. *IEEE Trans. Ind. Electron.* **2015**, *62*, 2021–2031. [CrossRef]
35. Hammad, E.; Farraj, A.; Kundur, D. Fundamental limits on communication latency for distributed control via electromechanical waves. *IEEE Int. Conf. Commun.* **2017**. [CrossRef]
36. Farraj, A.; Hammad, E.; Kundur, D. A systematic approach to delay: Adaptive control design for smart grids. *IEEE Int. Conf. Smart Grid Commun.* **2015**, 768–773. [CrossRef]
37. Guo, F. Comprehensive real-time simulation of the smart grid. *IEEE Trans. Ind. Appl.* **2013**, *49*, 899–908. [CrossRef]
38. Cai, Y.; Li, Y.; Cao, Y.; Li, W.; Zeng, X. Modeling and impact analysis of interdependent characteristics on cascading failures in smart grids. *Int. J. Electr. Power Energy Syst.* **2017**, *89*, 106–114. [CrossRef]
39. Khan, R.; McLaughlin, K.; Laverty, D.; Sezer, S. STRIDE-based threat modeling for cyber-physical systems. *IEEE Innov. Smart Grid Technol. Conf. Eur.* **2017**, 1–6. [CrossRef]

40. Khan, R.; McLaughlin, K.; Laverty, D.; Sezer, S. IEEE C37.118-2 synchrophasor communication framework: Overview, cyber vulnerabilities analysis and performance evaluation. *Proc. Int. Conf. Inf. Syst. Secur. Priv.* **2016**, 167–176. [CrossRef]
41. Wang, Y.; Gamage, T.T.; Hauser, C.H. Security implications of transport layer protocols in power grid synchrophasor data communication. *IEEE Trans. Smart Grid* **2016**, *7*, 807–816. [CrossRef]
42. Khan, R.; McLaughlin, K.; Laverty, D.; Sezer, S. Analysis of IEEE C37.118 and IEC 61850-90-5 synchrophasor communication frameworks. *IEEE Power Energy Soc. Gen. Meet.* **2016**. [CrossRef]
43. Liu, S.; Liu, P.X.; Wang, X. Effects of cyber attacks on islanded microgrid frequency control. *Proc. IEEE Int. Conf. Comput. Support. Coop. Work Des.* **2016**, 461–464. [CrossRef]
44. Zhang, H.; Meng, W.; Qi, J.; Wang, X.; Zheng, W.X. Distributed load sharing under false data injection attack in an inverter-based microgrid. *IEEE Trans. Ind. Electron.* **2019**, *66*, 1543–1551. [CrossRef]
45. Chlela, M.; Joos, G.; Kassouf, M.; Brissette, Y. Real-time testing platform for microgrid controllers against false data injection cybersecurity attacks. *IEEE Power Energy Soc. Gen. Meet.* **2016**. [CrossRef]
46. Nasirian, V.; Moayedi, S.; Davoudi, A.; Lewis, F.L. Distributed cooperative control of dc microgrids. *IEEE Trans. Power Electron.* **2015**, *30*, 2288–2303. [CrossRef]
47. Vu, T.V.; Nguyen, B.H.L.; Ngo, T.A.; Steurer, M.; Schoder, K.; Hovsapian, R. Distributed optimal dynamic state estimation for cyber intrusion detection in networked dc microgrids. In Proceedings of the IECON 45th Annual Conference of the IEEE Industrial Electronics Society 2019, Lisbon, Portugal, 14–17 October 2019. Available online: https://ieeexplore.ieee.org/document/8927045 (accessed on 25 June 2020).
48. Zhao, J.; Mili, L.; Wang, M. A generalized false data injection attacks against power system nonlinear state estimator and countermeasures. *IEEE Trans. Power Syst.* **2018**, *33*, 4868–4877. [CrossRef]
49. Sahoo, S.; Mishra, S.; Peng, J.C.H.; Dragicevic, T. A stealth cyber-attack detection strategy for dc microgrids. *IEEE Trans. Power Electron.* **2019**, *34*, 8162–8174. [CrossRef]
50. Alhelou, H.; Golshan, M.E.; Hatziargyriou, N.D. Deterministic dynamic state estimation-based optimal lfc for interconnected power systems using unknown input observer. *IEEE Trans. Smart Grid* **2020**, *11*, 1582–1592. [CrossRef]
51. Chaojun, G.; Jirutitijaroen, P.; Motani, M. Detecting false data injection attacks in ac state estimation. *IEEE Trans. Smart Grid* **2015**, *6*, 2476–2483. [CrossRef]
52. Liu, Y.; Ning, P.; Reiter, M.K. False data injection attacks against state estimation in electric power grids. In Proceedings of the 16th ACM Conference on Computer and Communications Security, Chicago, IL, USA, 9–13 November 2009.
53. Dán, G.; Sandberg, H. Stealth attacks and protection schemes for state estimators in power systems. *IEEE Int. Conf. Smart Grid Commun.* **2010**, 1–6. [CrossRef]
54. Hu, Q.; Fooladivanda, D.; Chang, Y.H.; Tomlin, C.J. Secure state estimation and control for cyber security of the nonlinear power systems. *IEEE Trans. Control Netw. Syst.* **2017**, *5*. [CrossRef]
55. Gallo, A.J.; Turan, M.S.; Nahata, P.; Boem, F.; Parisini, T.; Ferrari-Trecate, G. Distributed cyber-attack detection in the secondary control of dc microgrids. In Proceedings of the European Control Conference, Limassol, Cyprus, 12–15 June 2018. Available online: https://zenodo.org/record/2590092#.XzYHZzURXIU (accessed on 3 July 2020).
56. Habibi, M.R.; Baghaee, H.R.; Dragicevic, T.; Blaabjerg, F. Detection of false data injection cyber-attacks in dc microgrids based on recurrent neural networks. *IEEE J. Emerg. Sel. Top. Power Electron.* **2020**. [CrossRef]
57. Sahoo, S.; Peng, J.C.H.; Devakumar, A.; Mishra, S.; Dragičević, T. On detection of false data in cooperative dc microgrids a discordant element approach. *IEEE Trans. Ind. Electron.* **2020**, *67*, 6562–6571. [CrossRef]
58. Beg, O.A.; Johnson, T.T.; Davoudi, A. Detection of false-data injection attacks in cyber-physical DC microgrids. *IEEE Trans. Ind. Inform.* **2017**, *13*, 2693–2703. [CrossRef]
59. Li, W.; Joós, G.; Bélanger, J. Real-time simulation of a wind turbine generator coupled with a battery supercapacitor energy storage system. *IEEE Trans. Ind. Electron.* **2010**, *57*, 1137–1145. [CrossRef]
60. Zhang, J.; Chu, Z.; Sankar, L.; Kosut, O. Can attackers with limited information exploit historical data to mount successful false data injection attacks on power systems? *IEEE Trans. Power Syst.* **2018**, *33*, 4775–4786. [CrossRef]
61. Hammad, E.; Ezeme, M.; Farraj, A. Implementation and development of an offline co-simulation testbed for studies of power systems cyber security and control verification. *Int. J. Electr. Power Energy Syst.* **2018**, *104*, 817–826. [CrossRef]

62. Kosek, A.M.; Lünsdorf, O.; Scherfke, S.; Gehrke, O.; Rohjans, S. Evaluation of smart grid control strategies in co-simulation: Integration of IPSYS and mosaic. In Proceedings of the 2014 Power Systems Computation Conference, Wroclaw, Poland, 18–22 August 2014.
63. The History of Making the Grid Smart Engineering and Technology History Wiki. Available online: https://ethw.org/The_History_of_Making_the_Grid_Smart (accessed on 18 June 2020).
64. Uribe-Pérez, N.; Hernández, L.; de la Vega, D.; Angulo, I. State of the art and trends review of smart metering in electricity grids. *Appl. Sci.* **2016**, *6*, 68.
65. European Parliament and Council. Legislative acts, Directive 2012/27/EU of the European Parliament and of the Council of 25 October on energy efficiency, amending Directives 2009/125/EC and 2010/30/EU and repealing Directives 2004/8/EC and 2006/32/EC, 2012/27/EU. *Off. J. Eur. Union* **2012**, *12*, 1–56.
66. Avancini, D.B.; Rodrigues, J.J.P.C.; Martins, S.G.B.; Rabêlo, R.A.L.; Al-Muhtadi, J.; Solic, P. Energy meters evolution in smart grids: A review. *J. Clean. Prod.* **2019**, *217*, 702–715. [CrossRef]
67. Tellbach, D.; Li, Y.F. Cyber-attacks on smart meters in household nanogrid: Modeling, simulation and analysis. *Energies* **2018**, *11*, 316. [CrossRef]
68. Patil, Y.S.; Sankpal, S.V. Multi-Player Attack Detection Model for Smart Meter Security in Smart Grid Systems. *Int. J. Appl. Eng. Res.* **2019**, *7*, 1488–1492.
69. Hasse, F.; Von Perfall, A.; Hillebrand, T.; Smole, E.; Lay, M.; Charlet, L. Blockchain–an Opportunity for Energy Producers and Consumers? Available online: https://asian-power.com/sites/default/files/asianpower/print/AP_Novdec16_p44-45.pdf (accessed on 12 July 2020).
70. Andoni, M.; Valentin, R.; David, F.; Simone, A.; Dale, G.; David, J.; Peter, M.; Andrew, P. Blockchain technology in the energy sector: A systematic review of challenges and opportunities. *Renew. Sustain. Energy Rev.* **2018**, *100*, 143–174. [CrossRef]
71. Li, Z.; Bahramirad, S.; Paaso, A.; Yan, M.; Shahidehpour, M. Blockchain for decentralized transactive energy management system in networked microgrids. *Electr. J.* **2019**, *32*, 58–72. [CrossRef]

of litz wire, used to damp VFTO. Such winding meets the requirement of sufficient resistance to damp VFTO. Section 6 summarizes the damping effects of the damping busbar and the improved design. Finally, conclusions are drawn in Section 7.

2. The Suppression Mechanism of the New Method

Several studies have shown that VFTO is generated as a superposition of multiple different reflected electromagnetic waves with complex nonharmonic time dependence and covers a wide frequency range from 100 kHz to 100 MHz [27]. The wavefront steepness, amplitude, and high-frequency components of VFTO can be suppressed by inductance and resistance. Based on this, this research proposes a VFTO suppression method—the spiral tube damping busbar—located before the disconnector switch (DS), as shown in Figure 1b. At the rated frequency, the damping busbar transmits current and voltage waves like a normal busbar, but when VFTO passes through the damping busbar, it will be suppressed. Also, the energy of the travelling waves is consumed by the resistance. When the resistance and inductance of the damping busbar match each other, the biggest wave energy consumption can be obtained and the best VFT damping can be achieved.

(a) (b)

Figure 1. The new damping busbar installed in a gas insulated substations (GIS): (**a**) general view of the damping busbar with its components; (**b**) the placement of the damping busbar.

2.1. The Structure of the Damping Busbar

The conventional GIS busbar was hollowed out into a spiral tube, and the busbar conductor was changed into a series circuit with a multiturn hollow inductance coil and a multiturn gap; then, we paralleled the damping resistance with the spiral tube inductance circuit. This method increases the active losses of the busbar by changing the wave impedance of the busbar. As a result, the transient energy is consumed, and the amplitude of the VFTO is reduced. Thus, the main components of the spiral tube damping busbar are (a) a spiral tube damping busbar made by screwing out the conventional busbar to a spiral slotted solenoid busbar; (b) solid metal oxide resistors with an antipulse voltage function as the noninductive damping resistors, as shown in Figure 2; and (c) electrical connection tools.

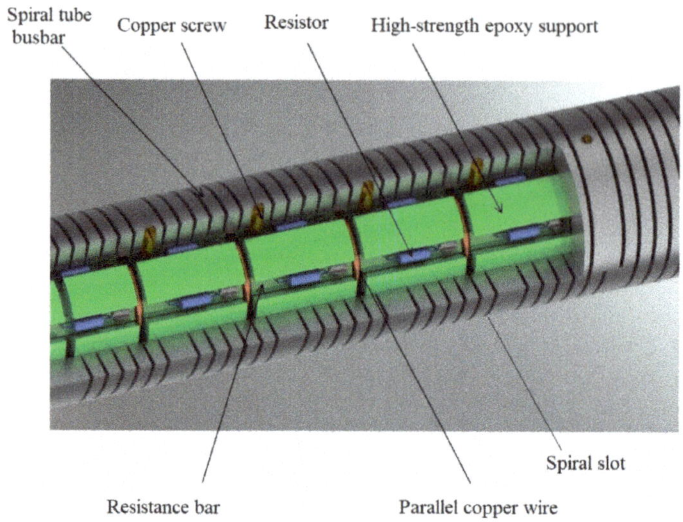

Figure 2. The structure of the damping busbar.

2.2. The Equivalent Circuit of the Damping Busbar

The equivalent circuit of the damping busbar is resistance and inductance connected in series with the GIS bus, as shown in Figure 3. $L_i(i = 0,1,2, \ldots)$ represents the inductance of each unit coil of the metal spiral tube conductor. R_i and $L_{Ri}(i = 0,1,2, \ldots)$ represent the noninductive resistance and its residual inductance of the parallel connection of each turn, respectively, and the value of the resistance can be adjusted to achieve the best damping effect. $g_i(i = 0,1,2, \ldots)$ is the hollowing gap of the damping busbar and $r_i(i = 0,1,2, \ldots)$ is the arcing resistance of the gap, which represents the losses formed by the discharge channel.

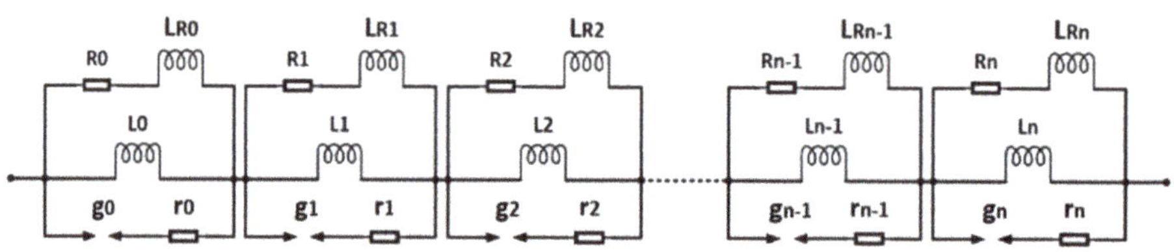

Figure 3. The distributed parameters of the equivalent circuit of the damping busbar.

Inductance Calculation of the Damping Busbar at High Frequency-Damping Busbar Parameters

The purpose of this section is to calculate a fairly precise value of the inductance. Determining the inductance was carried out by studying the effect of high-frequency fields with all the geometrical parameters of the busbar [28]. To this end, a calculation method for the self-inductance of the damping busbar at high frequency was developed and analyzed by the finite element method (FEM). In order to design the damping busbar, we needed to use a simulation method to obtain the VFTO waveform and then calculate the VFTO distributions. Thus, the mathematical expression of the VFTO waveform was theoretically calculated using the curve fitting method to get the fitting data of the VFTO waveform [29]. Then, the curve fitting method was applied to get the fitting data of the associated very fast transient current (VFTC) waveform for a single SF_6 gas discharge. Due to VFTC being assigned as current excitation for the simulation procedures, the VFTC equation was obtained for Fourier 8, with goodness of fit parameters R-square = 0.9768 and adjusted R-square = 0.9759.

$$VFTC(t) = a_0 + \sum_{i=1}^{8} a_i \cdot \cos(i\omega t) + \sum_{i=1}^{8} b_i . \sin(i\omega t) \quad (1)$$

Table 1 illustrates the coefficients of the VFTC equation.

Table 1. Coefficients of the very fast transient current (VFTC) equation.

	Values								
i	0	1	2	3	4	5	6	7	8
a_i	1025	403.3	−7.241	809.5	−28.8	46.59	23.15	−1149	1881
b_i	0	642.6	2540	−3334	−808.4	161.9	−859.5	1063	−1489
ω	1.372×10^7								

The aim of the simulation study was to calculate the equivalent inductance of the damping busbar at high frequency, so some assumptions were proposed for ease of simulation. In this paper, the simple geometry model was imported into ANSYS Maxwell, and the eddy current solution type was chosen. However, this design depends on many variables according to ANSYS analysis: material properties (aluminum); the geometry of the busbar (busbar body); and operating frequency and input current (the VFTC frequency is 2 MHz and the VFTC peak value is 8256 A—see Equation (1)—for singular switching opening operation, as shown in Figure 4). All of these can be included in design sweeps to study how the impedance matrix—and, therefore, the inductance—is affected by changes in these properties (frequency sweep setup: set the frequency field from 1 to 2 MHz with 0.1 MHz step size).

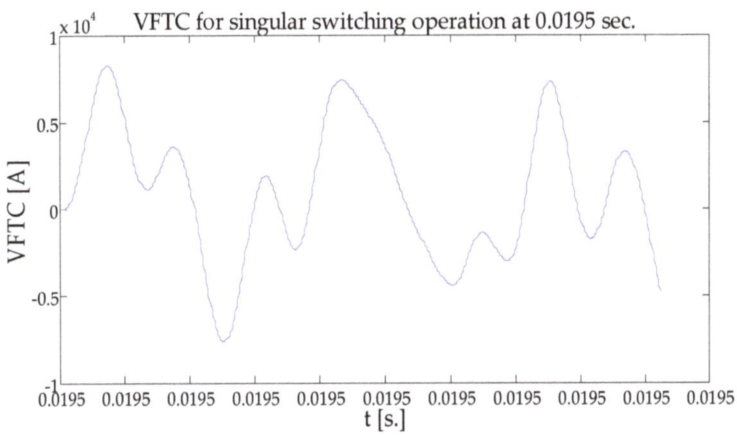

Figure 4. VFTC for a single breakdown at 0.0195 s.

The simulation results show a decrease in the inductance value of the damping busbar taking place for different frequencies, as illustrated in Figure 5. Notably, however, Figure 5 shows that the inductance value of the damping busbar is 0.33656 mH at 2 MHz. As a consequence, the damping busbar parameters are illustrated in Table 2.

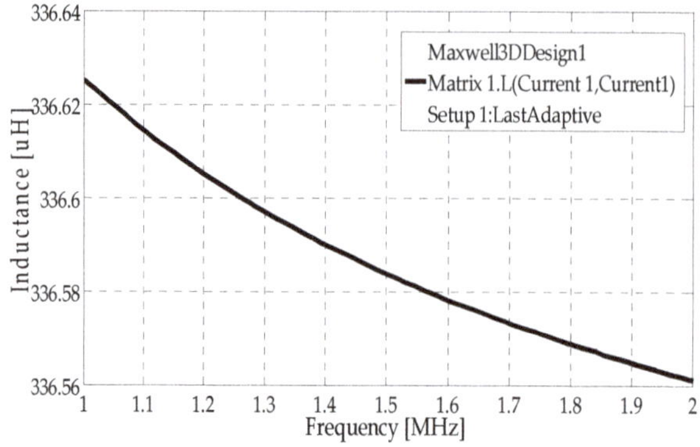

Figure 5. Inductance calculation of the damping busbar at high frequency.

Table 2. Damping busbar parameters.

	Parameters			
	Number of Turns	**Units**	**Resistance per Unit [Ω]**	**Inductance [mH]**
Value	28	14	6 Resistors in parallel Req./unit = 20 Ω	0.33656

3. Simulation Results

3.1. Modelling of a 1000 kV GIS

The equivalent circuit diagram of a 1000 kV GIS was simulated by using EMTP (Electromagnetic Transients Program) simulation in order to study the VFTO with and without installing the damping busbar [30]. However, the high-frequency characteristics of VFTO led to simulating GIS components as capacitances dominating the other parameters. In order to model the GIS bus duct, distributed parameters and lumped elements can be utilized. Surge impedance and wave velocity could be calculated for a GIS section of any length by using the physical dimensions of the bus duct [31,32]. In the following equations, C and L are self-capacitance and inductance, Z is surge impedance, and V is wave velocity in the GIS [33,34]:

$$C = \frac{2\pi\varepsilon_0\varepsilon_r}{\ln R/r}, \tag{2}$$

$$L = \frac{\mu_0\mu_r}{2\pi}\ln\frac{2R}{r}, \tag{3}$$

$$Z = \sqrt{\frac{L}{C}} = \frac{1}{2\pi}\sqrt{\frac{\mu_0\mu_r}{\varepsilon_0\varepsilon_r}}\ln\frac{R}{r}, \tag{4}$$

$$V = \frac{1}{\sqrt{LC}}. \tag{5}$$

Figure 6 shows the equivalent circuit of a 1000 kV GIS using EMTP simulation after installing the equivalent circuit of the damping busbar, as illustrated in Table 2.

Let r be the inner radius of the GIS shell and R be the inner radius of the GIS bus. In addition, $\varepsilon_0 = 8.85418782 \times 10^{-12}$ F/m, $\varepsilon_r = 1.0024$ and $\mu_0 = 4\pi \times 10^{-7}$ H/m. According to the dimensions of the GIS busbar, the calculations showed that the propagation velocity is 295 m/μs and the surge impedance is about 93.382 ohm/m.

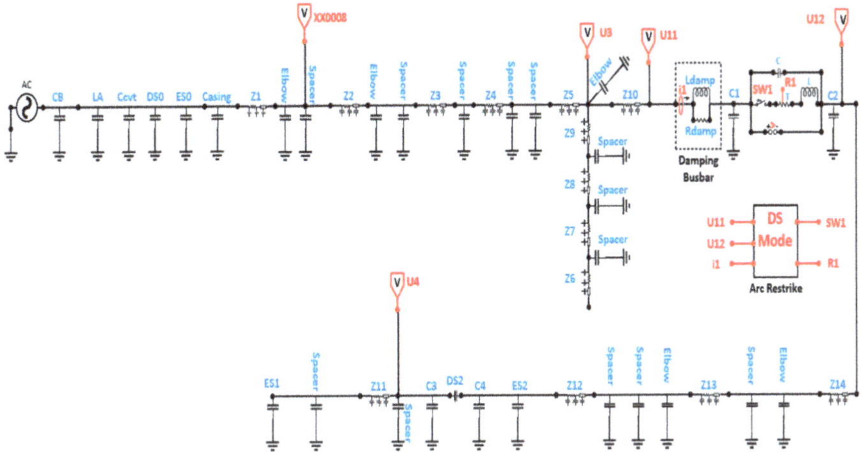

Figure 6. The equivalent circuit of a 1000 kV GIS after installing the damping busbar.

3.2. Discussion

As shown in Figure 7a, when the disconnector is opened, the VFTO for the whole process at the power supply side appears as a superposition of high-frequency transients and a sine wave. After SF_6 gas discharges, the disconnector gap restores the insulation state; then, the current disappears, the high-frequency transient components gradually decay, and, finally, the voltage wave continues as a sinusoidal wave with the supply voltage.

In addition, the VFTO for the full process at the load side has an approximate step shape, which is a superposition of high-frequency transients and a stepped wave. Each step of the ladder wave corresponds to SF_6 gas breakdown, the step magnitude is the breakdown voltage, and the narrow pulse at the ladder edge is the highest frequency of the VFTO, as shown in Figure 7c.

Moreover, the VFTC for the whole process is the superposition of high-frequency transients with zero current amplitude. When the critical breakdown voltage (BV) of the SF_6 gas is equal to the transient recovery voltage of the gap, the SF_6 gas collapses and the gap transfers from the insulation state into a conductive state; then, the circuit produces a high-frequency transient current, as shown in Figure 7e. Additionally, when the critical breakdown voltage of SF_6 gas exceeds the transient recovery voltage of the gap, the current disappears during discharge of the SF_6 gas.

The simulation results show that the maximum amplitude of the VFTO decreased from 1.554 to 1.13 p.u., which means that the damping busbar functioned as a VFTO suppressor. The damping effect of the new design is evident from the simulation results, and the mitigation effect was about 27%. Table 3 illustrates the transient characteristics of the VFTO waveform under opening operation before and after installing the damping busbar. A singular switching operation (opening) of the disconnector was simulated in EMTP-ATP (Alternative Transients Program) software.

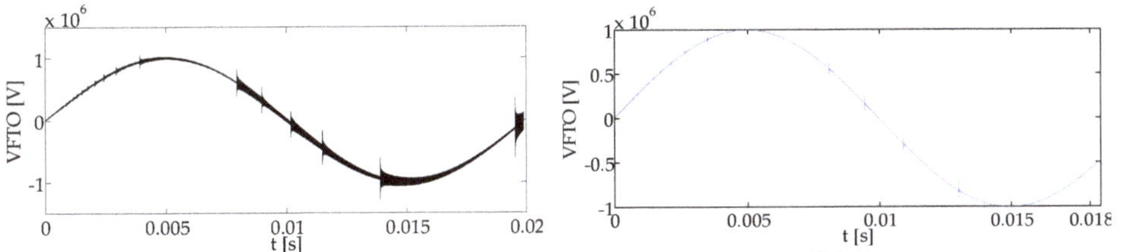

(a) (b)

Figure 7. *Cont.*

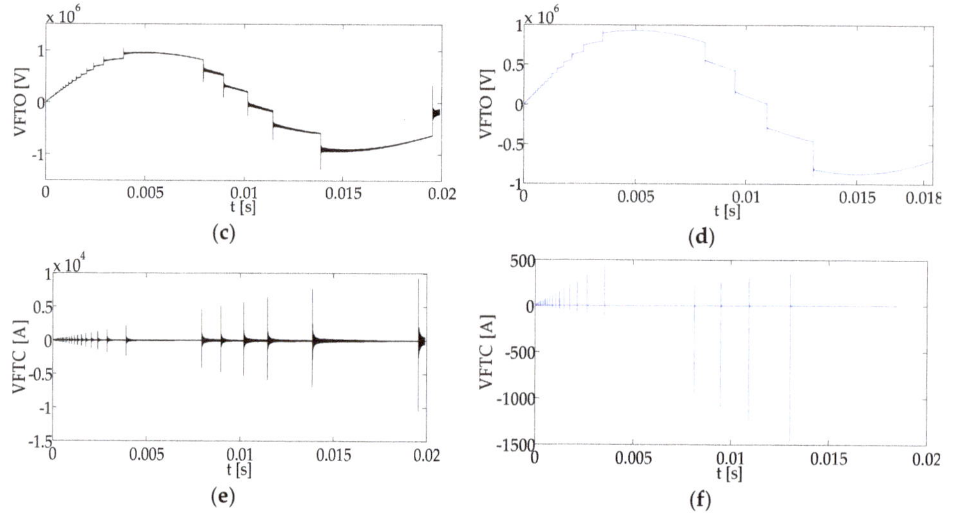

Figure 7. Very fast transient overvoltage (VFTO) and VFTC waveforms: VFTO at the source side (**a**) before installing the damping busbar and (**b**) after installing it; VFTO at the load side (**c**) before installing the damping busbar and (**d**) after installing it; VFTC (**e**) before installing the damping busbar and (**f**) after installing it.

Table 3. The transient characteristics of VFTO under an opening operation at the load side before and after installing the damping busbar.

VFTO Characteristic Parameters	Simulation Results	
	Before Installing the Damping Busbar	After Installing the Damping Busbar
Rise time [ns]	10	30
Average breakdown steepness [kV/ns]	71.53	48.725
Maximum amplitude of VFTO [p.u.]	1.554	1.13
Average voltage overshoot coefficient	0.487	0.244
Delay time [μs]	24.4	47.2

4. Inductance Effect on Suppressing VFTO

When VFTO passes through the inductance, the wavefront is smoothed, its steepness is reduced, and the amplitude is decreased. The inductance L_i increases the round-trip time of the travelling wave, which in turn leads to a reduction in the high-frequency components of the VFTO. The travelling wave leads to a higher voltage on the inductance of the small circuit, which causes the parallel resistance R_i to absorb the travelling wave energy, increase the consumption of active power, and reduce the amplitude of the VFTO.

The inductance value of the damping busbar is an important factor for an obvious VFTO suppression effect. The limitation on the steepness of the transients is a result of the inductance effect of the damping busbar. Figure 8 describes the influence of the inductance value on VFTO suppression at the load side during transient phenomena.

The more considerable value of the equivalent inductance of the damping busbar leads to an obvious suppression effect. Table 4 illustrates the transient characteristics of the VFTO waveform under an opening operation for different values of equivalent inductance of the damping busbar.

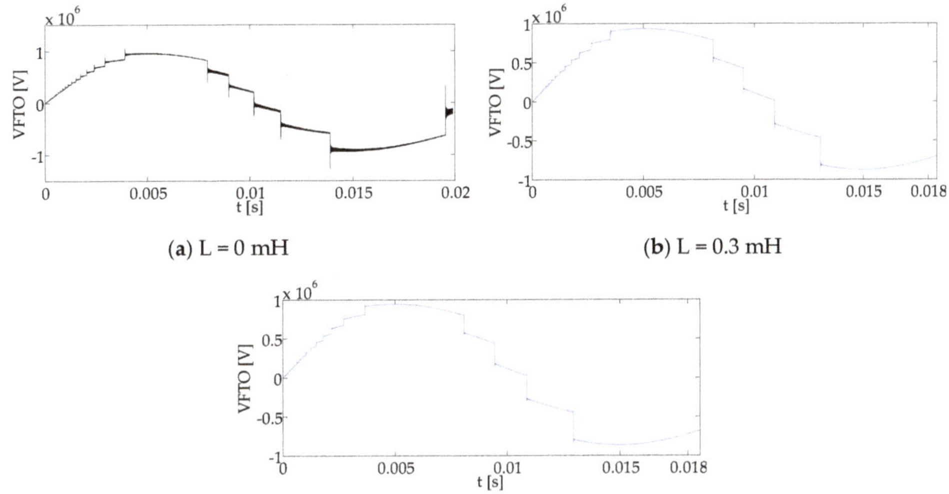

(**a**) L = 0 mH (**b**) L = 0.3 mH (**c**) L = 3 mH

Figure 8. Influence of the inductance value on VFTO suppression at the load side of a 1000 kV GIS: at inductance values (**a**) L = 0 mH, (**b**) L = 0.3 mH, (**c**) L = 3 mH.

Table 4. Influence of the equivalent inductance of the damping busbar on VFTO suppression.

VFTO Characteristics Parameters	Simulation Results		
	L = 0 mH	L = 0.3 mH	L = 3 mH
Rise time [ns]	10	30	35
Maximum amplitude of VFTO [p.u.]	1.554	1.13	1.0

It can be observed that a higher damping effect was achieved by increasing the equivalent inductance of the damping busbar. Furthermore, increasing the inductance value can damp the VFT waveform's front. It is also helpful for absorbing the sharp spikes of the VFT since surge arresters do not act quickly enough to prevent steep-fronted switching transients. Therefore, installing an additional spiral coil with the damping busbar is highly valuable to attenuating the effects of VFTs. For this reason, a spiral coil of resistive litz wire was designed to be installed in the helical groove and connected in series with the damping busbar.

5. The Improved Design of the Damping Busbar

A further enhancement of the suppression effect was investigated by designing a spiral litz coil connected in series with the busbar, which increases the total inductance value. In order to design the spiral coil, an algorithm based on air-gap calculation was developed. This algorithm designs a spiral coil for a specific inductance value, which was determined as an input value. Other parameters, such as VFTC characteristics and the number of turns of the damping busbar, were also inserted in the input data.

Furthermore, litz wire was chosen for the improved design because it is mostly used for the frequency range 10 kHz to 2 MHz. The main advantage of this configuration is to minimize the power losses due to "skin and proximity effect", and it is desirable in high-frequency applications as well. Sullivan, C.R. and Zhang, R.Y applied a simple method for a suitable litz wire design [35]. The flowchart of the procedure used to calculate the spiral coil design parameters is shown in Figure 9. By determining the inductance value and the number of turns, then magnetic flux density could be calculated. After that, the eddy power losses in the litz coil were calculated for the optimal design.

However, we can divide the strand-level proximity effect into the internal proximity effect (the effect of other currents within the bundle) and the external proximity effect, but the total proximity effect may be considered a result of the total field at any given strand [36–38]:

$$P_{eddy,strand} = \frac{\pi\omega^2\overline{\hat{B}^2}{d_s}^4 nl}{128\rho_c}\left(1 + \frac{\pi^2 n{d_s}^2}{4K_a p^2}\right) \quad (6)$$

where ω is the angular frequency, l is the length of the bundle, ρ_c is the resistivity of copper, p is the pitch of the twisting, n is the number of strands in a bundle, d_s is the diameter of each strand, $\overline{\hat{B}^2}$ is the peak magnetic flux density, and K_a is the strand packing factor. Thus, the AC resistance factor, $F_r = R_{ac}/R_{dc}$, for a litz-wire winding for arbitrary waveforms and 2-D or 3-D field geometries can be expressed as follows:

$$F_r = \frac{R_{ac}}{R_{dc}} = 1 + \frac{(\pi n N_s)^2{d_s}^6}{192\delta^4 b^2}. \quad (7)$$

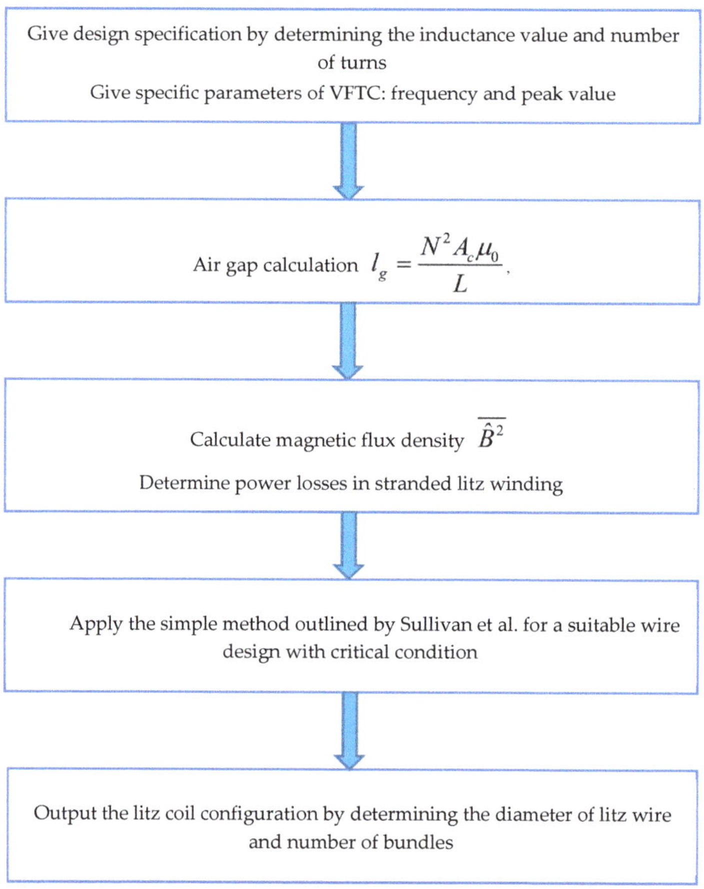

Figure 9. The flowchart of the procedure to calculate the litz coil design.

The power loss in a stranded-wire winding is derived from $P = F_r P_r = F_r \frac{I^2_{rms}.l}{nA_s}$, where A_s is the cross-sectional area of a strand. Consequently, by inserting the operating frequency, the AWG (American wire gauge) can be chosen (i.e., when frequency range is from 1.4 MHz to 2.8 MHz, then the best choice is AWG = 48), which means that about 70% loss reduction is achieved. Eddy power losses were calculated, and the final design was determined according to the critical factor (diameter of litz wire no more than the width of the spiral slot on the damping busbar surface).

The aim of calculating the air gap was to determine the magnetic flux density in order to calculate the power losses in the litz wire for the optimal design of the coil. Thus, for a typical design, we completed the following:

1. Determine A_c of the core, which was an aluminum cylinder (busbar body) in our study.
2. Determine the number of turns which was commensurate with the design.
3. Determine the inductance value which achieves an effective design to avoid saturation.
4. Calculate the air gap length as a first step to calculate the magnetic flux density in order to calculate the power loss in the litz wire for the optimal design of the coil. Thus, the length of the air gap was calculated by using Ampere's law, as shown below:

$$l_g = \frac{N^2 A_c \mu_0}{L}. \tag{8}$$

Many different combinations of strand diameter and number of strands could be designed and manufactured for any given cost. Thus, the number of bundles and, finally, the litz coil diameter could be obtained using a critical condition (diameter of the litz wire no greater than the width of the spiral slot on the damping busbar surface). The previously described technique was used to calculate the litz coil design. Simulations were performed using MATLAB software by keeping the length of the coil, number of turns, and transient current parameters constant. The results are listed in Table 5.

Table 5. Simulation results for the litz coil design.

Inductance Value [mH]	Simulation Results for Litz Coil Design			
	Number of Bundles	Number of Strands	Strand Diameter [mm]	Construction Type
0.1	5	140	0.899	Stranded litz wire, type 2
0.3	5	140	0.838	
3.0	5	140	0.702	

It is worth noting that the inductance value of both the damping busbar and the spiral litz coil was calculated at 2 MHz. An accurate choice of construction type of the litz design leads to a higher capability for greater current carrying capacities. The larger Type 2 and 3 litz designs have this essential feature for high-frequency power supply, inverter, and grounding applications. A Type 2 litz construction is distinguished by bundles of twisted wires together, while Type 3 features individually insulated bundles of Type 2 litz wire [39]. As a consequence, in this study we aimed to add a spiral litz coil located in the sweeping spiral on the surface of the damping busbar in order to improve its performance in VFTO mitigation. EMTP simulation was carried out to study the VFTO with and without installing the improved damping busbar. $L_{coil} = 0.1$ mH is the equivalent inductance of the spiral litz coil and $L_{busbar} = 0.33656$ mH is the equivalent inductance of the damping busbar. Thus, $L_{eq.} = 0.43656$ mH is the equivalent inductance of the improved damping busbar. The simulation was set up based on $L_{damp} = 0.43656$ mH, as presented in Figure 6. Thus, Figure 10b shows the higher damping effect of the improved damping busbar. In order to clarify the damping effect, Figure 10d illustrates the damping effect of the improved damping busbar for a singular switching operation at approximately the same moment (0.013 s.). In order to clarify the higher damping effect shown in Figure 10b, further details about the damping effect of the improved design are discussed in the next section.

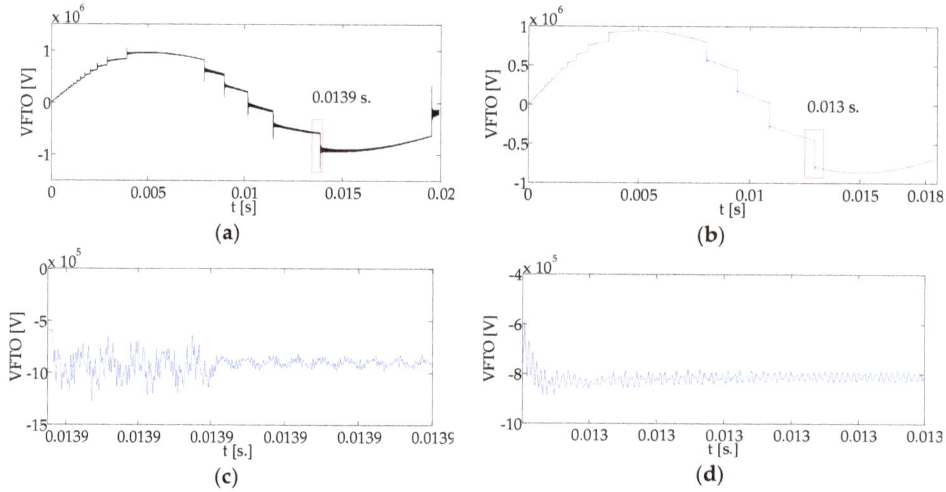

Figure 10. VFTO waveform at the load side: (**a**) before installing the damping busbar, (**b**) after installing the improved design of the damping busbar; VFTO at the load side for one breakdown at 0.013 s: (**c**) before installing the damping busbar, (**d**) after installing the improved design of the damping busbar.

6. Summary

In this work, a higher damping effect was investigated by designing a spiral litz coil connected in series to the damping busbar, which increased the total inductance value. The simulation results show that there was a considerable increase in the damping effect, as shown in Figure 10. Furthermore, improving the design and material type could be useful to achieving an optimal suppressing effect. The main advantages of this improved design are the higher damping effect, accuracy, simplicity, and ability to be used in any existing GIS without any change in the GIS design or construction. The transient characteristics of the VFTO after adding a 0.1 mH litz coil are illustrated in Table 6.

Table 6. The transient characteristics of VFTO under an opening operation at load side with and without installing the spiral litz coil (0.1 mH).

VFTO Characteristics Parameters	Simulation Results		
	Without Damping Busbar	Damping Busbar Only	Spiral Litz Coil in Series with the Damping Busbar
Rise time [ns]	10	29	31
Maximum amplitude of VFTO [p.u.]	1.554	1.13	1.081

Another obvious advantage of this improved design is no magnetic saturation problem and mitigation of the overvoltage magnitude due to dissipation of the transient energy in the damping resistance. On the other hand, litz wire has been classified as a cost-effective choice to reduce eddy current loss in high-frequency transformer and inductor windings. Some solutions could be successful in increasing the dielectric strength for a high voltage level or for safety requirements, like adding thermoplastic insulation such as PTFE, which could be used for bundle insulation on litz wire in some applications [36].

The proposed improved damping busbar has limitations when thermoplastic insulation is added due to the extra cost. So, in future work, it will be fundamental to develop the design to improve the estimation quality to meet the industry's requirements for low costs. However, compared with other existing expensive VFTO suppression methods that require changes in GIS design and construction, the proposed method can be easily implemented in any existing GIS without any change in its structure, making it very cost-effective. Therefore, it is necessary to use simulation calculations to optimize the insulation reliability of the uniform electric field and increase the insulation of the busbar. Table 7 shows the mitigation effect of different damping methods, including the new method.

Table 7. Comparison of different mitigation methods with the new damping busbar.

Method	Mitigation Effect
Damping resistor	Up to 25%
Slow-acting disconnector (low trapped charge voltage)	(15–25)%
High-frequency resonator	Up to 20%
Ferrite rings	Up to (10–30)%
Damping busbar	(22.22–52.94)%

7. Conclusions

A new method to suppress VFTO was proposed, called a damping busbar. The mechanism of this new method was analyzed. A calculation method for the self-inductance of the damping busbar at high frequency was presented. Moreover, the inductance effect of the damping busbar on VFTO suppression was studied as well. As a result, the amplitude of the VFTO was reduced by approximately 27% after applying the damping busbar. A further enhancement of the suppression effect was investigated by designing a spiral litz coil connected in series with the busbar, which increased the total inductance value. Consequently, a good damping effect was achieved by utilizing the inductive and resistance arrangement which was formed by the improved design of the spiral tube damping busbar. The conclusions are as follows.

1. We proposed a new design which has a significantly better damping effect when compared with other existing suppression methods. The proposed damping busbar can be easily implemented in any existing GIS without any change in its structure and is therefore very cost-effective.
2. We investigated a calculation method for the self-inductance at high frequency.
3. We improved the performance of the damping busbar by adding a spiral litz coil connected in series with the busbar in order to enhance the suppression effect, and we developed an algorithm based on air gap calculation to design the litz coil.
4. We compared the damping effect with and without the damping busbar and the improved design.

However, research on improving the proposed design to increase the equivalent inductance value by increasing the length of the busbar and the number of turns has to be considered. Many different adaptations, tests, and experiments should be investigated in future work (i.e., experimental study of a 1000 kV GIS station VFTO after installing the improved damping busbar).

Author Contributions: Conceptualization, R.A.A., Y.-X.S., and W.X.; formal analysis, R.A.A.; investigation, R.A.A., Y.-X.S., and W.X.; writing—original draft preparation, R.A.A. and W.X.; supervision, Y.-X.S., and W.X.

Conflicts of Interest: The authors declare no conflict of interest.

References

1. Povh, D.; Pyc, I.; Retzmann, D.; Weinhold, M.G. Future Developments in Power Industry. In Proceedings of the 14th IERE General Meeting and IERE Central and Eastern Europe Forum, Krakow, Poland, 17–21 October 2004; pp. 1–12.
2. Riechert, U.; Holaus, W. Ultra High-Voltage Gas-Insulated Switchgear—A technology milestone. *Eur. Trans. Electr. Power* **2012**, *22*, 60–82. [CrossRef]
3. Chauhan, M.; Joshi, U.; Asija, U.D. Features and Design Concepts of Gas Insulated Switch Gear. *Int. J. Adv. Res. Sci. Eng.* **2014**, *8354*, 60–67.

4. Mikes, E.; Ponchon, P.; Doumer, P. GIS Substation Extensions and Upgrades. In Proceedings of the CEPSI Paper T1-068 Conference, Manila, Philippines, 23–27 October 2000; pp. 1–8.
5. Nagarsheth, R.; Sushant, S. Study of Gas Insulated Substation and its Comparison with Air Insulated Substation. *Int. J. Electr. Power Energy Syst.* **2014**, *55*, 481–485. [CrossRef]
6. Singha, J.T.M.; Thomas, J.; Naidu, M.S. Study of Fast Transients Generated in a GIS Bus Duct in Compressed SF_6-N_2 Gas Mixtures Containing Lower Percentages of SF_6. In Proceedings of the Dielectric Materials, Measurements and Applications Conference Publication, Bengaluru, India, 17–21 September 2000; pp. 118–122.
7. Ullah, R.; Ullah, Z.; Haider, A.; Amin, S.; Khan, F. Dielectric Properties of Tetrafluoroethane (R134) Gas and its Mixtures with N_2 and Air as a Sustainable Alternative to SF_6 in High Voltage Applications. *Electr. Power Syst. Res.* **2018**, *163*, 532–537. [CrossRef]
8. Carsimamovic, S.; Bajramovic, Z.; Ljevak, M.; Veledar, M. Very Fast Electromagnetic Transients in Air Insulated Substations and Gas Insulated Substations due to Disconnector Switching. In Proceedings of the International Symposium on Electromagnetic Compatibility (EMC), Chicago, IL, USA, 8–12 August 2005; pp. 382–387.
9. Tavakoli, A.; Gholami, A.; Nouri, H.; Negnevitsky, M. Comparison Between Suppressing Approaches of Very Fast Transients in Gas-Insulated Substations (GIS). *IEEE Trans. Power Deliv.* **2013**, *28*, 303–310. [CrossRef]
10. Dhaware, A.M.; Potdar, M.S. Review on Mitigation Methods of Very Fast Transient Overvoltage in Gas-Insulated Switchgear. *Resincap Int. J. Sci. Eng.* **2017**, *1*, 20–25.
11. Yamagata, Y.; Tanaka, K.; Nishiwaki, S.; Takahashi, N.; Kokumai, T.; Miwa, I.; Komukai, T.; Imai, K. Suppression of VFT in 1100kV GIS by Adopting Resistor-Fitted Disconnector. *IEEE Trans. Power Deliv.* **1996**, *11*, 872–880. [CrossRef]
12. Szewczyk, M.; Piasecki, W.; Stosur, M.; Riechert, U.; Kostovic, J. Impact of Disconnector Design on Very Fast Transient Overvoltages in Gas-Insulated UHV Switchgear. In Proceedings of the International Symposium on High Voltage Engineering (ISH), Hannover, Germany, 22–26 August 2011; pp. 35–38.
13. Burow, S.; Straumann, U.; Kohler, W.; Tenbohlen, S. New Methods of Damping Very Fast Transient Overvoltages in Gas-Insulated Switchgear. *IEEE Trans. Power Deliv.* **2014**, *29*, 2332–2339. [CrossRef]
14. Babaeilaktarashani, M. A New Cost Effective Approach to Suppress Very Fast Transients on Power Transformers Connected to Gas Insulated Substations. Master's Thesis, Curtin University, Perth, Australia, 2016.
15. Smajic, J.; Shoory, A.; Burow, S.; Holaus, W.; Riechert, U.; Tenbohlen, S. Simulation Based Design of HF-Resonators for Damping of Very Fast Transients in GIS. *IEEE Trans. Power Deliv.* **2014**, *29*, 2528–2533. [CrossRef]
16. Riechert, U.; Bösch, M.; Smajic, J.; Shoory, A.; Szewczyk, M.; Piasecki, W.; Burow, S.; Tenbohlen, S. Mitigation of Very Fast Transient Overvoltages in Gas Insulated UHV Substations. In Proceedings of the CIGRÉ Report A3-110, 44th CIGRÉ Session 2012, Palais des Congrès de Paris (CIGRÉ), Paris, France, 26–31 August 2012; pp. 26–31.
17. Szewczyk, M.; Pawłowski, J.; Kutorasiński, K.; Piasecki, W.; Florkowski, M.; Straumann, U. High-Frequency Model of Magnetic Rings for Simulation of VFTO Damping in Gas-Insulated Switchgear with Full-Scale Validation. *IEEE Trans. Power Deliv.* **2015**, *30*, 2331–2338. [CrossRef]
18. He, J.; Guan, Y.; Liu, W.; Chen, W.; Li, Z. Design Optimization of Ferrite Rings for VFTO Mitigation. *IEEE Trans. Power Deliv.* **2015**, *32*, 1181–1186. [CrossRef]
19. Guan, Y.; Yue, G.; Chen, W.; Li, Z.; Liu, W. Experimental Research on Suppressing VFTO in GIS by Magnetic Rings. *IEEE Trans. Power Deliv.* **2013**, *28*, 2558–2565. [CrossRef]
20. Reddy, P.R.; Amarnath, J. Nanocrystalline to Suppress VFTO and VFTC for 245kV Gas Insulated Substations. *Int. J. Comput. Appl.* **2013**, *70*, 1–6.
21. Szewczyk, M.; Kuniewski, M.; Piasecki, W.; Florkowski, M.; Straumann, U. Determination of Breakdown Voltage Characteristics of 1'100 kV Disconnector for Modeling of VFTO in Gas-Insulated Switchgear. *IEEE Trans. Power Deliv.* **2016**, *31*, 2151–2158. [CrossRef]
22. Szewczyk, M.; Kuniewski, M. Controlled Voltage Breakdown in Disconnector Contact System for VFTO Mitigation in Gas-Insulated Switchgear (GIS). *IEEE Trans. Power Deliv.* **2017**, *32*, 2360–2366. [CrossRef]

23. Yamagata, Y.; Nakada, Y.; Nojima, K.; Kosakada, M.; Ozawa, J.; Ishigaki, I. Very Fast Transients in 1000 kV Gas Insulated Switchgear. In Proceedings of the 1999 IEEE Transmission and Distribution Conference (Cat. No. 99CH36333), New Orleans, LA, USA, 11–16 April 1999; pp. 501–508.
24. Chen, W.J.; Wang, H.; Han, B.; Wang, L.; Ma, G.M.; Yue, G.C.; Li, Z.; Hu, H. Study on the Influence of Disconnector Characteristics on Very Fast Transient Overvoltages in 1100 kV Gas-Insulated Switchgear. *IEEE Trans. Power App. Syst.* **2015**, *30*, 2037–2044. [CrossRef]
25. Yadav, D.N.; Gorayan, R.S. Investigation of VFTO's in GIS Using EMTP. In Proceedings of the 13th National Power Systems Conference NPSC, Madras, India, 27–30 December 2004; pp. 619–623.
26. Rao, J.R.; Amarnath, J.; Kamakshaiah, S. Simulation and Measurement of Very Fast Transient Over Voltages in a 245kV GIS and Research on Suppressing Method Using Ferrite Rings. *ARPN J. Eng. Appl. Sci.* **2010**, *5*, 88–95.
27. Li, Y.; Shang, Y.; Zhang, L.; Shi, R.; Shi, W. Analysis of Very Fast Transient Overvoltages (VFTO) from Onsite Measurements on 800 kV GIS. *IEEE Trans. Dielectr. Electr. Insul.* **2012**, *19*, 2102–2110.
28. Coffin, J.G. The Influence of Frequency upon the Self-Inductance of Coils. *Proc. Am. Acad. Arts Sci.* **1906**, *41*, 275–296. [CrossRef]
29. Almenweer, R.A.; Yixin, S.; Xixiu, W. Research on the Fitting Method to Describe the Mathematical Expression of VFTO in GIS. *IET. J. Eng.* **2019**, *2019*, 2053–2057. [CrossRef]
30. Rodríguez-Medina, B.; Santiago-Luna, M. *Quick Guide ATP/EMTP*; Electric Power Engineering Group: Mayagüez, Puerto Rico, 18 June 2002; pp. 1–34.
31. Ram, A.R.; Kumar, K.S. Analysis of Very Fast Transients in EHV Gas Insulated Substations. *Int. J. Eng. Sci. Innov. Technol.* **2012**, *1*, 18–26.
32. Prakasam, K.; Prabhavathi, D.; Kalavathi, M.S.; Reddy, B.R. Analysis of Very Fast Transient Over Voltages in Gas Insulated Substation. *Int. J. Electr. Eng. Technol.* **2015**, *6*, 51–59.
33. Hu, R.; Cui, X.; Chen, W.; Sun, Z.; Qi, L.; Zhang, W.; Li, Z.; Dai, M. Simulation of Very Fast Transient Overvoltage on UHV 1000kV GIS Test Circuit. In Proceedings of the 6th International Conference on Electromagnetic Field Problems and Applications-ICEF, Dalian, China, 19–21 June 2012; pp. 1–4.
34. Hosseini, S.M.H.; Bagheri, Z.; Niafenderi, S.D. Modeling of 420 kV GIS Substation and Research on Very Fast Transient Overvoltage (VFTO). *Res. J. Appl. Sci. Eng. Technol.* **2013**, *5*, 2656–2660. [CrossRef]
35. Sullivan, C.R.; Zhang, R.Y. Simplified Design Method for Litz Wire. In Proceedings of the IEEE Applied Power Electronics Conference and Exposition—APEC, Fort Worth, TX, USA, 16–20 March 2014; pp. 2667–2674.
36. Tang, X.; Sullivan, C.R. Optimization of Stranded-Wire Windings and Comparison with Litz Wire on the Basis of Cost and Loss. In Proceedings of the IEEE 35th Annual Power Electronics Specialists Conference (IEEE Cat. No.04CH37551), Aachen, Germany, 20–25 June 2004; pp. 854–860.
37. Murgatroyd, P.N. Calculation of Proximity Losses in Multi-stranded Conductor Bunches. *IEE Proc. Sci. Meas. Technol.* **1989**, *136*, 115–120.
38. Wojda, R.P. Winding Resistance and Winding Power Loss of High-Frequency Power Inductors. Ph.D. Thesis, Wright State University, Dayton, OH, USA, 2012.
39. Reese, B.A.; Joseph, R.; Sullivan, C.R. Improved Litz-Wire Designs for the MHz Range. In Proceedings of the IEEE 19th Workshop on Control and Modeling for Power Electronics (COMPEL), Padua, Italy, 25–28 June 2018; pp. 1–8.

applied sciences

MDPI

Article

Detecting Grounding Grid Orientation: Transient Electromagnetic Approach

Aamir Qamar [1,*], Inzamam Ul Haq [2], Majed Alhaisoni [3] and Nadia Nawaz Qadri [1]

1 Department of Electrical and Computer Engineering, COMSATS University Islamabad, Wah Campus, Wah 47040, Pakistan; drnadia@ciitwah.edu.pk

2 Department of High Voltage and Insulation Technology, Chongqing University, Chongqing 400044, China; inzimam.324@gmail.com

3 College of Computer Science and Engineering, University of Ha'il, Ha'il 2440, Saudi Arabia; majed.alhaisoni@gmail.com

* Correspondence: aamirqamar@ciitwah.edu.pk; Tel.: +92-3339869637

Received: 7 November 2019; Accepted: 2 December 2019; Published: 4 December 2019

Abstract: The configuration is essential to diagnose the status of the grounding grid, but the orientation of the unknown grounding grid is ultimately required to diagnose its configuration explicitly. This paper presents a transient electromagnetic method (TEM) to determine grounding grid orientation without excavation. Unlike the existing pathological solutions, TEM does not enhance the surrounding electromagnetic environment. A secondary magnetic field as a consequence of induced eddy currents is subjected to inversion calculation. The orientation of the grounding grid is diagnosed from the equivalent resistivity distribution against the circle perimeter. High equivalent resistivity at a point on the circle implies the grounding grid conductor and vice versa. Furthermore, various mesh configurations including the presence of a diagonal branch and unequal mesh spacing are taken into account. Simulations are performed using COMSOL Multiphysics and MATLAB to verify the usefulness of the proposed method.

Keywords: grounding grid; magnetic field; orientation; transient electromagnetic method (TEM)

---

1. Introduction

The grounding grid ensures the safety of personnel and power equipment in the substation facility. It also provides stable voltages to the equipment without disturbing the continuity of supply. The main aim of the grounding grid is to provide a low impedance path to fault currents caused by lightening strikes, short circuits, and switching surges [1–4].

The grounding grid is a lattice of horizontal bare conductors extending across the entire area of the substation. It is mainly made up of steel, galvanized steel, copper, copper clade steel, etc. Practically, the mesh size of a grounding grid varies from 3 m to 7 m with a depth from 0.7 m to 1 m [5]. As the grounding grid is hidden inside Earth, vertical conductors are the only access points from the Earth's surface. Moreover, fault currents are effectively dissipated into the Earth via vertical grounding rods that connect the grounding grid with low resistivity soil. Based on the substation configuration, the grounding grid mesh may be of equal or unequal spacing and may have a diagonal branch. The optimized configuration of the grounding grid plays a vital role in improving its efficiency. Therefore, the configuration is frequently modified, which is achieved by changing the mesh and adding vertical grounding rods [6,7]. A typical grounding grid is shown in Figure 1.

Grounding grid resistance (Rg), ground potential rise (GPR), maximum touch voltage, and maximum step voltage are the key parameters for measuring the performance of the grounding grid [5]. The step and touch voltages are influenced by the GPR and the configuration of the equipment in a substation. In case of a lightning strike, the step and touch voltages are the significant factors to improve the safety of the grounding grid [8,9]. After years of operation, grounding conductors corrode and even break. Corrosion occurs due to the presence of water particles and air gaps in soil. Corrosion and breakpoints reduce the efficiency of the grounding grid, which can cause serious damage to the equipment, as well as personnel. Therefore, the stable operation of the grounding grid requires regular diagnostic tests.

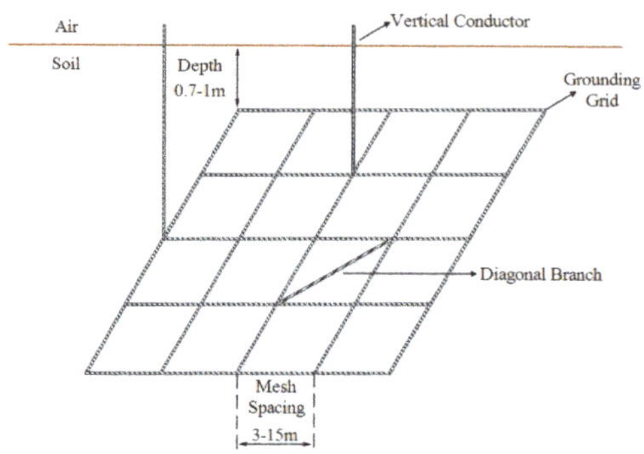

Figure 1. Grounding grid and its characteristics.

In essence, the performance analysis of unknown grounding grids is comprised of the following stages: fault diagnosis, configuration detection, and orientation detection; the former being dependent on the latter. Out of the three stages, orientation detection is the least addressed stage in spite of being very basic to the performance analysis of the unknown grounding grid. This is because the existing literature has considered the orientation of unknown grounding grids as parallel to the substation boundary, which practically may differ and leads the existing methods of configuration detection to ultimately fail [10]. Therefore, this paper proposes the transient electromagnetic method (TEM) to diagnose the orientation of the unknown grounding grid. Furthermore, the proposed method is validated for different mesh configurations of the grounding grid.

2. Related Work

There has been growing interest in fault diagnosis as excavation is resource intensive both in terms of time and effort. Recent studies on fault diagnosis of the grounding grid can be categorized into electric network methods [11–14], electromagnetic methods [15–18], electrochemical detection methods [19], and transient electromagnetic methods [16,20]. Electrical network methods are based on the surface potential difference and port resistance. These methods have low accuracy as the surface potential difference and port resistance are very small even if the grid is broken. Electromagnetic methods are based on processing the surface magnetic intensity once the current is injected in the grid. The accuracy of these methods depends on soil resistivity. Once the soil condition is changed, re-measurements are required. Grounding grid corrosion level is easily detected by the electrochemical method by measuring the electrochemical properties between grounding conductors and soil. However, this method fails to diagnose breakpoints in the grounding grid. In the transient electromagnetic method, equivalent resistivity is calculated by performing fast inversion calculations on the secondary magnetic field. Faults in the grounding grid are diagnosed from the distribution of equivalent resistivity.

The configuration or topology plays a vital role in the performance of the grounding grid. It is also an essential requirement for fault diagnosis. Although the drawing layout of the grounding grid shows its complete configuration, it is prone to human error, leading to spoilage or loss. Research on the configuration detection of the grounding grid is limited. The derivative method was used by [21,22] to measure the grounding grid configuration. The drawback associated with the derivative method is the occurrence of false peaks due to the surrounding electromagnetic environment. The transient electromagnetic method (TEM) was used by [23] to determine the grounding grid configuration. Measuring points with high equivalent resistivity and low magnetic intensity showed the presence of the grounding conductor. Furthermore, the wavelet edge based detection technique was utilized by [24] to image the configuration of the grounding grid.

Currently employed methods of configuration detection have assumed grounding grid orientation parallel to the substation boundary. This makes the grounding grid orientation parallel in the plane of the Earth. Practically, the orientation of unknown grounding grids is not known. In such a scenario, existing configuration detection methods fail to deliver accurate results [10]. Although [25] utilized magnetic detection electrical impedance tomography (MDEIT) to measure grounding grid configuration irrespective of its orientation, this method requires numerous measurements. Figure 2 illustrates the parallel and non-parallel orientation of the grounding grid with respect to the substation boundary.

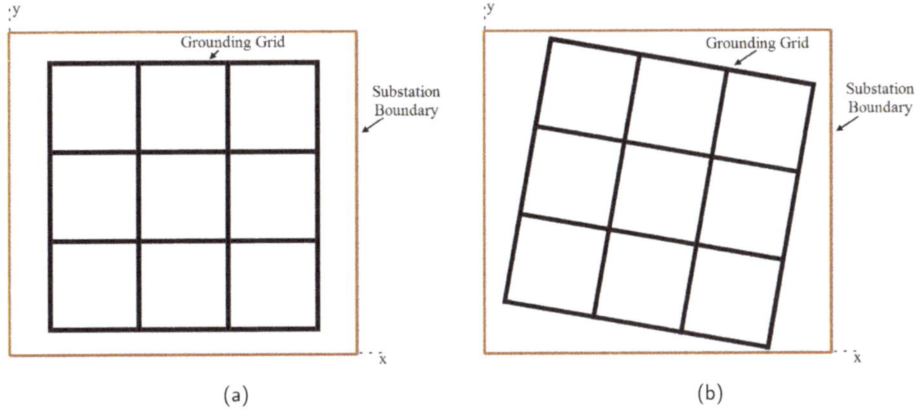

Figure 2. Grounding grid orientation with respect to the substation boundary. (**a**) Grounding grid oriented parallel along the substation boundary. (**b**) Grounding grid with non-parallel orientation along the substation boundary.

The literature on grounding grid orientation detection is extremely limited. Existing methods regarding orientation detection only include the derivative method [10,26]. This method is based on the derivative of the surface magnetic flux density and the concept of locating the geometrical object in the polar plane. The derivative method [10,26] performs well only when the substation electromagnetic environment (EME) is ignored, otherwise the method collapses. The effect of EME on the derivative method is illustrated in Figure 3. This figure is comprised of the following: Figure 3a is the original signal (magnetic flux density $\vec{B_z}$) from the grid [26]; Figure 3b is the surrounding EME signal; Figure 3c is the magnetic flux density $\vec{M_z}$ when $\vec{B_z}$ and EME signals are combined; and Figure 3d is the derivative of $\vec{M_z}$. False peaks along true peaks are originated, distorting the result completely. The incorrect resulting consequences come from the EME enhancement due to the derivative.

This paper employs the transient electromagnetic method (TEM) to diagnose the grounding grid orientation without soil excavation. Unlike the derivative method, TEM is independent of the current injection that brings the disturbing inhomogeneity of the surface magnetic flux density. Furthermore, it does not enhance the effect of surrounding EME. The feasibility of the proposed method is also tested for various complex mesh configurations. This incorporates the presence of the diagonal branch and unequally spaced grid configuration.

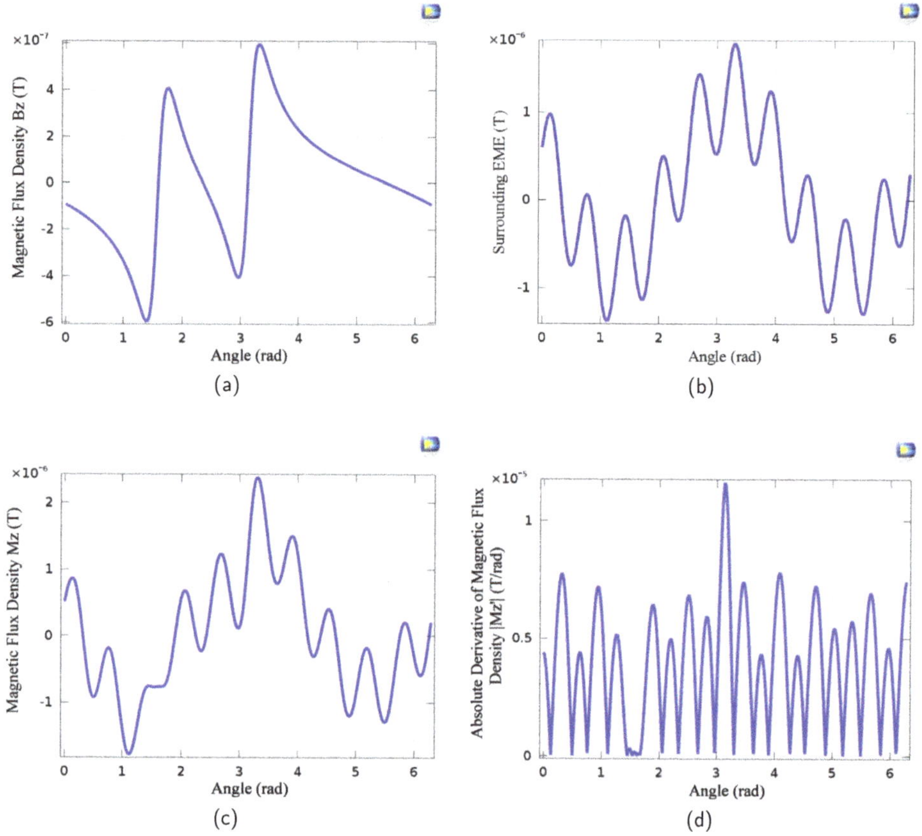

Figure 3. Influence of the surrounding electromagnetic environment (EME) on the derivative method. (**a**) Surface magnetic flux density $\vec{B}_z$ pertaining to the grounding grid in [26]. (**b**) Surrounding EME. (**c**) Mixed signal $\vec{M}_z$ of magnetic flux density $\vec{B}_z$ and the surrounding EME. (**d**) Derivative of mixed signal $\vec{M}_z$. This signal contains fake peaks due to the presence of EME, which causes the identification of true peaks to be impossible.

3. Transient Electromagnetic Method

The transient electromagnetic method (TEM) is widely used for geological exploration of underground minerals [27–29]. It is an effective method for determining the electrical resistivity of underground layers [30], as well as the fault diagnosis of the grounding grid [16,20].

Illustrated in Figure 4 is a typical TEM system that includes a transmitter-receiver pair to transmit and receive magnetic fields. The primary magnetic field is produced by injecting the pulse current of the ramp wave in the transmitter coil. The time varying primary magnetic field induces eddy currents in the grounding conductors. The secondary magnetic field due to the induced eddy currents is recorded above the surface by the receiver coil. This coil is located in the center of the transmitter coil. Inversion calculation of the induced electromotive force (emf) in the secondary coil is obtained utilizing equivalent resistivity imaging equations based on the smoke ring concept [31]. The location of grounding conductors is determined from the equivalent resistivity and magnetic field distribution. High equivalent resistivity and low magnetic field indicate the presence of the grounding conductor and vice versa.

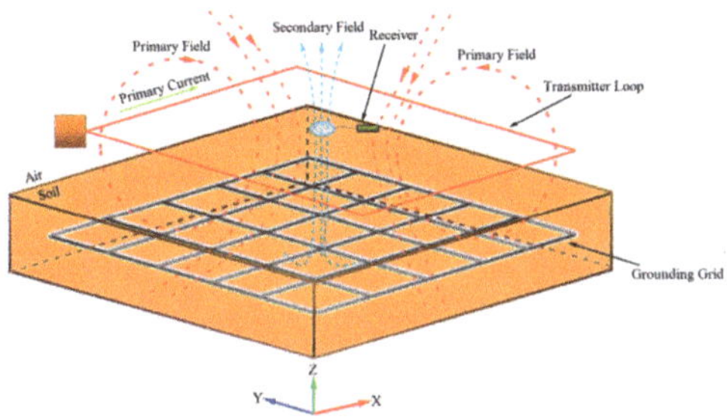

Figure 4. A typical transient electromagnetic method (TEM) system probing the underground grid. The primary magnetic field due to the transmitter coil interacts with the grid buried in the soil and induces eddy currents. Induced eddy currents produce a secondary magnetic field that travels upward to the Earth's surface and collected by the receiver coil placed in the center of the transmitter coil.

The vertical component of the secondary magnetic field in the center of transmitter coil is expressed as [32]:

$$H_z = \frac{I_{tc}}{2r_{tc}}\left[\frac{3}{\sqrt{\pi}u}e^{-u} + (1 - \frac{3}{2u^2})erf(u)\right] \tag{1}$$

where r_{tc} is the radius of the transmitter coil, I_{tc} is the magnitude of transmitter current, u is the transient magnetic field parameter, and $erf(u)$ is the error function expressed as:

$$erf(u) = \frac{2}{\sqrt{\pi}}\int_0^u e^{-u^2}dt \tag{2}$$

The induced electromotive force $E(t)$ is obtained as [32]:

$$E(t) = \frac{I_{tc}}{\sigma r_{tc}^3}\left[3erf(u) - \frac{2}{\sqrt{\pi}}u(3 + 2u^2)e^{-u^2}\right] \tag{3}$$

where σ is the conductivity of the underground medium. The transient magnetic field parameter u is expressed as:

$$u = \sqrt{\frac{\mu\sigma r_{tc}^2}{4t}} \tag{4}$$

The conductivity σ is obtained from (4) as:

$$\sigma = \frac{4u^2 t}{\mu r_{tc}^2} \tag{5}$$

Inserting σ in (3), $E(t)$ becomes:

$$E(t) = \frac{I_{tc}\mu}{4u^2 t r_{tc}}\left[3erf(u) - \frac{2}{\sqrt{\pi}}u(3+2u^2)e^{-u^2}\right] \tag{6}$$

A function $F(u)$ is setup using (6):

$$3erf(u) - \frac{2}{\sqrt{\pi}}u(3+2u^2)e^{-u^2} - \frac{4u^2 r_{tc} t E(t)}{\mu I_{tc}} = 0 \tag{7}$$

$$F(u) = 3erf(u) - \frac{2}{\sqrt{\pi}}u(3+2u^2)e^{-u^2} - \frac{4u^2 r_{tc} t E(t)}{\mu I_{tc}} \tag{8}$$

As resistivity ρ is reciprocal to conductivity σ, so the apparent resistivity in terms of u is given by:

$$\rho(t) = \frac{\mu r_{tc}^2}{4u^2 t} \tag{9}$$

Employing the iterative method in (8), the transient magnetic field parameter u is determined.

The vertical depth $d(m)$ and downward velocity $v(m/s)$ of the induced eddy currents can be calculated as [31]:

$$d = \frac{4}{\sqrt{\pi}}\sqrt{\frac{t\rho}{\mu}} \tag{10}$$

$$v = \frac{2}{\sqrt{\pi}}\sqrt{\frac{\rho}{t\mu}} \tag{11}$$

where t is the sampling time and μ is the permeability of the medium. Downward velocity $v(m/s)$ between two consecutive time samples is expressed as:

$$v = \frac{d_{i+1} - d_i}{t_{i+1} - t_i} \tag{12}$$

where t_i and t_{i+1} are the two consecutive time samples and d_i and d_{i+1} are the corresponding vertical depths. Comparing (11) and (12) yields:

$$\frac{d_{i+1} - d_i}{t_{i+1} - t_i} = \frac{2}{\sqrt{\pi}}\sqrt{\frac{\rho}{t\mu}} \tag{13}$$

$$\rho_r = \left(\frac{\pi t\mu}{4}\right)\frac{(d_{i+1} - d_i)^2}{(t_{i+1} - t_i)^2} \tag{14}$$

where ρ_r is the equivalent resistivity. Taking two consecutive time samples t_i and t_{i+1} into account, (10) is expressed as:

$$d_{i+1} - d_i = \frac{4}{\sqrt{\pi\mu}}\left[\sqrt{t_{i+1}\rho_{i+1}} - \sqrt{t_i\rho_i}\right] \tag{15}$$

Inserting (15) into (14), the equivalent resistivity ρ_r is equal to:

$$\rho_r = 4t\left[\frac{\sqrt{t_{i+1}\rho_{i+1}} - \sqrt{t_i\rho_i}}{t_{i+1} - t_i}\right]^2 \tag{16}$$

where $t = \frac{t_{i+1}+t_i}{2}$ is the average of two consecutive time samples and ρ_r equals:

$$\rho_r = 4\left[\frac{\sqrt{t_{i+1}\rho_{i+1}} - \sqrt{t_i\rho_i}}{t_{i+1} - t_i}\right]^2\left[\frac{t_{i+1} + t_i}{2}\right] \tag{17}$$

where ρ_i is the apparent resistivity at the ith time sample.

4. Performance Evaluation and Results' Analysis

In this section, a performance study to demonstrate the viability of the proposed method for orientation detection of the grounding grid is conducted. The evaluation study was performed through simulations. Simulations were performed using Comsol Multiphysics 5.0, a Finite Element Method (FEM) based tool. Furthermore, the inversion calculations of the recorded magnetic field above the Earth's surface were performed in MATLAB, and the results of the calculations are presented graphically.

4.1. Simulation Model

The simulation model shown in Figure 5 features a square grid of dimensions 4 m × 4 m. The conductors are labeled C_1 to C_{12} and arranged such that the mesh dimensions are 2 m × 2 m. The conductors were cylindrical steel rods of radius 0.01 m and conductivity 4.032 $\times 10^6$ S/m. The soil considered was homogeneous with electrical resistivity equal to 5 Ωm. The grounding grid was buried 0.5 m under the Earth's surface. The transmitter coil of radius 0.15 m was excited with a 16 A pulse current of a trapezoidal wave. The secondary magnetic field was recorded 0.05 m above the surface at the center of the transmitter coil after each 10 μs for a total of 100 time samples after the transmitter coil current was turned off. The transmitter coil was moved in a circle of radius 1 m along points P_1 to P_8 such that the angular displacement between adjacent points was 0.785 rad. The circle was centered at Node 5 acting as pole, which in general must be identified from the position of the vertical conductor. According to IEEE std 80-2013 [5], the length of the grounding grid branches varies from 3 m to 7 m. Therefore, the radius of the circle must be constrained to be between 0 m and 3 m.

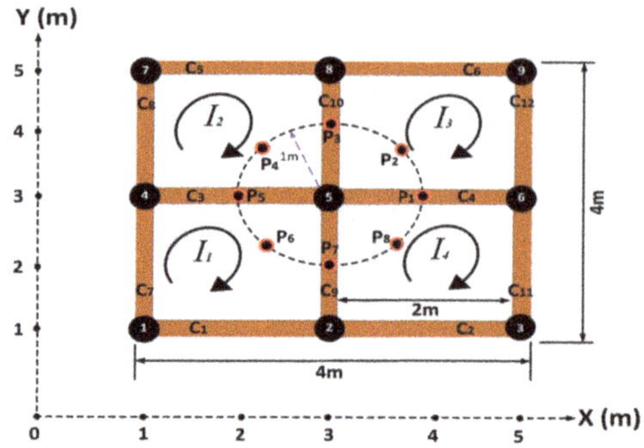

Figure 5. Simulation model featuring the square grounding grid of dimensions 4 m × 4 m and mesh spacing 2 m. Conductors are labeled C_1 to C_{12} and nodes 1 to 9. I_1 to I_4 are the induced eddy currents whose direction of flow is indicated by arrows. The TEM system is moved 0.05 m above the surface along circle of radius 1 m from point P_1 to P_8.

Employing the inversion calculation for the secondary magnetic field of Figure 5, the corresponding equivalent resistivity ρ_r is plotted in Figure 6. Here, ρ_r is high at points P_1, P_3, P_5, and P_7, illustrating the presence of conductors C_4, C_{10}, C_3, and C_9 at analogous points. For instance, the equivalent resistivity at P_1 was high due to the opposite flow of eddy currents I_3 and I_4 in C_4. Keeping in view the rectangular geometry of a typical grounding grid and the characteristics of the polar coordinate system, it is inferred from Figure 6 that the grid was oriented parallel in the plane of the Earth (parallel to the substation boundary). Furthermore, the secondary magnetic field along P_1 to P_8 is shown in Figure 7. Here, the average value of the magnetic field is plotted against each point. Due to the fact that the magnetic field from a medium is inversely proportional to its resistivity, Figure 7 shows an inverse relation with Figure 6. Therefore, $\overline{H_z}$ was low at P_1, P_3, P_5, and P_7, confirming C_4, C_{10}, C_3, and C_9.

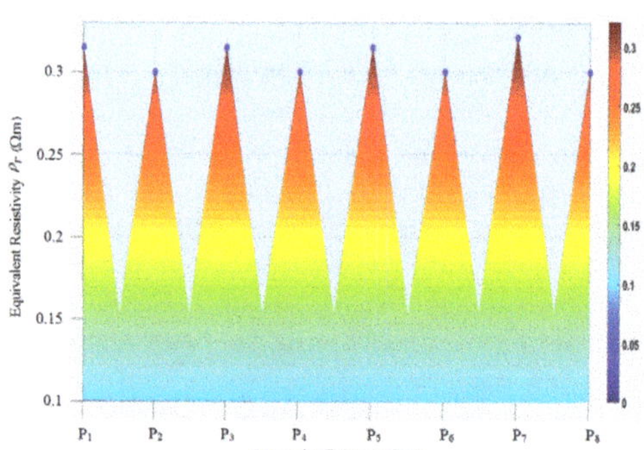

Figure 6. Equivalent resistivity ρ_r along the circle from P_1 to P_8. High ρ_r at P_1, P_3, P_5, and P_7 corresponds to the presence of conductors C_4, C_{10}, C_3, and C_9. C_4 at 0 rad, C_{10} at 1.57 rad, C_3 at 3.14 rad, and C_9 at 4.71 rad along the circle showed the parallel orientation of the grid in the plane.

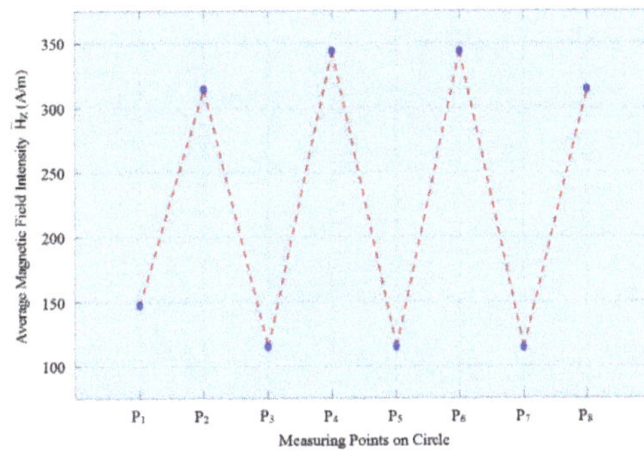

Figure 7. Average magnetic field intensity $\overline{H_z}$ along the circle from P_1 to P_8. $\overline{H_z}$ is low at P_1, P_3, P_5, and P_7, confirming the presence of C_4, C_{10}, C_3 and C_9, and the parallel orientation of the grid in the plane of the Earth.

4.2. Grounding Grid with a Diagonal Branch

Grounding grids exist in different configurations depending on the substation layout. A diagonal branch often exits in grounding grids. To check the feasibility of the proposed method for the orientation detection of the grounding grid with a diagonal branch, conductor C_{13} was added to Figure 5. The model with a diagonal branch is shown in Figure 8. C_{13} connected Nodes 1 and 5. TEM was applied by moving the transmitter-receiver pair in a circle from P_1 to P_8, and the result of equivalent resistivity ρ_r is shown in Figure 9. This time, ρ_r was high at P_6 as eddy currents I_1 and I_2 opposed each other in C_{13}, validating the presence of diagonal conductor C_{13}. Contrarily, ρ_r at P_5 and P_7 decreased although conductors C_5 and C_7 existed beneath them. This was due to unequal magnetic coupling as the mesh size had changed due to the presence of diagonal conductor C_{13}. Moreover, the average magnetic field $\overline{H_z}$ graph related to Figure 8 is demonstrated in Figure 10. Diagonal conductor C_{13} was represented by low $\overline{H_z}$ at P_6.

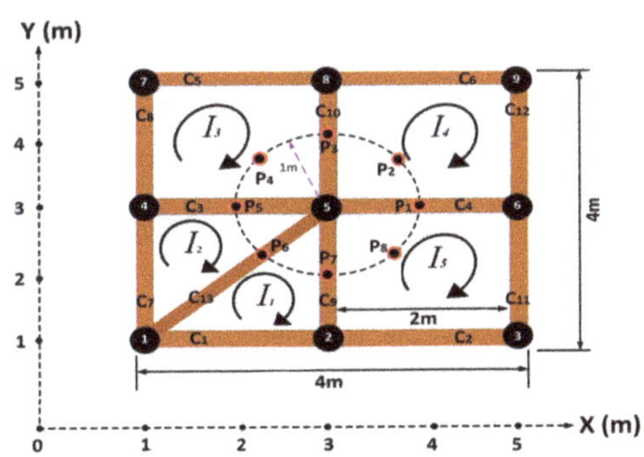

Figure 8. Grounding grid with diagonal conductor C_{13}. C_{13} connects Nodes 1 and 5 while carrying $I_1 - I_2$.

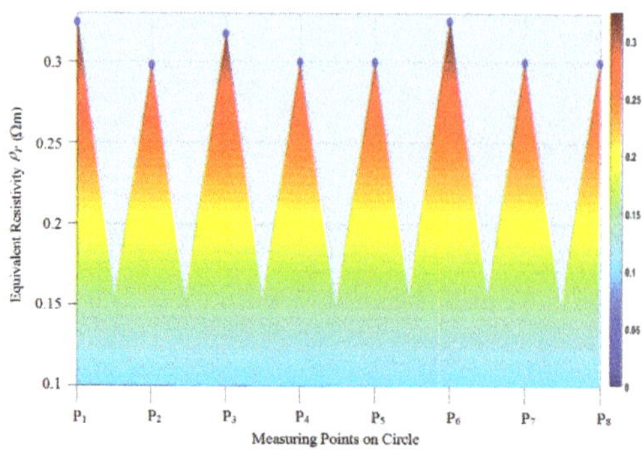

Figure 9. Equivalent resistivity ρ_r of Figure 8. High ρ_r at P_6 validates the presence of diagonal conductor C_{13}. Unequal magnetic coupling due to an unequal mesh size results in low ρ_r at P_5 and P_7.

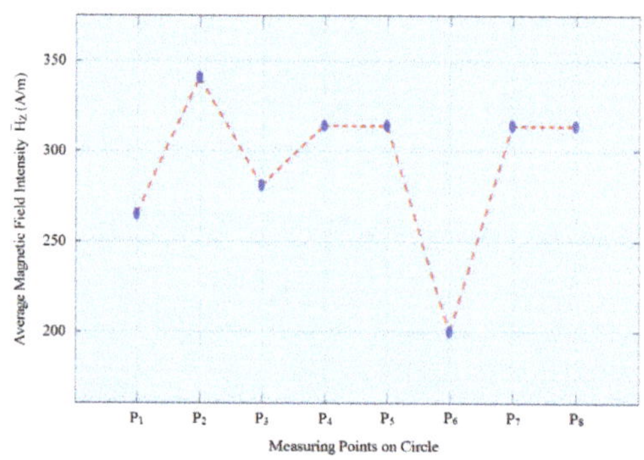

Figure 10. Average magnetic field intensity $\overline{H_z}$ along P_1 to P_8 related to Figure 8. Here, diagonal conductor C_{13} is indicated by low $\overline{H_z}$ at P_6.

4.3. Grounding Grid with Unequal Mesh Spacing

Demonstrating the feasibility of TEM for orientation detection of an unequally spaced grounding grid, Figure 11 is taken into account. In this figure, conductors are labeled C_1 to C_{12} and nodes 1 to 9. The dimensions of meshes M_1 and M_2 were 2.5 m × 2 m, and those of M_3 and M_4 were 1.5 m × 2 m. Consider Node 5 as a pole and moving the transmitter coil along the circle from point P_1 to P_8.

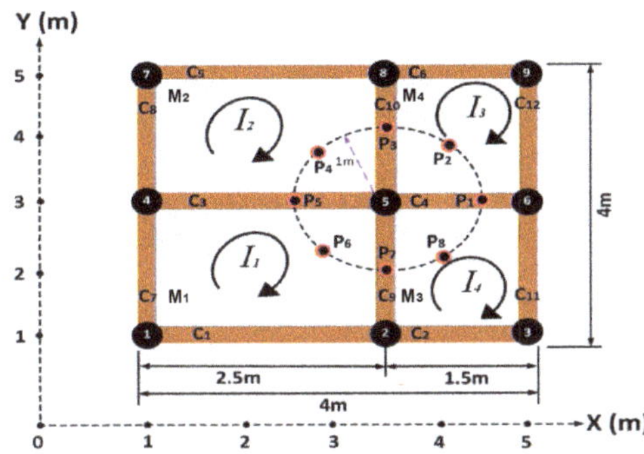

Figure 11. Grounding grid with an unequal mesh configuration. The dimensions of meshes M_1 and M_2 are 2.5 m × 2 m and M_3 and M_4 are 1.5 m × 2 m.

In Figure 11, when the transmitter loop was at measuring point P_1 with an angle of 0° and coordinates (4.5,3), currents I_3 and I_4 flowed in conductor C_4 in the opposite direction and canceled each other out. Thus, the current was less and so was the recorded magnetic field at measuring point P_1. At points P_2 and P_8 with angles 45° and 315°, respectively, the downward electromagnetic signal coupled extensively with mesh M_4 and M_3, thus inducing large eddy currents in the meshes. The recorded magnetic field at measuring points P_2 and P_8 was almost equal and higher than at measuring point P_1. Furthermore, when the transmitter loop was at measuring point P_3 with an angle of 90°, currents I_3 and I_2 flowed in conductor C_{10} with an unequal magnitude and did not cancel each other completely. This was because of unequal magnetic coupling in meshes M_4 and M_2. Thus, the recorded magnetic field at measuring point P_3 was less than at point P_2, but higher than at point P_4. Similarly, the recorded magnetic field at measuring point P_7 was less than at point P_8, but higher than at point P_6.

Figure 12 and Table 1 illustrate that due to weak magnetic coupling, the recorded magnetic field at measuring points P_4 and P_6 was almost equal and smaller than at measuring points P_2 and P_8. Thus, the size of grounding grid meshes M_1 and M_2 was greater than the size of meshes M_3 and M_4.

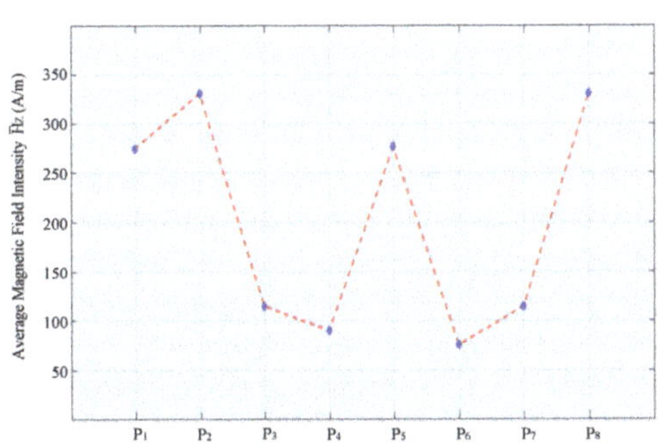

Figure 12. Average magnetic field intensity $\overline{H_z}$ along P_1 to P_8 related to Figure 11. The large size of meshes M_1 and M_2 results in weak magnetic coupling, and therefore, $\overline{H_z}$ at P_4 and P_6 is less than $\overline{H_z}$ at P_2 and P_8.

Figure 13 displays the distribution of equivalent resistivity ρ_r calculated against the circle perimeter in Figure 11. At point P_1, ρ_r was high since currents I_3 and I_4 flowed in conductor C_4 in the opposite direction. Thus, the current was less, and the equivalent resistivity ρ_r was high. As depicted in Table 2, ρ_r at points P_2 and P_8 in Figure 11 was low as compared to Figure 5. This was due to the small size of meshes M_4 and M_3 and, therefore, the strong magnetic coupling in these meshes. On the contrary, ρ_r was high at P_4 and P_6 due to the large size of meshes M_1 and M_2 and, therefore, the weak magnetic coupling. When the transmitter loop was at measuring point P_3 with an angle of 90°, currents I_2 and I_3 flowed in conductor C_{10} with unequal magnitude and did not cancel each other. Thus, the calculated equivalent resistivity ρ_r at measuring point P_3 was higher than at point P_2, but smaller than at point P_4. Similarly, ρ_r at measuring point P_7 was higher than at point P_8, but smaller than at point P_6.

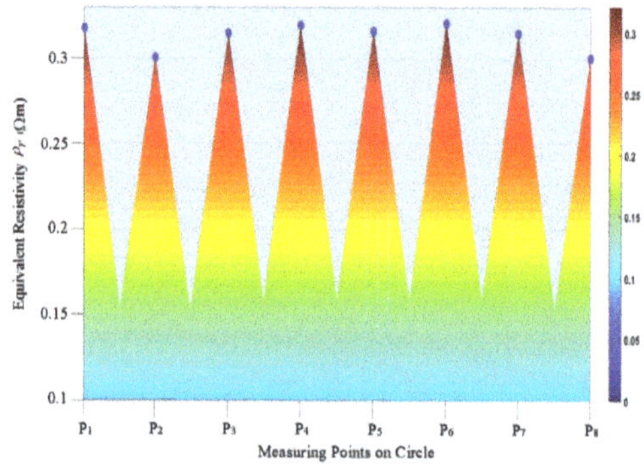

Figure 13. Equivalent resistivity ρ_r related to Figure 11. The large size of meshes M_1 and M_2 results in weak magnetic coupling and, therefore, ρ_r at P_4 and P_6 is higher than ρ_r at P_2 and P_8.

Table 1. Recorded magnetic field based on TEM.

Measuring Point	Average Magnetic Field Intensity (A/m)		
	Equal Mesh Spacing	Equal Mesh Spacing and Diagonal Branch	Unequal Mesh Spacing
P_1	147.0995	264.8785	274.8785
P_2	313.9748	340.748	330.748
P_3	114.8244	280.854	114.8244
P_4	343.3893	313.9748	90.345
P_5	114.8244	313.9748	276.74
P_6	343.3893	200.051	75.632
P_7	114.2897	313.9748	114.8244
P_8	313.9748	313.9748	330.748

Table 2. Equivalent resistivity calculated based on TEM.

Measuring Point	Average Equivalent Resistivity (Ωm)		
	Equal Mesh spacing	Equal Mesh Spacing and Diagonal Branch	Unequal Mesh Spacing
P_1	0.2889	0.2976	0.2916
P_2	0.2752	0.2736	0.2758
P_3	0.2889	0.2911	0.2889
P_4	0.2752	0.2752	0.2933
P_5	0.2944	0.2752	0.2900
P_6	0.2752	0.2982	0.2944
P_7	0.2889	0.2752	0.2889
P_8	0.2752	0.2752	0.2758

Graphs of the equivalent resistivity ρ_r and average magnetic field intensity $\overline{H_z}$ illustrated the presence of conductors C_3 and C_4 along the x-axis and conductors C_9 and C_{10} along the y-axis. It was deduced that the grounding grid was oriented parallel to the plane of the Earth.

5. Conclusions and Future Work

Grounding grid drawings are often lost and mishandled, altering the status of the grid from known to unknown. In this paper, a new method to measure the orientation of the grounding grid was presented. The method was not only independent of the current injection that brought the disturbing inhomogeneity of the surface magnetic flux density, but also did not enhance the effect of the surrounding EME. The transmitter-receiver pair of the TEM system was moved along a circle above the surface such that the vertical conductor acted as the pole of the circle. According to the mesh spacing of the grounding grid between 3 m to 7 m, the radius of the circle was constrained between 0 m to 3 m. Once the equivalent resistivity ρ_r was determined from the secondary magnetic field H_z, high ρ_r and low $\overline{H_z}$ at a point on the circle laid the basis for orientation detection of the grounding grid. Moreover, the proposed method was also investigated for complex mesh configurations including the presence of a diagonal branch and an unequally spaced mesh configuration. As an application, the paper used TEM to measure the orientation of the grounding grid. Simulation results showed that the diagnosis was feasible.

There is a great need for further research to detect grounding grid orientation. This includes a grounding grid with an unequal mesh spacing and a diagonal branch. Furthermore, the depth of the grounding grid must also be considered in future research because it is critical for fault diagnosis.

Author Contributions: A.Q. proposed the idea and drafted the manuscript. I.U.H. performed the simulations and composed the figures. M.A. and N.N.Q. arranged the funding.

Funding: This research received no external funding.

Acknowledgments: This research is funded by the Deanship of Scientific Research at the University of Ha'il.

Conflicts of Interest: The authors declare no conflict of interest.

References

1. Aamir, Q.; Wilayat, K.; Syed, R.N.; Farooq, A.O. Non-destructive depth detection approach for substation grounding grids based on magnetostatics. *IET Electron. Lett.* **2019**, *55*, 1121–1123. [CrossRef]
2. Gouda, O.E.; Dessouky, S.S.; Kalas, A.E.; Hamdy, M.A. Current distribution in grounding grid conductors with and without vertical rods buried in two-layer soil model. *IEEJ Trans. Electr. Electron. Eng.* **2018**, *13*, 1276–1284. [CrossRef]
3. Bo, Z.; Wu, J.; He, J.; Zeng, R. Analysis of transient performance of grounding system considering soil ionization by time domain method. *IEEE Trans. Magn.* **2013**, *49*, 1837–1840. [CrossRef]
4. Hideki, M. Electromagnetic transient response of buried bare wire and ground grid. *IEEE Trans. Power Deliv.* **2007**, *22*, 1673–1679. [CrossRef]
5. *IEEE Std 80-2013 (Revision of IEEE Std 80-2000/Incorporates IEEE Std 80-2013/Cor 1-2015). IEEE Guide for Safety in AC Substation Grounding*; IEEE: Piscataway, NJ, USA, 2015.
6. El-Refaie El-Sayed, M.; Elmasry, S.E.; Elrahman MKAbd Abdo, M.H. Achievement of the best design for unequally spaced grounding grids. *Ain Shams Eng. J.* **2018**, *6*, 171–179. [CrossRef]
7. Ferrante, N. A new evolutionary method for designing grounding grids by touch voltage control. In Proceedings of the IEEE International Symposium on Industrial Electronics, Ajaccio, France, 4–7 May 2004; Volume 2, pp. 1501–1505. [CrossRef]
8. Farkoush, S.G.; Wadood, A.; Khurshaid, T.; Kim, C.-H.; Irfan, M.; Rhee, S.-B. Reducing the Effect of Lightning on Step and Touch Voltages in a Grounding Grid Using a Nature-Inspired Genetic Algorithm With ATP-EMTP. *IEEE Access* **2019**, *7*, 81903–81910. [CrossRef]
9. Farkoush, S.G.; Khurshaid, T.; Wadood, A.; Kim, C.-H.; Kharal, K.H.; Kim, K.-H.; Cho, N.; Rhee, S.-B. Investigation and optimization of grounding grid based on lightning response by using ATP-EMTP and genetic algorithm. *Complexity* **2018**, *2018*, 8261413. [CrossRef]
10. Aamir, Q.; Muhammad, U.; Fan, Y.; Muhammad, U.; Zeeshan, K. Derivative Method Based Orientation Detection of Substation Grounding Grid. *Energies* **2018**, *11*, 1873. [CrossRef]
11. Li, X.; Yang, F.; Ming, J.; Jadoon, A.; Han, S. Imaging the Corrosion in Grounding Grid Branch with Inner-Source Electrical Impedance Tomography. *Energies* **2018**, *11*, 1739. [CrossRef]
12. Yuan, J.; Yang, H.; Zhang, L.; Cui, X.; Ma, X. Simulation of substation grounding grids with unequal-potential. *IEEE Trans. Magn.* **2000**, *36*, 1468–1471. [CrossRef]
13. Yang, F.; Wang, Y.; Dong, M.; Kou, X.; Yao, D.; Li, X.; Gao, B.; Ullah, I. A Cycle Voltage Measurement Method and Application in Grounding Grids Fault Location. *Energies* **2017**, *10*, 1929. [CrossRef]
14. Liu, K.; Yang, F.; Wang, X.; Gao, B.; Kou, X.; Dong, M.; Ammad, J. A novel resistance network node potential measurement method and application in grounding grids corrosion diagnosis. *Prog. Electromagn. Res. M* **2016**, *52*, 9–20. [CrossRef]
15. Liu, Y.; Cui, X.; Zhao, Z. A magnetic detecting and evaluation method of substations grounding grids with break and corrosion. *Front. Electr. Electron. Eng. China* **2010**, *5*, 501–504. [CrossRef]
16. Yu, C.; Fu, Z.; Wang, Q.; Tai, H.M.; Qin, S. A novel method for fault diagnosis of grounding grids. *IEEE Trans. Ind. Appl.* **2015**, *51*, 5182–5188. [CrossRef]
17. Aamir, Q.; Nadir, S.; Zeeshan, K.; Zahoor, U.; Farooq, A.O. Breakpoint Diagnosis of Substation Grounding Grid Using Derivative Method. *Prog. Electromagn. Res. M* **2017**, *57*, 73–80. [CrossRef]
18. Dawalibi, F. Electromagnetic fields generated by overhead and buried short conductors, Part 2—Ground networks. *IEEE Trans. Power Deliv.* **1986**, *1*, 112–119. [CrossRef]
19. Zhang, X.-L.; Zhao, X.-H.; Wang, Y.-G.; Mo, N. Development of an electrochemical in situ detection sensor for grounding grid corrosion. *Corrosion* **2010**, *66*, 076001. [CrossRef]
20. Yu, C.; Fu, Z.; Hou, X.; Tai, H.M.; Su, X. Break-point diagnosis of grounding grids using transient electromagnetic apparent resistivity imaging. *IEEE Trans. Power Deliv.* **2015**, *30*, 2485–2491. [CrossRef]
21. Li, C.; He, W.; Yao, D.; Yang, F.; Kou, X.; Wang, X. Topological measurement and characterization of substation grounding grids based on derivative method. *Int. J. Electr. Power Energy Syst.* **2014**, *63*, 158–164. [CrossRef]

22. Aamir, Q.; Fan, Y.; Wei, H.; Ammad, J.; Muhammad, Z.K.; Xu, N. Topology measurement of substation's grounding grid by using electromagnetic and derivative method. *Prog. Electromagn. Res. B* **2016**, *67*, 71–90. [CrossRef]
23. Yu, C.; Fu, Z.; Wu, G.; Zhou, L.; Zhu, X.; Bao, M. Configuration detection of substation grounding grid using transient electromagnetic method. *IEEE Trans. Ind. Electron.* **2017**, *64*, 6475–6483. [CrossRef]
24. Fu, Z.; Song, S.; Wang, X.; Li, J.; Tai, H.M. Imaging the topology of grounding grids based on wavelet edge detection. *IEEE Trans. Magn.* **2018**, *54*, 1–8. [CrossRef]
25. Liu, K.; Yang, F.; Zhang, S.; Zhu, L.; Hu, J.; Wang, X.; Irfan, U. Research on Grounding Grids Imaging Reconstruction Based on Magnetic Detection Electrical Impedance Tomography. *IEEE Trans. Magn.* **2018**, *54*, 1–4. [CrossRef]
26. Qamar, A.; Yang, F.; Xu, N.; Syed, A.S. Solution to the inverse problem regarding the location of substations grounding grid by using the derivative method. *Int. J. Appl. Electromagn. Mech.* **2018**, *56*, 549–558. [CrossRef]
27. Mohamed, M.; Gad, E.-Q.; Usama, M.; Abeer, E.-K.; Jun, M.; Naseer, A.-A. Integrated geoelectrical survey for groundwater and shallow subsurface evaluation: Case study at siliyin spring, el-fayoum, egypt. *Int. J. Earth Sci.* **2010**, *99*, 1427–1436.
28. Xue, G.; Yan, Y.; Li, X.; Di, Q. Transient electromagnetic s-inversion in tunnel prediction. *Geophys. Res. Lett.* **2007**, *34*. [CrossRef]
29. Adel, K.M.; Maxwell, A.M.; Sergio, L.F. Deep structure of the northeastern margin of the parnaiba basin, brazil, from magnetotelluric imaging. *Geophys. Prospect.* **2008**, *50*, 589–602. [CrossRef]
30. Yu, C.; Fu, Z.; Zhang, H.; Tai, H.M.; Zhu, X. Transient process and optimal design of receiver coil for small-loop transient electromagnetics. *Geophys. Prospect.* **2014**, *62*, 377–384. [CrossRef]
31. Misac, N.N. Quasi-static transient response of a conducting half-spacean approximate representation. *Geophysics* **1979**, *44*, 1700–1705. [CrossRef]
32. Stanley, H.W.; Gerald, W.H.; Misac, N. Electromagnetic theory for geophysical applications. *Electromagn. Methods Appl. Geophys.* **1988**, *1*, 131–311. [CrossRef]

applied sciences

MDPI

Article

Development of Railway Protective Relay Simulator for Real-Time Applications

Joorak Kim, Gyu-Jung Cho and Jaewon Kim *

Korea Railroad Research Institute, Uiwang 16105, Korea; jrkim@krri.re.kr (J.K.); gjcho@krri.re.kr (G.-J.C.)
* Correspondence: youngkjw@krri.re.kr; Tel.: +82-31-460-5549

Received: 20 November 2019; Accepted: 17 December 2019; Published: 25 December 2019

Abstract: Electric railways use a single-phase system, with the line comprising a trolley wire (TF) that supplies power to the load with a neutral wire and an autotransformer (AF) feeder to absorb the return current of the rail. Testing the performance of the protective relay that detects the fault of the traction power-supply system (TPSS) and operates the circuit breaker is very important. Until now, the performance test of protective relays for the TPSS has been conducted via a simple-steady test or using an expensive real-time simulator. However, under a fast-moving environment in which the load consumes a large amount of power, the protective relay must always detect faults and operate properly. This paper proposes a digital simulator that enables the dynamic testing of protective relays without using any steady test and expensive real-time simulators. This simulator includes both external waveform import and internal waveform generation functions. Users can test the operation of the protective relay by entering the waveform generated externally or internally into the protective relay. Additionally, it has the ability to monitor the operating protection elements and pickup time when the protective relay detects a fault and orders the circuit breaker trip.

Keywords: traction power-supply system; protective relay; digital simulator; monitoring and controlling system; railway

1. Introduction

The fault of the traction power-supply system (TPSS) directly results in the operation disturbance of railway trains because electric railways use electrical energy as an energy source. Therefore, system faults must be quickly identified and responded to, which is achieved by protective relays. In other words, protective relays are designed to monitor breakdowns and prevent breakdown impacts. Thus, they are very important for the stable operation of the system [1–4].

The autotransformer (AT) feeding method, which is applied to electric railways in many countries, supplies power to the line through an AT with twice the rated voltage of the load (i.e., train). In other words, it comprises a trolley wire (TF) that connects the neutral point of an AT with the rail to supply power to the load, and a feeder (AF) to absorb the return current of the load in the rail [5–8]. Unlike the conventional electric power system, substations that serve as a power source have a disconnected wire form instead of a network form; thus, short circuit fault at a distance from the power source can be confused with the load current. Additionally, because the load of the TPSS has a large capacity of approximately 15 [MVA] in terms of high-speed trains and rapid operation, the load current has great variability. Accordingly, the AC TPSS is characterized by asymmetrical single phases due to moving loads (such as trains); the TPSS characteristics differ significantly from conventional balanced three-phase power systems.

In addition, it is essential to establish performance verification procedures at the development stage because it is essential to ensure the reliability of the protective relay operation on railways where safety is important. However, many studies have been conducted on the fault analysis of railway systems, but little research has been conducted on the performance evaluation of relays in railways [9–12].

Generally, in power systems, the performance test of protective relay can be divided into steady and dynamic tests. A steady test is a method of injecting a specific voltage and current into a protective relay to determine whether it operates. A dynamic test tests the response characteristics of the protective relay by injecting a waveform that reflects the operating characteristics or environment of the system wherein the protective relay is equipped. A steady characteristic test of protective relay for railroads can be performed using general commercial testing devices including an amplifier that generates waveforms. However, a dynamic characteristic test can be performed through hardware in the loop simulation (HILS) in connection with the relay by simulating system operation environments through an expensive simulator such as digital real-time simulator (DRTS) [13–16]. However, DRTSs are not easy to construct, as they are expensive and require more effort than electromagnetic transients (EMT) simulators for transient state analysis. Additionally, the best way to test the protective relay with the correct system fault waveform is to obtain voltage and current waveforms from the protective relay installed in the actual TPSS as a common format for transient data exchange for power systems (COMTRADE) file and then inject it into the protective relay that needs to be tested.

Therefore, this paper proposes a new digital simulator for the railways called protective relay digital simulator for railway (PREDIS-R), which enables the accurate and efficient testing of protective relay performance at a low cost at the laboratory level. The proposed digital simulator has the following configurations and features. First, it provides an environment where users can create waveforms or import them from outside. In other words, users can create the desired test waveform on a digital simulator or import a fault waveform implemented by measurement or an external EMT simulator. The waveform generated is amplified to the potential transformer (PT) and current transformer (CT) levels of the train substation via a real-time amplifier. The amplified voltage and current waveform are injected into the protective relay, and it can perform an operational performance test. Additionally, the operating test of the protective relay can monitor the pickup time and operating protective elements of the protective relay to review the correct operation of the protective relay on the users' intended fault waveform.

2. Development of Protective Relay Digital Simulator for Railways

2.1. Performance Scheme

The main role of the protective relay is to analyze the waveform of the input voltage and current to determine any fault and provide output for the operation of the circuit breaker (CB). To test the performance of these protective relays, the simulator must enter a waveform that accurately reflects the characteristics of the faulty and steady states of the system and read the status information of the protective relay operating according to the input waveform.

Figure 1 shows the performance scheme of the PREDIS-R proposed herein. It was developed to be able to use fault data obtained from the field or enter waveform data obtained through presimulation based on an EMT simulation tool (i.e., power system computer aided design/electromagnetic transients with DC (PSCAD/EMTDC) [17], electro magnetic transients program (EMTP), etc.) into the protective relay simulators in the international standard COMTRADE format to simulate the fault of the system. Similarly, when external faulty waveform data input is loaded into the digital protective relay simulator used to test the performance of the protective relay, PREDIS-R generates the same waveform as the protective relay with the input waveform and outputs it.

Moreover, a function that can create a separate waveform is applied to the protective relay simulator to reflect the characteristics of the TPSS, which does not have the EMT simulation result waveform or cannot be implemented by EMT simulation. Thus, when creating a waveform internally, PREDIS-R generates the exact waveform that the user intends and outputs it to a protective relay.

The proposed PREDIS-R comprises hardware for loading external test waveforms or generating test waveforms directly showing motion information to users and software setting up test waveforms, controlling simulators, and storing and outputting operating information.

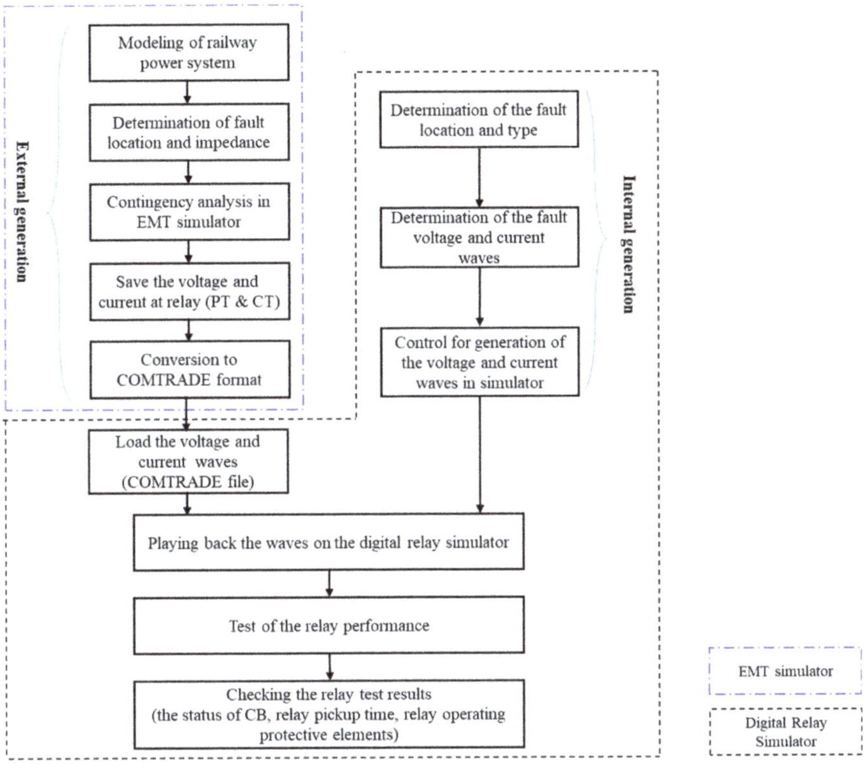

Figure 1. Protective relay digital simulator for railway (PREDIS-R) performance scheme.

2.2. Basic Formula for Relay Setting

The Zone 1 distance relay setting can be configured as follows in Equation (1), considering the protection range of 90%:

$$Z_1 = 0.9 \cdot Z_{55} \cdot L_1 \cdot CF \quad (1)$$

where Z_1 is the relay input impedance for Zone 1, Z_{55} is the unit line impedance per length based on 55 kV [Ω/km], L_1 is the line length between the substation (SS) and sectioning post (SP) [km], and CF is the conversion factor considering the CT and PT ratios.

The Zone 2 distance relay setting can be configured as follows in Equation (2), considering the protection range of 120%:

$$Z_2 = 1.2 \cdot Z_{55} \cdot L_2 \cdot CF \quad (2)$$

where Z_2 is the relay input impedance for Zone 2, and L_2 is the length between two substations [km].

The input impedance of the distance relay between Zone 1 and Zone 2 is 33.83 Ω and 66 Ω, respectively.

2.3. Configuration

Figure 2 shows the data flow in the hardware of the developed digital simulator. The control system of the simulator is designed to output the waveform set by the user accurately through software.

Additionally, operation information of the relay is obtained by communication between the protective relay and the simulator, and the status change information of the CB is output to the display device.

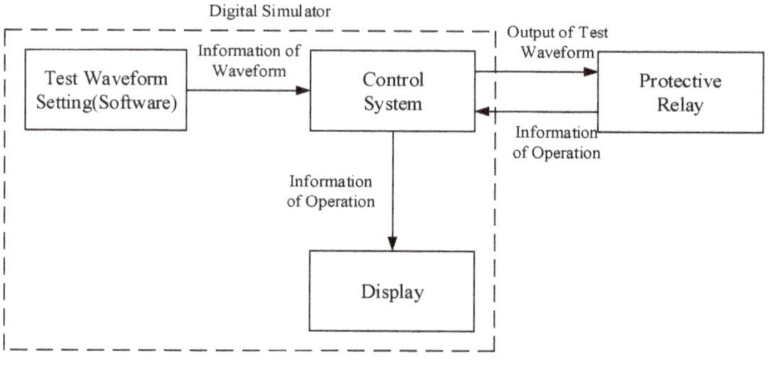

Figure 2. Data flow of the PREDIS-R.

The control system of the simulator comprises an amplifier, signal generator, and signal receiver. The signal generator and receiver print out the test waveform and obtain the output information of the protective relay. For this purpose, it is configured to enable digital and analog input and output. Figure 3 shows the role of each input and output terminal. The digital input (DI) part checks the status of the protective relay, and the analog input (AI) part acquires the monitored voltage and current signals from the protective relay. The digital output (DO) part creates the signal to operate the CB and the analog output (AO) part injects the test waveform into the protective relay. The waveforms from AO are small signals of 10 Vp.

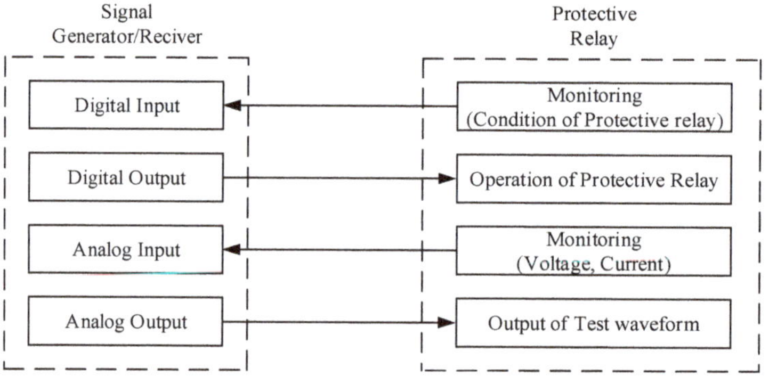

Figure 3. Signal generation and each function of the collecting part.

Figure 4 shows the hardware configuration diagram of the PREDIS-R. The amplifier was applied to generate the waveform output with the voltage and current levels of the protective relay installed in the real environment of the TPSS. The protective relay converts the voltage and current of the system through the PT (55,000 V: 110 V) and CT (800 A: 5 A) to monitor the fault occurrence. However, because the waveform output from the signal generation and reception parts is small (10 Vp), amplification of the signal at the same level as the output from the actual switchgear in the substation is necessary. To amplify the level of the installed PT and CT secondary side, the amplification rate of the voltage and current was set to be 18 and 1.34 times, respectively, and the signal could be amplified up to 180 Vp and 9 Ap, respectively.

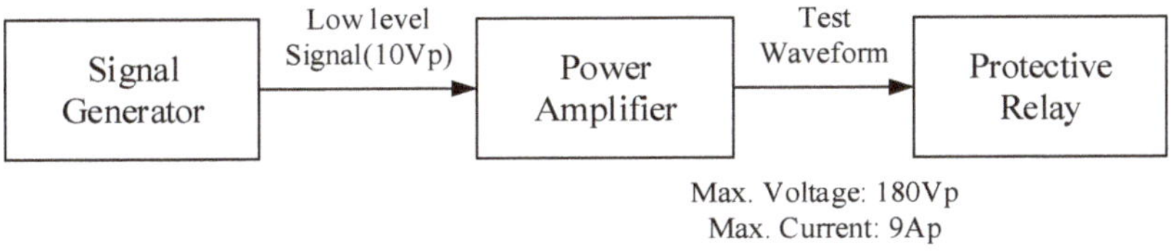

Figure 4. Hardware configuration diagram of the PREDIS-R.

The operating information of the protective relay and the change in the status of the CB according to the signal sent to the CB can be checked by the user in real time through the monitor. Figure 5 shows the fabricated PREDIS-R.

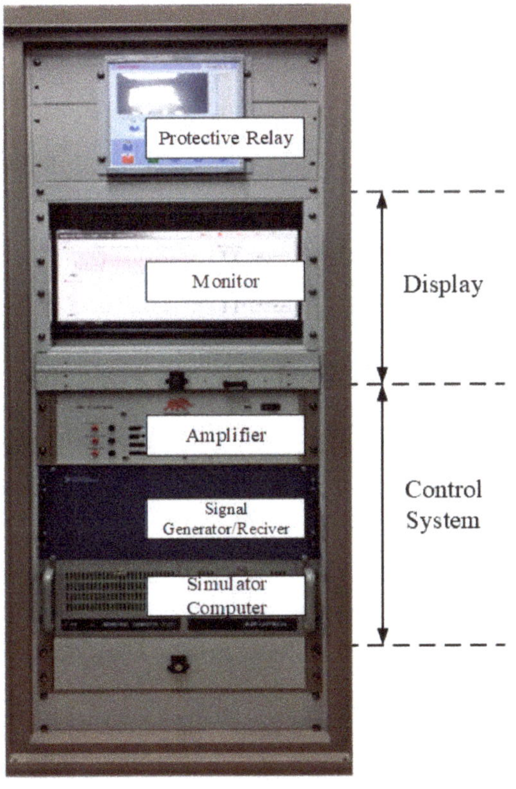

Figure 5. The fabricated PREDIS-R.

The simulator performance scheme in Figure 1 has been implemented to allow users to intuitively use it through the user-based main interface shown in Figure 6. The elements and functions that constitute the main interface are listed in Table 1.

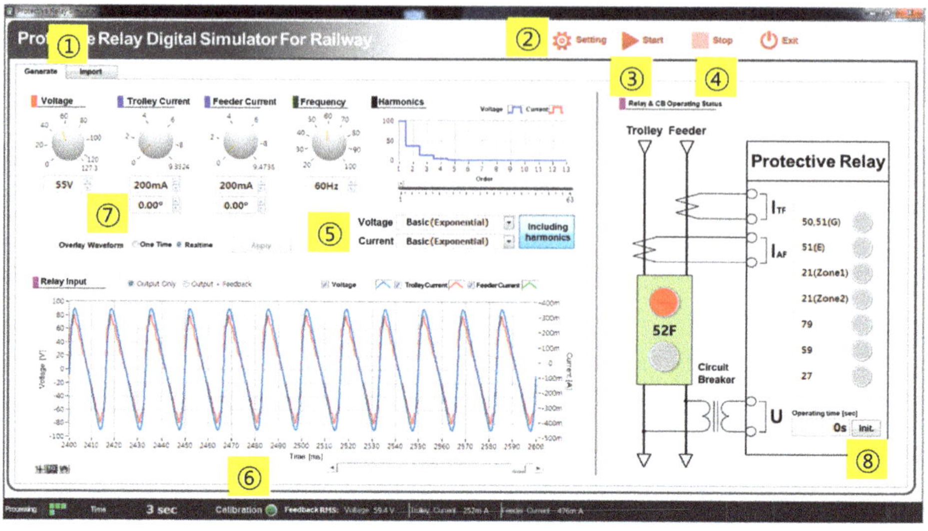

Figure 6. Main interface of the PREDIS-R.

Table 1. Main interface configuration and functions.

Number	Interface	Function
1	Selection Mode	Select one of the following modes - Import waveform - Generate waveform
2	Setting	- Set I/O channel - Input harmonics
3	Start	Reproduce simulation waveform
4	Stop	Stop reproducing simulation waveform
5	Harmonics	Load and apply input harmonics
6	Calibration	The error between input data and output waveform is larger than the threshold; the waveform will be expressed as a red line. At this time, recalibration is recommended
7	Phase Control	Change the phase of each current
8	Initialize	Reset operation time to 0

Various protective components in integrated relay are applied to the TPSS. The distance relay is used as the main protective relay to protect the TPSS from ground faults and short circuit faults. Further, the high resistance ground fault current is difficult to distinguish; thus, an overcurrent relay is used for back-up protection. Additionally, a reclosing relay to reclose the CB and an under-voltage relay for alarm and CB opening in the event of a power outage is also used. To test the performance of these various relay elements, various types of waveforms must be generated by software. For this, PREDIS-R developed and applied a function that can receive external input or generate the waveform for performance testing.

2.3.1. External Input Waveform

Obtain the input waveform that can occur from the TPSS implemented externally and use it to test the performance of the protective relay. For example, the waveform of each condition must be entered to examine the performance of the protective relay under steady-state load current and fault occurrence. As shown in Figure 7, the simulation results of PSCAD/EMTDC can be converted into COMTRADE format and inputted into the protective relay. This enables the performance test of protective relays in various ways, such as the location of fault, type of fault, and load current. In addition, the test waveform reliability can be improved because the output is based on simulation results.

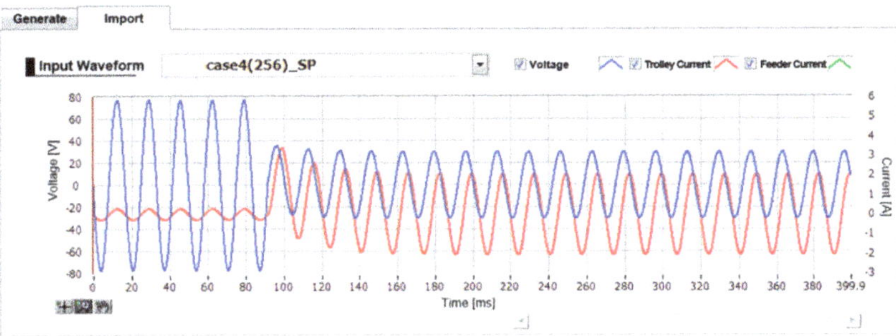

Figure 7. External (PSCAD/EMTDC COMTRADE) input waveform.

2.3.2. Internal Waveform Generation

Test waveforms can be implemented according to users' needs. For this purpose, PREDIS-R is designed to allow voltage, current, and phase control. Considering the variation ratio of substation PT and CT, it can be adjusted within Max 120 V, Max 9 A, and 0–180°, respectively, and implemented for continuous inputs.

The AC TPSS can easily distinguish between steady and faulty states because the supply voltage level is high and the difference between the load and fault currents is large. However, the load current includes harmonics, because it converts the power through the inverter to control the speed of the train.

Figure 8 shows the use of the harmonic generation function to test whether the protective relay is malfunctioning because of harmonics.

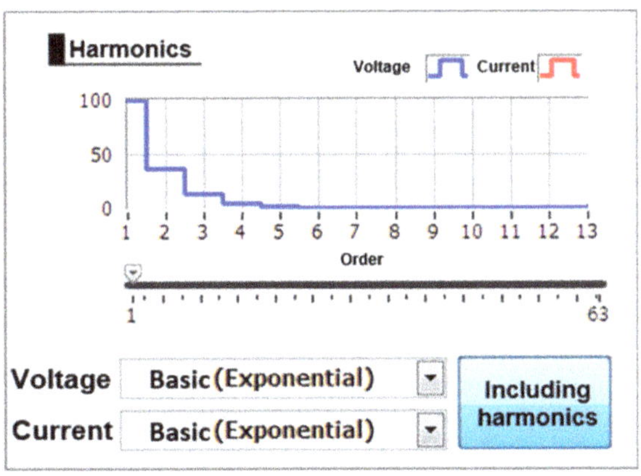

Figure 8. Internal waveform generation including harmonics.

PREDIS-R also includes the capability to monitor test progress and test results in real time. The signals obtained from the waveform input, output, and protective relay on the simulator are printed on the screen to provide an intuitive understanding to users, as shown in Figure 9. This screen shows the operating protective elements and operating time during the performance test of the protective relay.

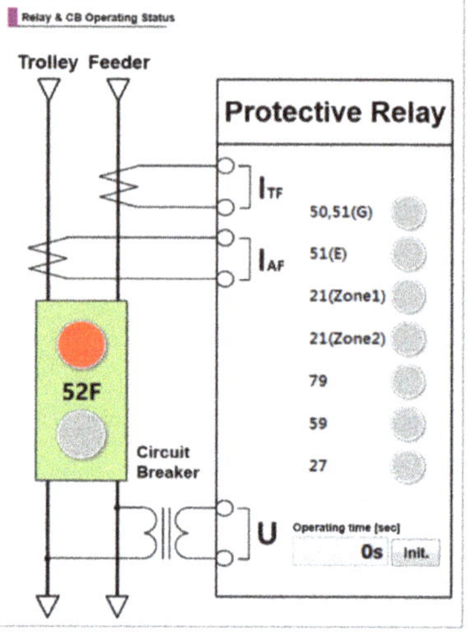

Figure 9. Output of protection element information for protective relay.

3. Verification of the PREDIS-R

PREDIS-R verification is divided into two main categories. The first is output performance verification, which checks whether the waveform output to the protective relay is implemented correctly when the waveform is received from the external input and outputs the desired waveform from the simulator. The second is output performance verification for waveforms, which is generated by the user.

First, to check the basic output performance, the voltage and current amplification rate of the amplifier that amplifies according to the operating voltage and current of the relay against the voltage signal generated by the simulator was verified. For this purpose, various sizes of sinusoidal wave voltage were inputted into the amplifier.

Table 2 shows the amplitude of the input and output currents for the current amplifier. The amplification rate of the current amplifier is 1.34 times, showing an average error rate of 0.9% in the amplification range. Table 3 shows the amplitude of the input and output voltages for the voltage amplifier. The amplification rate of the voltage amplifier is 18 times, showing an average error rate of 0.07% in the amplification range. As the error rate averages are very small, they are confirmed to be amplified according to their design conditions. The output performance under a continuously changing amplification waveform was verified against the external input waveform and internally generated waveform.

To verify that the external input waveform is accurately implemented with the protective relay, the COMTRADE file generated using PSCAD/EMTDC was entered into the simulator, and then the output waveform was checked, as shown in Figure 10. Upon comparing, the maximum value of the COMTRADE input waveform was 77.06 V, while the maximum value of the output waveform of the simulator was 78.4 V, showing that it is possible to output accurate waveforms with an error rate of 1.73%.

Table 2. Verification result for current amplifier.

Input (A_p)	Output (A_p)	I/O Ratio	Error Rate (%)
1	1.33	1.33	0.74
2	2.69	1.34	0.37
3	4.04	1.35	0.49
4	5.42	1.35	1.11
5	6.77	1.35	1.04
6	8.14	1.36	1.24
7	9.52	1.36	1.49

Table 3. Verification result for voltage amplifier.

Input (V_p)	Output (V_p)	I/O Ratio	Error Rate (%)
1	18.01	18.01	0.05
2	36.08	18.02	0.22
3	54.05	18.02	0.09
4	72.06	18.01	0.08
5	89.97	17.99	0.03
6	107.93	17.99	0.06
7	125.92	17.99	0.06
8	143.95	17.99	0.03
9	161.96	18.00	0.02
10	179.86	17.99	0.07

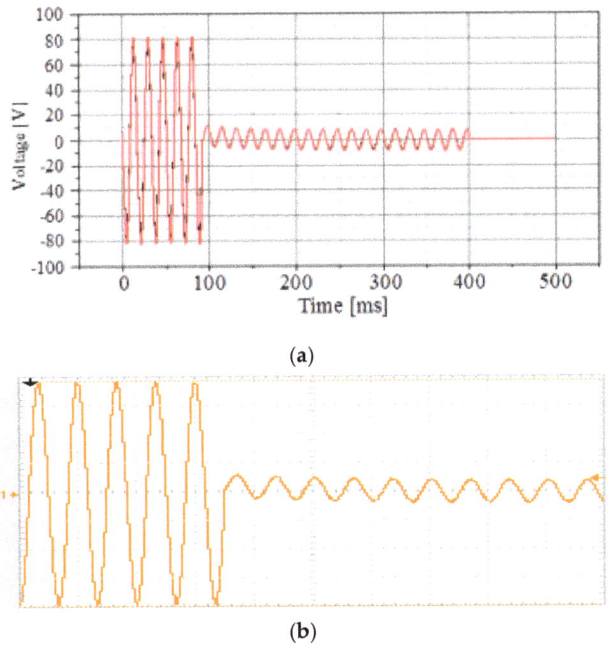

Figure 10. COMTRADE Output waveform test: (**a**) COMTRADE input waveform; (**b**) Simulator output waveform (voltage: 20 V/Div, time: 25 ms/Div).

To ensure that the correct output is produced when generating the desired waveform, simulator input/output waveform comparison was performed using a waveform generation function, as shown in Figure 11. The CT ratio used to measure the output waveform was 400 mV/A. The maximum value of the waveform set on the simulator was 1.4 A, and the maximum value of the output waveform was 1.5 A; thus, it was confirmed that accurate waveforms can be generated and output.

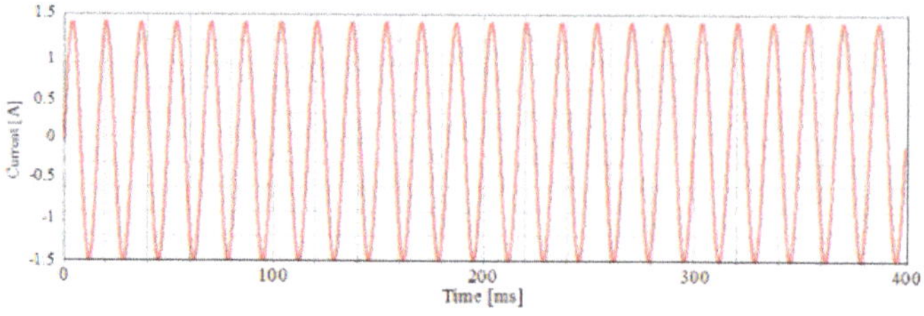

(a)

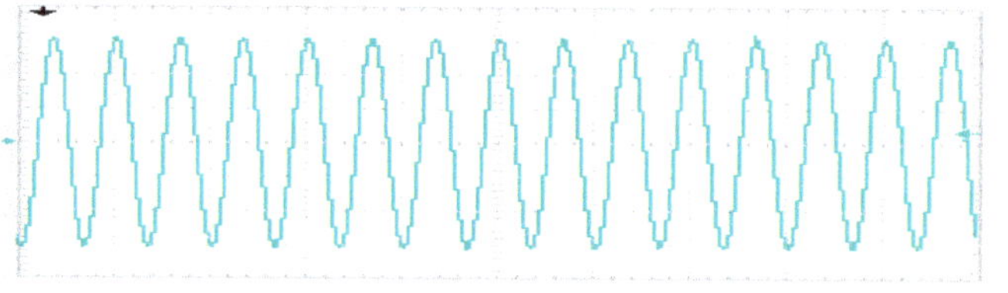

(b)

Figure 11. Customized waveform output test: (a) Simulator input waveform; (b) Simulator output waveform (voltage: 200 mV/div, time: 25 ms/div).

4. Case Study

As PREDIS-R is a simulator for testing the performance of the protective relay for the railway system, a case study was conducted according to the scenarios given in Table 4 to validate the performance test results of protective relay. The input waveform used in the case study consists of both the external input waveform and internal waveform generated by the user.

Table 4. Simulation scenario. SP: sectioning post, SS: substation.

Scenario	Waveform Type	Description
Case 1	COMTRADE	Adjacent to SS, trolley line fault
Case 2	COMTRADE	Adjacent to SP, trolley line fault
Case 3	COMTRADE	Adjacent to SS, feeder line fault
Case 4	COMTRADE	Adjacent to SS, feeder line fault
Case 5	Simulator product	Load current including harmonics

First, using PSCAD/EMTDC, the TPSS was modeled to reflect the actual domestic data, and then the contingency was simulated. In this case, the fault voltage and current waveform were converted into COMTRADE format and entered into the digital simulator interconnected to the protective relay; then, it was checked whether the relay worked with the simulator. To facilitate checking the operation of the relay, the fault resistance was assumed to be 0 Ω. Each protective relay element must operate in coordination depending on the voltage and current measurement position, such as the incoming panel and distributing panel of the substation and the location of the relay. However, this study does not consider them, as it focuses on verifying the performance of a simulator that injects abnormal waveforms into the relay and monitors the operating status of the relay.

The TPSS model using PSCAD/EMTDC in the simulation of Case 1 through Case 4 is shown in Figure 12, and the detailed parameters of the system are shown in Table 5.

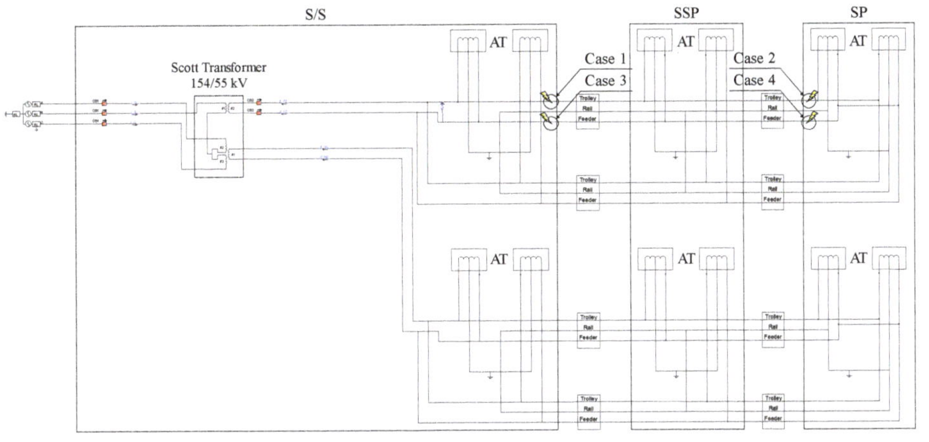

Figure 12. TPSS for case studies.

Table 5. Parameters of the traction power-supply system (TPSS) model.

Items	Parameters	
%Z of the SS	Z_1 = 1.1128 + j7.8827 [%] at 100 MVA base Z_0 = 3.0157 + j17.8050 [%] at 100 MVA base	
Scott transformer	%Z = 10 [%] at 30 MVA base Primary side: 154 [kV] Secondary side: 55 [kV]	
Auto transformer	Z = j 0.45 [Ω] Primary side: 55 [kV] Secondary side: 27.5 [kV]	
Impedance of the catenary system	Trolley Rail Feeder	0.1076 + j0.2645 [Ω/km] 0.1052 + j0.4699 [Ω/km] 0.1180 + j0.4548 [Ω/km]
Length of the each section	SS to SSP SSP to SP	10 [km] 10 [km]

It was modeled using the actual data of the TPSS, and checking the location of the fault simulated for each case is possible. The system is mainly divided into the substation (SS), sub-sectioning post (SSP), and SP. Each division has an AT to compensate for voltage drop. In SS, a Scott transformer is in operation, which can receive power from the utility and transform the three-phase voltage level of 154 kV to 55 kV on a single phase. This is a characteristic of the autotransformer feeding system and is intended to output twice the voltage applied to the electric train load in the substation. The autotransformers are installed at SS, SSP, and SP—one each at the up and down tracks to supply power. Each division is about 10 km apart, and the reason for operating the autotransformer is to compensate the voltage drop. In this feeding scheme, the installation interval of the substation can be extended. The neutral point of the autotransformer is commonly grounded together with rails and protective wires at 2.8-km intervals within the catenary section, and it causes an offset of the communication inductive interference.

To check the operation of the protective relay software, the distance relay protecting the TPSS was modeled using PSCAD/EMTDC. Figure 13 shows the relay operating trajectories for each case, and a detailed operating time comparison with simulators is shown in Table 6.

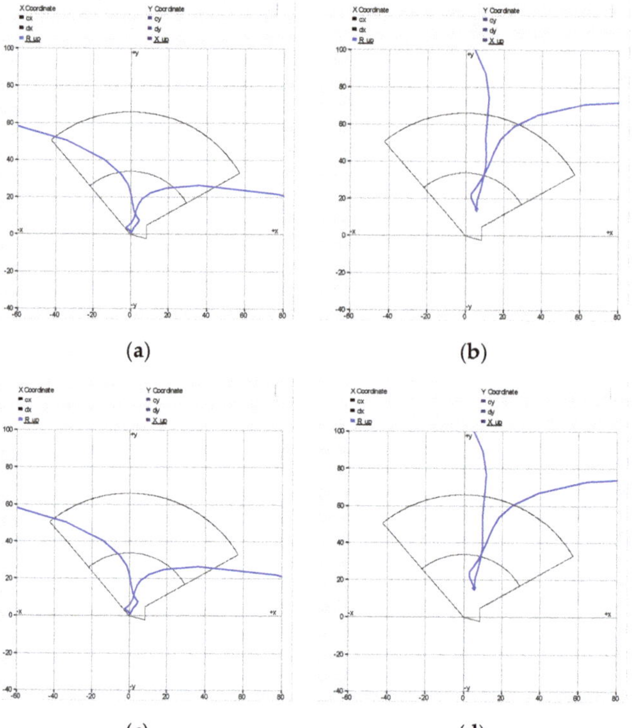

Figure 13. Simulation results using PSCAD/EMTDC: (**a**) Case 1; (**b**) Case 2; (**c**) Case 3; and (**d**) Case 4.

Table 6. Simulation results—relay operation time (s).

Scenario	PSCAD/EMTDC	Digital Simulator
Case 1	0.1266	0.1267
Case 2	0.1292	0.1293
Case 3	0.1287	0.1288
Case 4	0.1329	0.1327

Figure 14 represents the results for Case 1. The distance relay element and overcurrent element of the integrated protective relay were operated within one cycle owing to trolley-rail short circuit fault near the SS. Each relay was used as the main protective relay and the back-up protective relay in the TPSS. Figure 15 shows the input and output waveform information and relay operation information displayed on the simulator interface. It is possible to intuitively identify the relay elements that operate in response to a simulated fault.

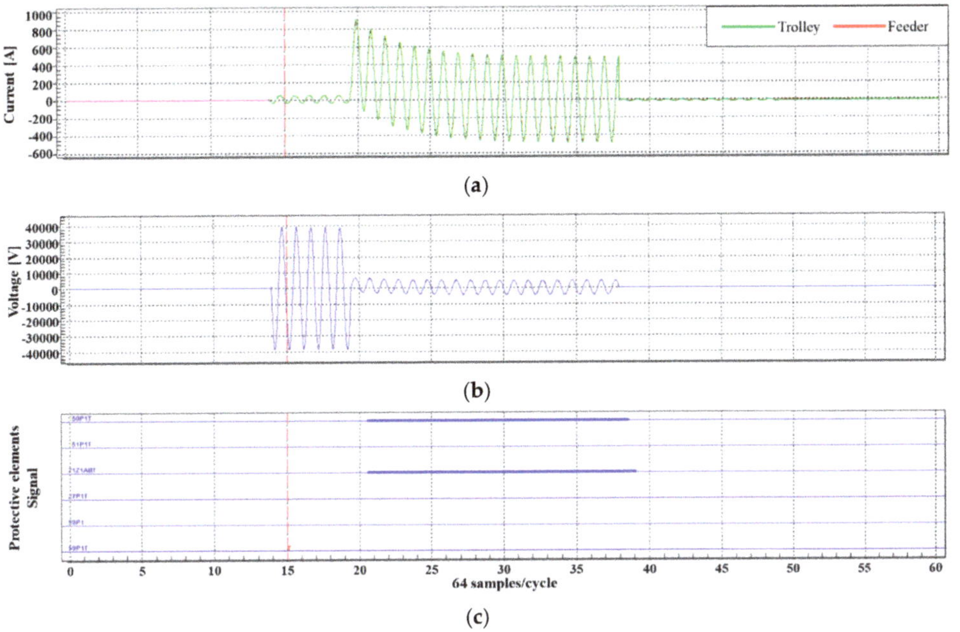

Figure 14. Simulation results for Case 1: (**a**) Input current; (**b**) Input voltage; and (**c**) Protective relay signal.

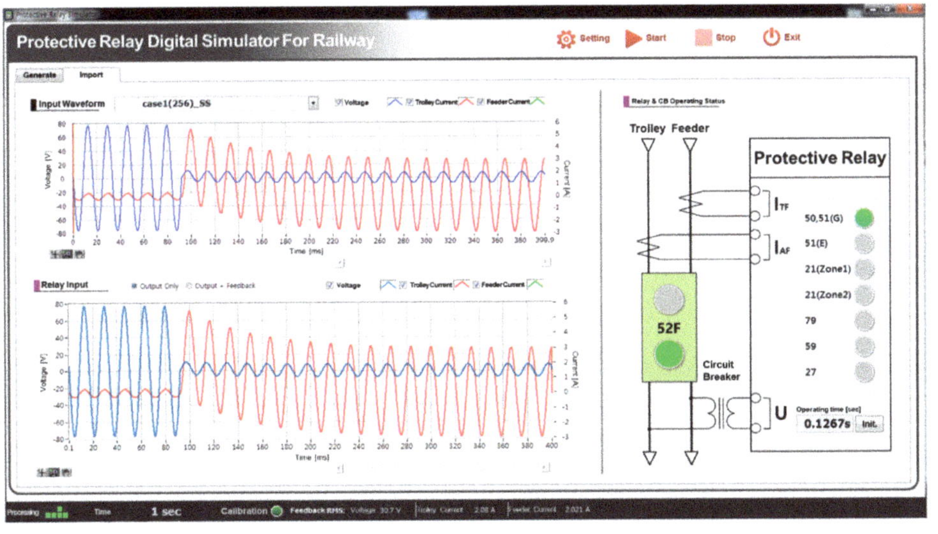

Figure 15. Simulation results in the digital simulator interface.

Figure 16 shows the result for Case 2 and the simulator interface output screen. Similar to Case 1, it is confirmed that the proposed simulator imports the COMTRADE file correctly and displays the results appropriately after checking the protective relay behavior.

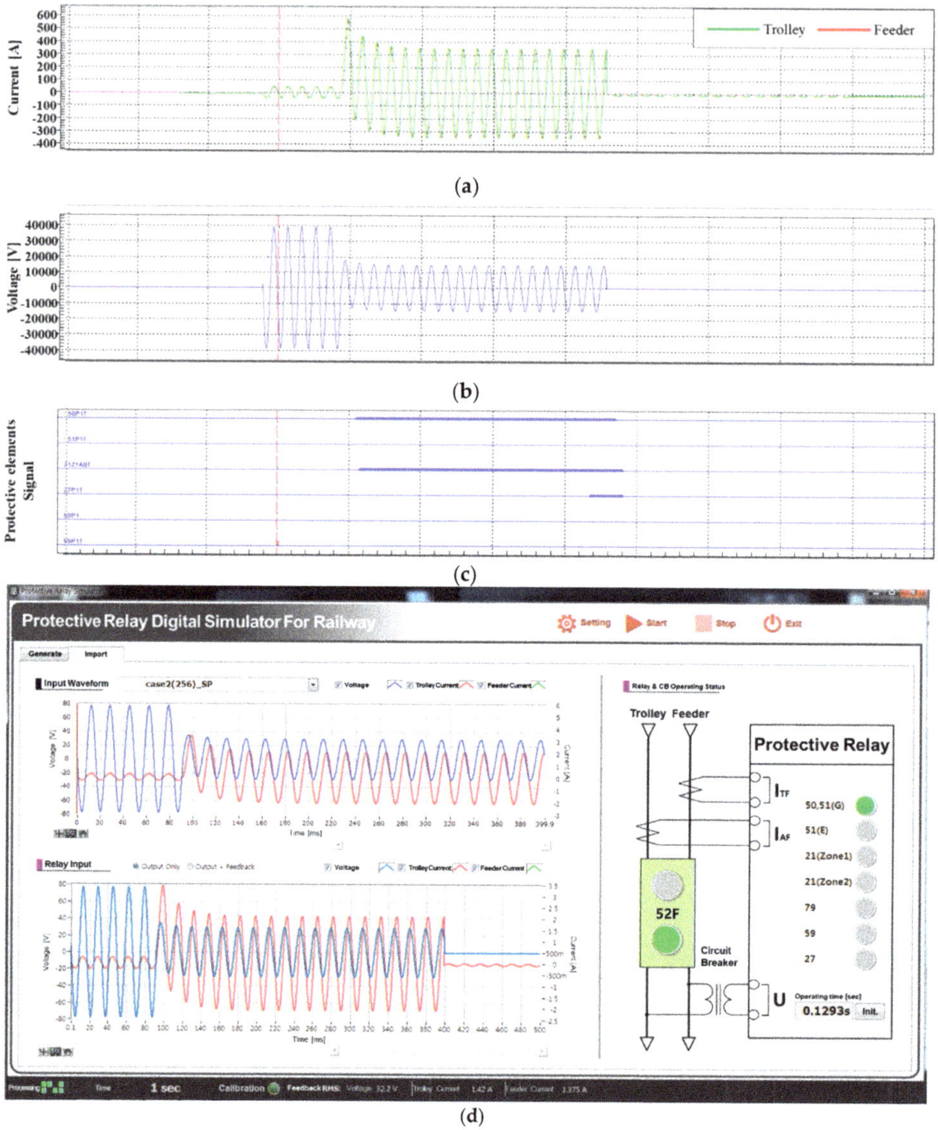

Figure 16. Simulation results for Case 2: (**a**) Input current; (**b**) Input voltage; (**c**) Protective relay signal; and (**d**) Digital simulator interface results.

Figures 17 and 18 represent the resulting waveforms corresponding to Cases 3 and 4, respectively, which are cases of feeder-rail short circuit fault near SS and SP, respectively. It was confirmed that identical to the previously simulated trolley-rail short circuit fault, the relay operated when an analog signal was input into the protective relay, and the result was accurately outputted from the simulator.

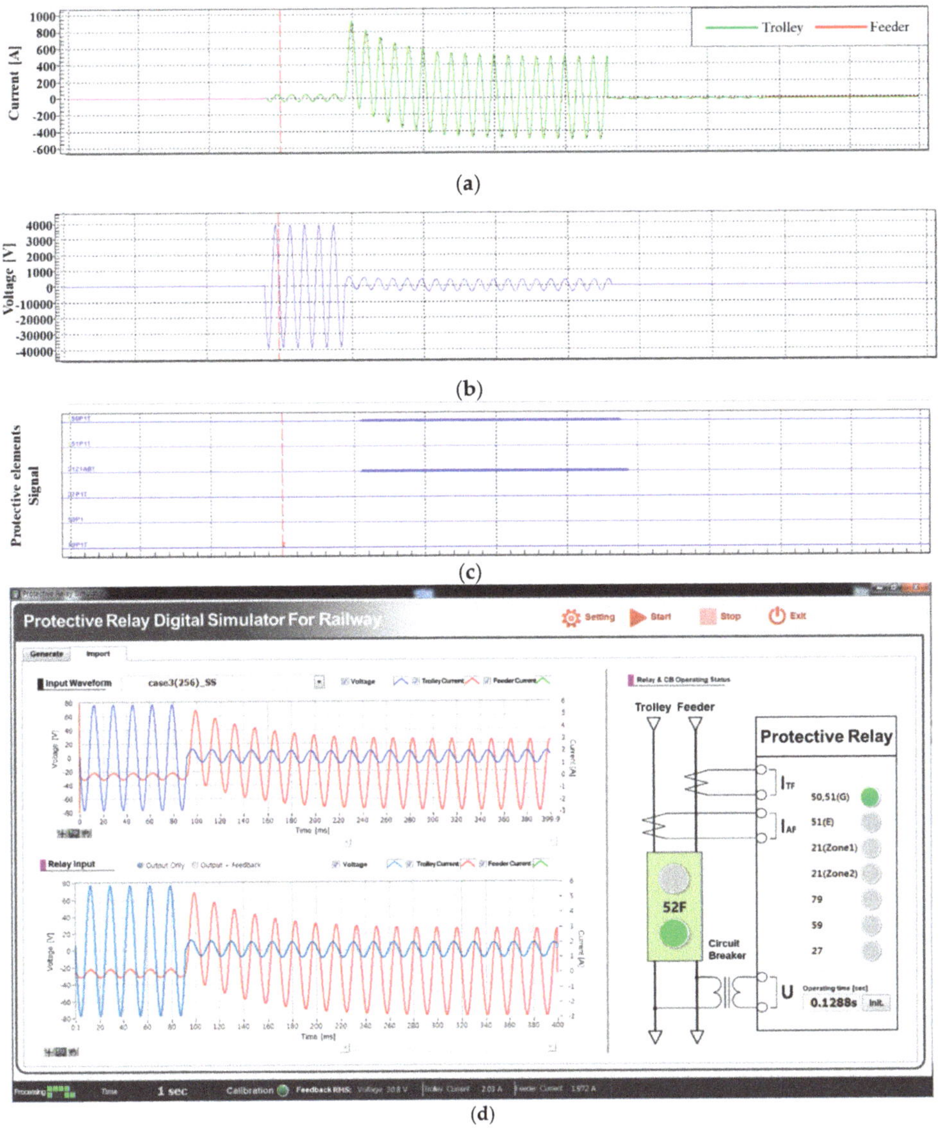

Figure 17. Simulation results for Case 3: (**a**) Input current; (**b**) Input voltage; (**c**) Protective relay signal; and (**d**) Digital simulator interface results.

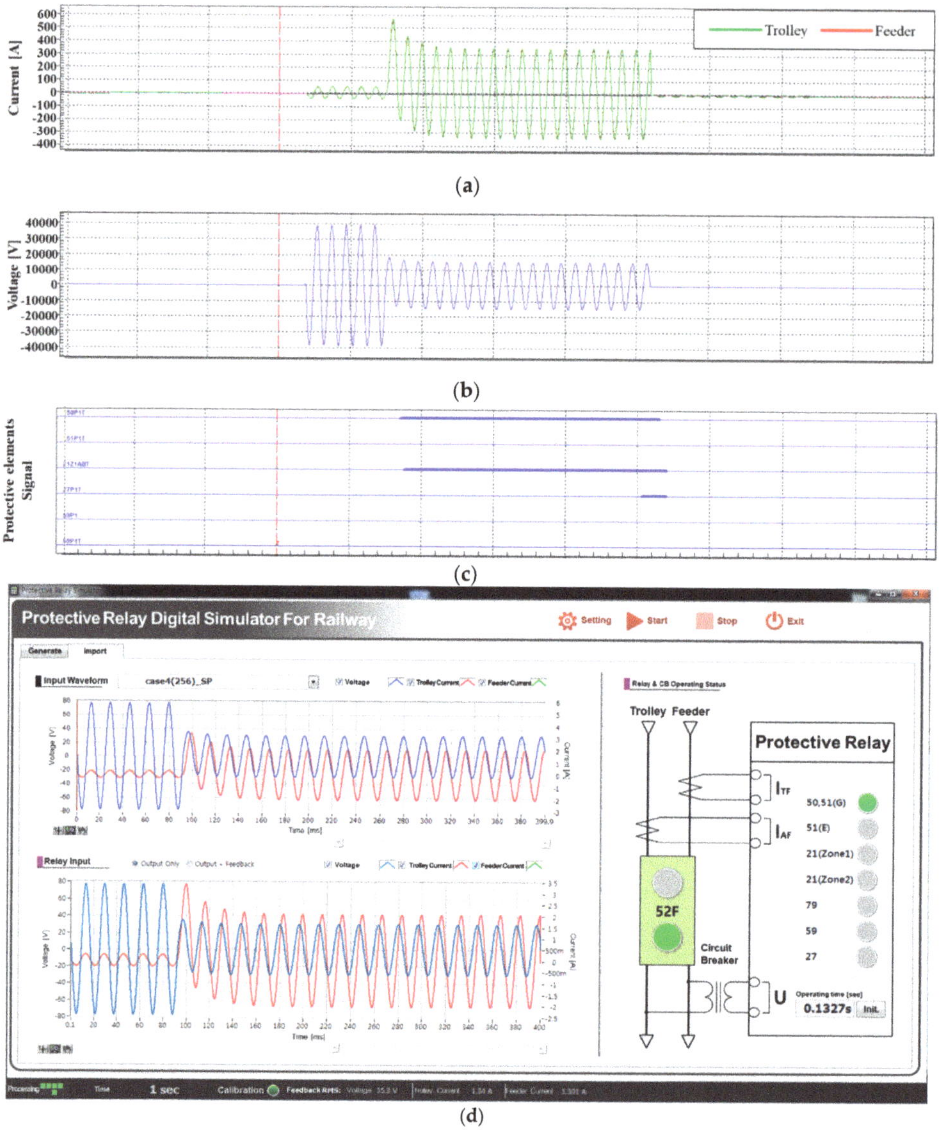

Figure 18. Simulation results for Case 4: (**a**) Input current; (**b**) Input voltage; (**c**) Protective relay signal; and (**d**) Digital simulator interface results.

Table 6 compares the operating time of the relay using the simulator with the relay operating time using the PSCAD/EMTDC. It is verified that both sides have been properly modeled with a smaller error in the operating time, and it is expected to be able to perform accurate simulations through cross-validation on both platforms when validating and testing the performance of protective relays later. The operating time of the relay shows the sequential operation of the protective elements, and the last operating protective element is time-overcurrent relay.

The current of the TPSS contains harmonics owing to the switching element of the train, and the protective relay must not operate on these currents including the inrush currents of the transformer.

Therefore, as a simulator for testing protective relays, including a function generating harmonics to enable flexible testing in various ways is necessary. Figure 19 shows the test waveform applied in Case 5, which is assumed to be the load current containing harmonics generated randomly by the simulator. The harmonic component included is 13.5% of the second harmonics and 4.9% of the third harmonics.

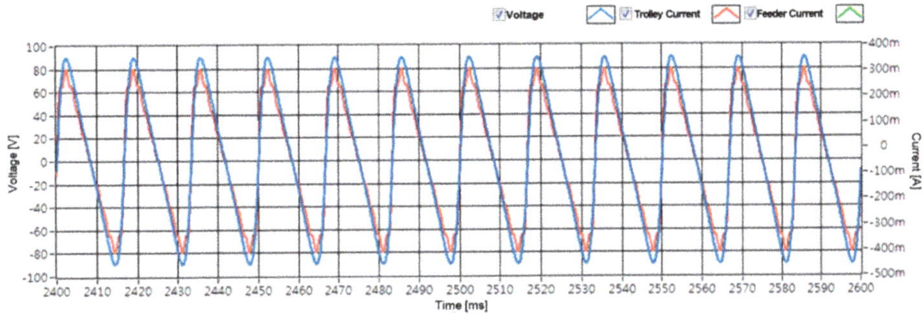

Figure 19. Current waveform with harmonics.

Figure 20 represents the results of a protective relay malfunction test for waveforms containing harmonics in Figure 19. The protective relay did not work, implying that the simulator injected the correct test waveform and displayed the relay's same operating information as the actual operation.

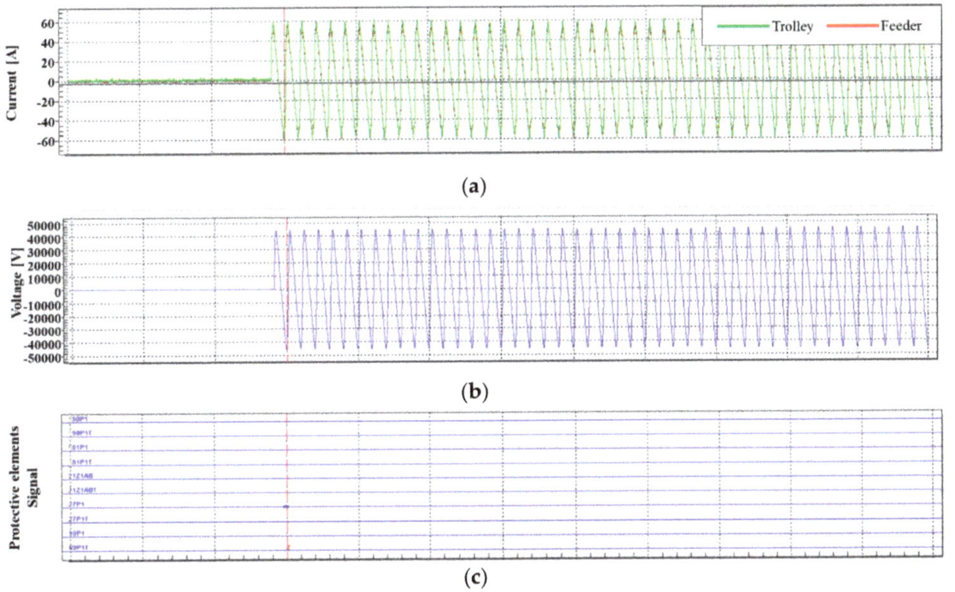

(**a**)

(**b**)

(**c**)

Figure 20. *Cont.*

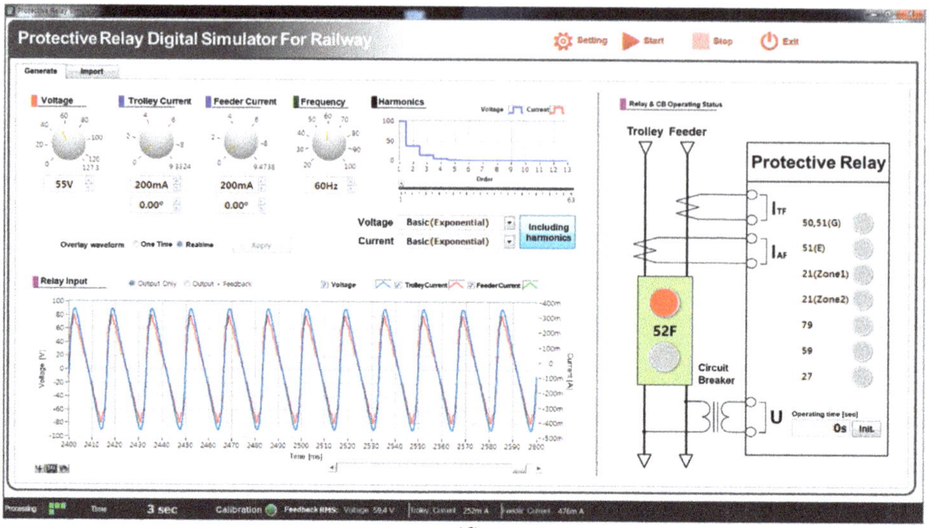

(**d**)

Figure 20. Simulation results for Case 5: (**a**) Input current; (**b**) Input voltage; (**c**) Protective relay signal; and (**d**) Digital simulator interface results.

5. Conclusions

Simulator development is essential when developing or testing new protective relays because it is time consuming and costly to verify performance on the real system; moreover, the test space is limited. To solve these real-world engineering problems, this paper presented and validated the modern testing simulator for protective relay in the TPSS. The developed simulator has the appropriate amplification rate to inject analog signals into the integrated relay and have a function to input external COMTRADE files and generate the simulator's self-waveform. It was also confirmed that the relay behaves the same as in the actual system under various contingency fault simulations, and the harmonics generation function also works accurately through the relay's malfunction test.

As a result, by using this proven digital simulator, it is possible to replicate in the laboratory electrical faults that have been detected in the real field without running the test on the real infrastructure, which can significantly speed up testing. The simulator developed in this study is expected to be used in the future to develop new protective relays or validate protective relay algorithms.

Author Contributions: J.K. (Joorak Kim) conceived and designed the main concept; J.K. (Joorak Kim), J.K. (Jaewon Kim), and G.-J.C. performed test and simulation; J.K. (Joorak Kim) and J.K. (Jaewon Kim) wrote the paper. All authors have read and agreed to the published version of the manuscript.

Funding: This research was funded by the Ministry of Land, Infrastructure and Transport's Railway Technology Research Project, "Development of Real Time Simulator and Analysis Model for Railway Power System (19RTRP-B146034-02)".

Acknowledgments: The authors are grateful for the support by Korea Agency for Infrastructure Technology Advancement and Korea Railroad Research Institute.

Conflicts of Interest: The authors declare no conflict of interest.

References

1. Kim, J.C. *Analysis of Power Supply System and Understanding of the Korean Electric Railway*; Kidari: Seoul, Korea, 2008.
2. Kim, B. *Electric Railway Power Engineering*; Kidari: Seoul, Korea, 2015.

3. Sezi, T.; Menter, F.E. Protection scheme for a new AC railway traction power system. In Proceedings of the 1999 IEEE Transmission and Distribution Conference (Cat. No. 99CH36333), New Orleans, LA, USA, 11–16 April 1999; pp. 388–393.
4. Fujie, H. Fault locator system in autotransformer feeding circuit of AC electric railway. *Quart. Rep. Railw. Tech. Res. Inst.* **1997**, *18*, 33–38.
5. Deng, Y.; Huang, K.; Su, D.; Liu, Z. Spatial Magnetic-Field Description Method Aimed at 2 × 25 kV Auto-Transformer Power Supply System in High-Speed Railway. *Appl. Sci.* **2018**, *8*, 997. [CrossRef]
6. Mariscotti, A.; Pozzobon, P.; Vanti, M. Distribution of the traction return current in AT electric railway systems. *IEEE Trans. Power Deliv.* **2005**, *20*, 2119–2128. [CrossRef]
7. Oura, Y.; Mochinaga, Y.; Nagasawa, H. Railway electric power feeding systems. *Jpn. Railw. Transp. Rev.* **1998**, *16*, 48–58.
8. Sibal, V. Traction Power Supply System for California High-Speed Train Project. Presented at AERMA. Available online: https://www.arema.org/files/library/2011_Conference_Proceedings/Traction_Power_Supply_System_for_California_High-Speed_Train_Project.pdf (accessed on 15 October 2019).
9. Lee, C.; Lee, H.; Yoon, D.-H.; Lee, H.; Song, J.; Jang, G.; Han, B. A novel fault location scheme on Korean electric railway system using the 9-conductor representation. *J. Elect. Eng. Technol.* **2010**, *5*, 220–227. [CrossRef]
10. Jung, H.; Kim, H.; Chang, S.-H.; Kim, J.; Min, M.-H.; An, T.-P.; Kwon, S.-I. A Study on a Catenary Impedance Estimation Technique using Boosting Current Compensation Based on Current Division Characteristics of an AT Feeding System. *J. Elect. Eng. Technol.* **2015**, *10*, 1370–1376. [CrossRef]
11. Cho, G.-J.; Kim, C.-H.; Kim, M.-S.; Kim, D.-H.; Heo, S.-H.; Kim, H.-D.; Min, M.-H.; An, T.-P. A Novel Fault-Location Algorithm for AC Parallel Autotransformer Feeding System. *IEEE Trans. Power Deliv.* **2019**, *34*, 475–485. [CrossRef]
12. Serrano, J.; Platero, C.; Lopez-Toledo, M.; Granizo, R. A new method of ground fault location in 2 × 25 kV railway power supply systems. *Energies* **2017**, *10*, 340. [CrossRef]
13. McLaren, P.G.; Kuffel, R.; Wierckx, R.; Giesbrecht, J.; Arendt, L. A real time digital simulator for testing relays. *IEEE Trans. Power Deliv.* **1992**, *7*, 207–213. [CrossRef]
14. Avalos, A.; Zamora, A.; Escamilla, O.; Paternina, M.R.A. Real-time Hardware-in-the-loop Implementation for Power Systems Protection. In Proceedings of the 2018 IEEE PES Transmission & Distribution Conference and Exhibition—Latin America (T&D-LA), Lima, Peru, 18–21 September 2018.
15. Souza, F.C.; Benemar, A. Souza Adaptive overcurrent adjustment settings: A case study using RTDS®. In Proceedings of the 2013 IEEE PES Conference on Innovative Smart Grid Technologies (ISGT Latin America), Sao Paulo, Brazil, 15–17 April 2013.
16. Wang, B.; Dong, X.; Bo, Z. Anthony Perks RTDS Environment Development of Ultra-High-Voltage Power System and Relay Protection Test. *IEEE Trans. Power Deliv.* **2008**, *23*, 618–623. [CrossRef]
17. PSCAD Homepage. Available online: https://hvdc.ca/ (accessed on 16 December 2019).

applied sciences

MDPI

Article

Comparison of Various Mother Wavelets for Fault Classification in Electrical Systems

Chaichan Pothisarn [1], Jittiphong Klomjit [1], Atthapol Ngaopitakkul [1,*], Chaiyan Jettanasen [1], Dimas Anton Asfani [2] and I Made Yulistya Negara [2,*]

1 Department of Electrical Engineering, Faculty of Engineering, King Mongkut's Institute of Technology Ladkrabang, Bangkok 10520, Thailand; chaichan.po@yahoo.com (C.P.); tor_1_4@hotmail.com (J.K.); chaiyan.je@kmitl.ac.th (C.J.)

2 Department of Electrical Engineering, Institut Teknologi Sepuluh Nopember, Surabaya, East Java 60111, Indonesia; anton@ee.its.ac.id

* Correspondence: knatthap@kmitl.ac.th (A.N.); yulistya@ee.its.ac.id (I.M.Y.N.)

Received: 7 January 2020; Accepted: 7 February 2020; Published: 11 February 2020

Abstract: This paper presents a comparative study on mother wavelets using a fault type classification algorithm in a power system. The study aims to evaluate the performance of the protection algorithm by implementing different mother wavelets for signal analysis and determines a suitable mother wavelet for power system protection applications. The factors that influence the fault signal, such as the fault location, fault type, and inception angle, have been considered during testing. The algorithm operates by applying the discrete wavelet transform (DWT) to the three-phase current and zero-sequence signal obtained from the experimental setup. The DWT extracts high-frequency components from the signals during both the normal and fault states. The coefficients at scales 1–3 have been decomposed using different mother wavelets, such as Daubechies (db), symlets (sym), biorthogonal (bior), and Coiflets (coif). The results reveal different coefficient values for the different mother wavelets even though the behaviors are similar. The coefficient for any mother wavelet has the same behavior but does not have the same value. Therefore, this finding has shown that the mother wavelet has a significant impact on the accuracy of the fault classification algorithm.

Keywords: transformer; transmission line; discrete wavelet transform; mother wavelet; fault classification

---

1. Introduction

Disturbance in the power system is a significant concern for many electric utilities due to its impact on the operation and reliability of the overall system. Besides, it may cause damage to vital equipment in the power system such as a generator, transmission line, and transformer in case the operator does not address the fault with speed and accuracy, which results in large scale electric outage, economic losses, and the possibility of loss of life. Thus, the protection system with the ability to accurately identify, locate, and classify faults in the power system needs to be implemented into the system to ensure the quick and correct response to fault clearing.

The algorithm to detect and classify faults in power transmission lines has been widely developed based on different types of methodologies. The wavelet transform is one of the popular methods for fault detection and discrimination due to its ability to analyze high-frequency components [1–6]. The discrete wavelet transform (DWT)-based methodology has been used for the detection of transmission line faults [1]. Three-phase voltage signals are decomposed by the Daubechies wavelet (db4). By considering the results, the coefficient details at level 4 clearly distinguish the different types of faults in the power system. The new differential relaying scheme based on the transient energy extracted using the DWT in the current signals of each line distinguishes between external and internal faults under all operating conditions [2]. This paper proposed an adaptive technique to detect low-impedance faults (LIFs) and high-impedance faults (HIFs) and classified LIFs in transmission systems depending on the DWT. The current of each phase is analyzed by using the DWT (db1), and faults are classified through comparison with the current approximation coefficient (Sg) and current detail coefficients (Da, Db, and Dc) [3]. This reveals that the wavelet transform is significant in the detection and classification of faults in power systems. However, the algorithm can be improved by using artificial intelligence (AI).

AI has been combined with wavelets to improve the accuracy of fault classification algorithms in electrical systems [7–12]. AI enhances accuracy and reduces the time to classify faults. Wavelets with neural networks (NNs) have been used to detect and identify fault types in transmission line systems [7–11]. The algorithm makes use of wavelet transform-based approximate coefficients of three-phase voltage and current signals obtained over a quarter cycle to detect and classify faults. Fault detection and classification and fault location estimation are carried out using an artificial neural network (ANN), and the alienation coefficients of current signals are used as ANN input [7]. This paper presents an application of NNs and wavelet transforms for fault classification in transmission lines in comparison with particle swarm optimization–artificial neural network (PSO–ANN)-, back-propagation neural networks (BPNN)-, and support-vector machines (SVM)-based classification schemes. The PSO–ANN technique has a very high accuracy (99.912%) in the classification of power system faults [11]. This paper presents a survey and review of the research and developments in the field of fault detection, classification, and location in transmission networks [12].

In addition, wavelets are used to detect faults or abnormalities in transformers [13–22]. They are used to detect vibrations or electrical signals. Transformer vibration signals are decomposed into several empirical wavelet functions. The signals are calculated to construct the eigenvectors of the transformer vibration signals for classifying three different working conditions (normal conditions, winding axial deformation, and winding radial deformation) [13]. Most papers use wavelets to classify inrush and internal and external faults. The differential current is used as input for the wavelet transform to analyze faults [14–21]. The boundary wavelet transform is implemented in the differential protection of power transformers to distinguish internal faults from other disturbances. The method is designed for real-time applications and implemented in a digital signal processor for real-time analysis [14]. The algorithm distinguishes between internal faults and inrush currents in power transformers. Fault currents are analyzed by using the DWT to evaluate their remarkable characteristic values, and the highest values produced by the total wavelet correlation matrix are used to identify inrush and internal fault currents in power transformers. The results are validated with an experimental test setup [18]. Moreover, the algorithm can identify fault occurrence with the continuous wavelet transform (CWT) and improve conventional current differential protection methods in the presence of current transformer (CT) saturation [20]. A spectrum of wavelets has been used for the prediction of winding insulation defects in transformers [21].

From the literature review, it can be seen that the wavelet transform methodology has been widely used in power system fault analysis. However, many studies only focus on transmission lines or transformers, and few have combined the two components when analyzing systems. Most of the references use wavelets to detect and identify fault types in electrical systems, but they do not analyze the mother wavelet [1–17,20,21]. Additionally, there are research articles that compare mother wavelets [18,19,22,23]. They study the comparative use of 16 different wavelets for fault classification in overhead transmission line systems. However, it is revealed that the Db4 wavelet completely satisfies the fault classification algorithm [23].

Another point is that the result from a simulation that has been widely used may not correctly represent the actual system in the real world due to some parameter and factor that has been simplified, which might affect the result. In terms of wavelet transform application, the research using wavelets to detect and discriminate fault types on power systems might not take effects from different mother wavelets into consideration when evaluating the performance of the algorithm. Thus, this paper presents the effect of mother wavelets on the performance of fault classification algorithms in the power system. The result in terms of accuracy between different mother wavelets was evaluated by using an experimental setup modeled after the actual system with a transmission line connected to the power transformer.

The paper is divided into six sections as follows. The second section provides the fundamentals and theory of the DWT. The experimental setup and system characteristics are presented in Section 3. The details of the fault classification algorithm used in this study are contained in Section 4. The results and conclusions of the study are summarized in Sections 5 and 6, respectively.

2. Fundamentals and Theory

The wavelet transform is a method based on signal processing that was developed from Fourier and short-Fourier transforms for suitable specific applications. The time width and frequency can be adjusted for optimal analysis.

The wavelet theory based on mathematics integrates small signals into one signal. The small signal is the wavelet, which has a specific character because the wavelet is a wave-like oscillation with an amplitude that begins at zero, increases, and then decreases back to zero. Therefore, any signals in the wavelet theory are a combination of wavelet groups that are structured from the same function. This function is the origin wavelet called the mother wavelet. The signals are caused by the mother wavelet stretching (scaling: a) or shrinking (shifting: k) the wavelet itself, which occurs as a position change along the time axis.

DWT signal analysis can select many scales (m) and many positions (k) for integrated wavelets and can form a signal at the interest scale. When integrating the signal at all resolutions, the actual input signal is received. In multilevel analysis, the scale of analysis is reduced by two times. Hence, this is a DWT-type dyadic wavelet transform, as described in Equation (1) [24].

$$DWT(m,n) = \frac{1}{\sqrt{2^m}} \sum_k f(k)\psi\left[\frac{n-k2^m}{2^m}\right] \tag{1}$$

where

m, n, k = Integers
Ψ = Mother wavelet
n = Number of data points
m = Scaling
k = Shifting

Signals are decomposed by the wavelet. The mother wavelet is a filter signal that separates the following two channels:

- High-frequency component (detail coefficient: cD).
- Low-frequency component (approximation coefficient: cA).

The high-frequency component will be used to analyze signal during transient state.

The mother wavelets used in the DWT are db, sym, coif, and bior [24]. Each mother wavelet provides a differential coefficient because of its differential characteristic signal, as shown in Figure 1.

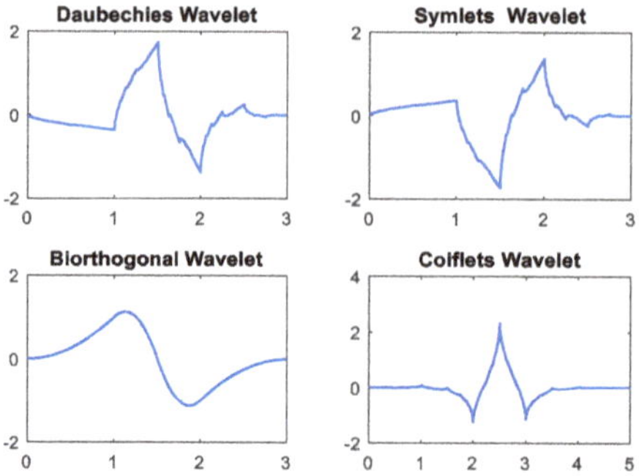

Figure 1. Characteristic signals of the mother wavelets.

The Daubechies (db) mother wavelet has asymmetric basic functions, while the symlets (Sym) mother wavelet has the least asymmetric basic functions. The biorthogonal (Bior) mother wavelet has symmetric basic functions, and the coif1 mother wavelet has nearly asymmetric basic functions. It is necessary to choose a suitable mother wavelet for increased efficiency.

3. Experimental

In this section, the faults that occur in transmission lines and transformers are tested by using the experimental setup. The difference current signals and transients generated from the experimental setup are detected and recorded for analysis with a wavelet. The experimental model is shown in Figure 2.

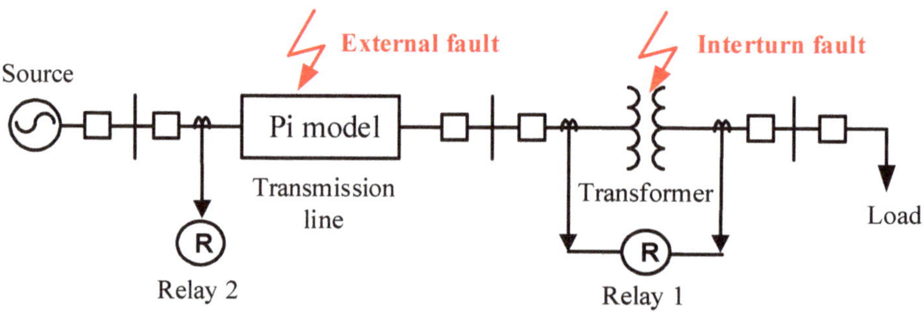

Figure 2. Characteristic signal of the mother wavelet.

Figure 2 shows the fault simulation circuit in the case of transmission lines connected to a transformer. The 40 km 115 kV transmission lines are calculated to be resistor, inductor and capacitor (RLC) parameters that are connected to a nominal pi model and a 15 kVA, 440/220 V transformer. The transformer is designed to accept internal short-circuiting. The design of the transformer permits a tap change every 10% of both the primary and secondary sides to test internal faults in the transformer. Moreover, the faults occurring in the transmission lines are tested by following these conditions:

- The fault types include a single line to the ground fault, a double line to the ground fault, a line to line fault, and a three-phase fault.
- The fault positions are designated to any phases of the transmission lines at lengths of 30%, 50%, and 70% of the length of the transmission line.

The experimental setup is shown in Figure 3.

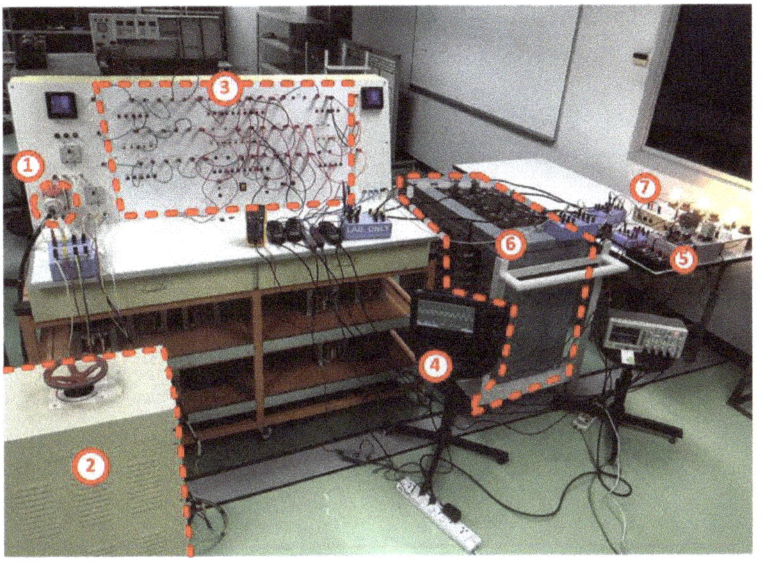

1. Supply
2. Variable-voltage transformer
3. Transmission line
4. Oscilloscope
5. Zero-crossing circuit
6. Transformer
7. Load

Figure 3. Experimental setup.

After recording the differential current from the transformer, the differential current is obtained through the wavelet transform method. The current signal is transformed using the DWT to extract high-frequency components at scales 1–3 with different mother wavelets, such as the Daubechies (db), symlets (sym), biorthogonal (bior), and Coiflets (coif) wavelets. The coefficient from the DWT is varied depending on various factors, and this coefficient is squared to emphasize the change in coefficient. The four mother wavelets are utilized to decompose the differential current signals. The DWT coefficients of the phase faults are larger than those of the phase unfaults.

The example of differential current obtained from the transformer (relay 1) in case of internal fault and the coefficient after using different mother wavelets is shown in Figure 4. From the figure, it can be seen that the DWT signals at scale 1 generate much noise compared with those at scales 2–3. The coefficients at scales 2–3 are notable when faults occur, and at scale 3, a lower frequency can be detected compared to that at scale 2. Thus, this research chose the coefficients at scale 2 for analysis of the fault types.

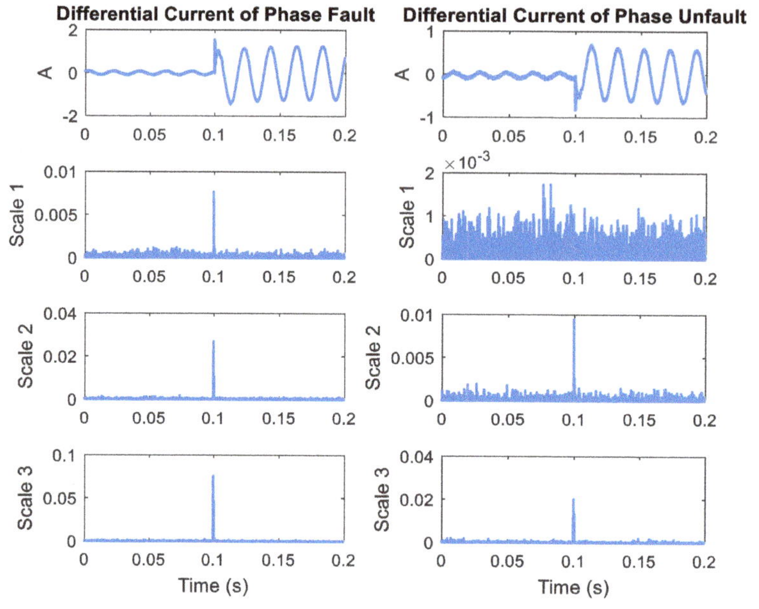

(a)

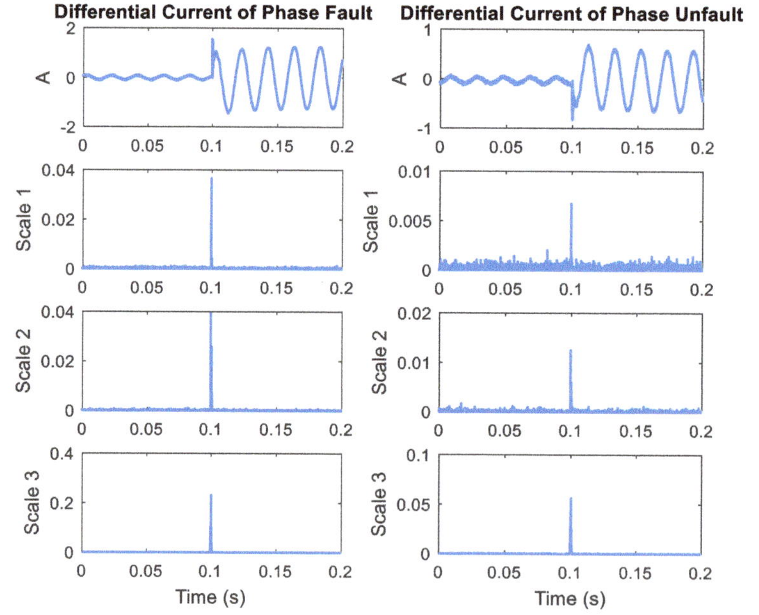

(b)

Figure 4. *Cont.*

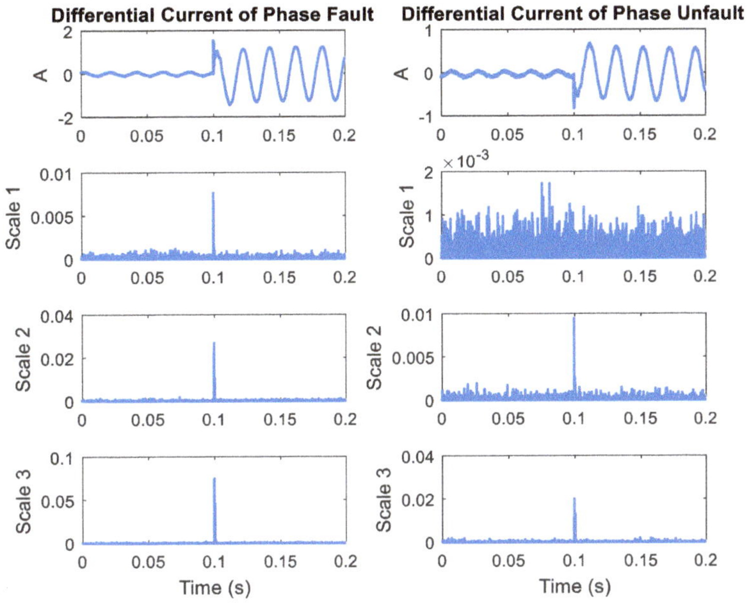

(c)

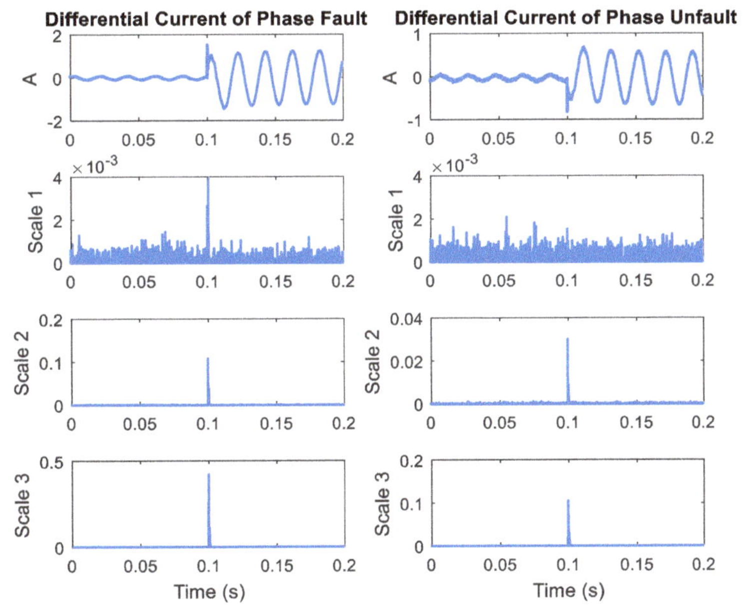

(d)

Figure 4. Differential current signals from the wavelet transform in the case of internal faults at phases in each mother wavelet: (**a**) Daubechies (db2); (**b**) symlets (sym2); (**c**) biorthogonal (bior3.1); (**d**) Coiflets (coif1).

4. Fault Classification

The fault classification algorithm using in this research was designed based on the DWT methodology as described in the flowchart in Figure 5. The operation of the algorithm detecting the fault on the power system consists of a transmission line and a transformer by first checking the status of relay 1 (at the transformer). Relay 1 obtains the differential current signals from the transformer and extracts the coefficient value using DWT. This algorithm considers the maximum coefficients at scale 2 every 5 ms (1/4 cycle). When the coefficients changes more than five times and attains values larger than 5×10^{-3}, an internal fault occurs. This status will signal the relay 1 to trip. Otherwise, it is an external fault and will send a signal for relay 2 (at the transmission line).

For the relay 2, the required signals from the transmission line are the positive sequence of the three-phase current signal. These signals will input through DWT to extract the coefficient value. The fault classification conditions in relay 2 are the maximum coefficients at scale 2 of the positive sequence every 5 ms (1/4 cycle). When the coefficients change more than 2 times and attain values larger than 1×10^{-2}, a fault occurs. This status will signal the relay 2 to trip.

The designed fault classification algorithm and testing of the proposed algorithm consists of testing of the experimental setup to obtain the signal from both relays 1 and 2. The number of case studies was varied with the different conditions to obtain a number of data points, as shown in Table 1. The data were divided into three sets with a total number of data points of 1776; the first set of data was used for algorithm design—50% (888 data points), the second set of data was used for algorithm testing—25% (444 data points), and the last set of data was used for a case study—25% (444 data points). The first set was used to design the condition within the algorithm with an accuracy of more than 95%. The second set of data, different from the first set, was used to test the accuracy of the algorithm. The third set of the data, the newer data differentiated from the first two sets, was used as the case study to evaluate the performance of the proposed algorithm.

Table 1. Number of data points used in the case study.

	Total Data Point	Algorithm Design	Algorithm Testing	Case Study
Transmission line	176	88	44	44
Transformer	1600	800	400	400
Total data point	1776	888	444	444

The four different mother wavelets were used to decompose the current signals to obtain the coefficient value from the high-frequency component. The extraction from the DWT of a differential current showed that the coefficients of the internal faults changed more than those of the external faults. However, the results of the four mother wavelets were similar, as shown in Figure 6. This characteristic was used to design the condition in a fault classification algorithm for relay 1 (transformer). External faults can be analyzed by using coefficients of transmission line systems. The coefficient characteristics between faulty and normal conditions when applied different mother wavelets are shown in Figure 7. The extraction from the DWT of three-phase and positive sequence signals showed a similar trend with the fault condition having a higher coefficient than the normal condition. However, there was a difference in the coefficient value between different mother wavelets. The above behavior was used to design algorithms for fault classification.

The algorithm was operated by using the maximum coefficients, as shown in Figure 8, for the transformer (relay 1) and Figure 9 for the transmission line (relay 2). From Figure 8 it can be seen that the result from DWT of differential current in scale 2 provided a higher maximum coefficient during the transient condition compared to the normal condition. In Figure 9, a similar trend was also shown with the maximum coefficient from scale 2 in the case of an external fault higher than the normal case. The normal case only consisted of noise due to no significant transient state in the signal. This characteristic was used to indicate fault status, and the different mother wavelet provided different maximum coefficient values, which could impact the design algorithm.

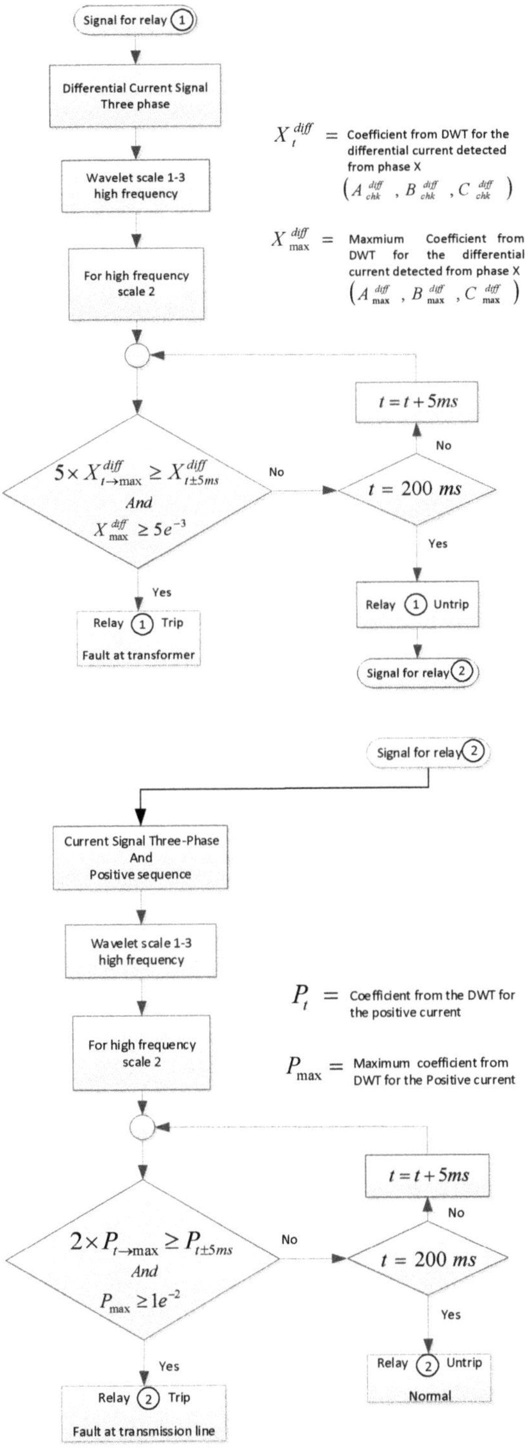

Figure 5. Algorithms for fault classification.

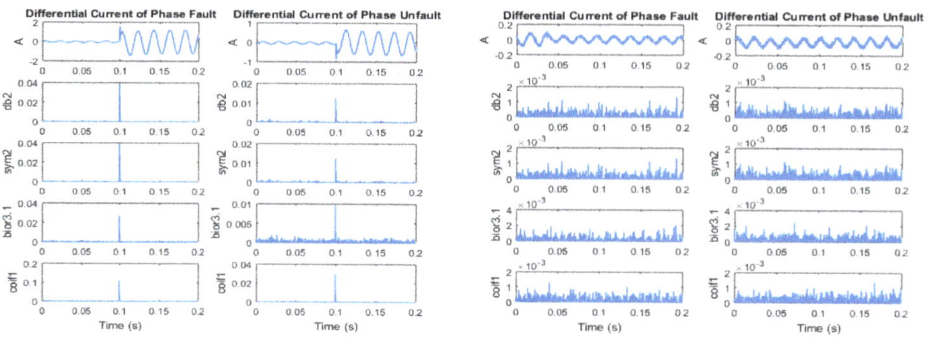

(**a**) (**b**)

Figure 6. Differential current signals from wavelet transform in scale 2 obtained from relay 1: (**a**) internal fault; (**b**) external fault.

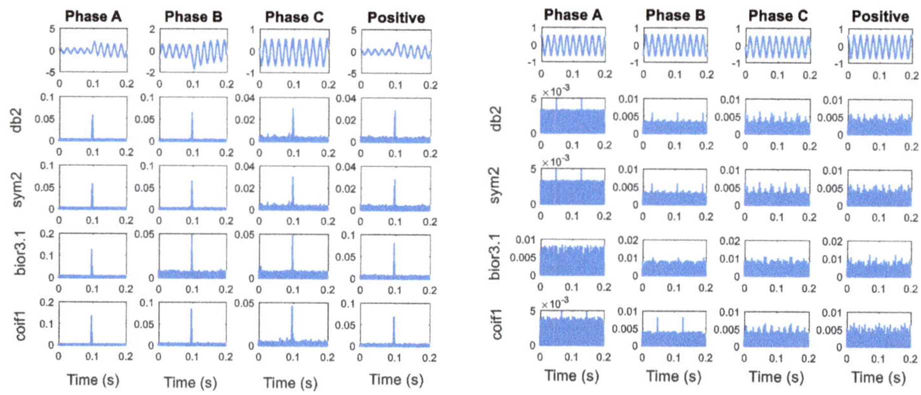

(**a**) (**b**)

Figure 7. Current signals from wavelet transform in scale 2 obtained from relay 2: (**a**) fault; (**b**) normal.

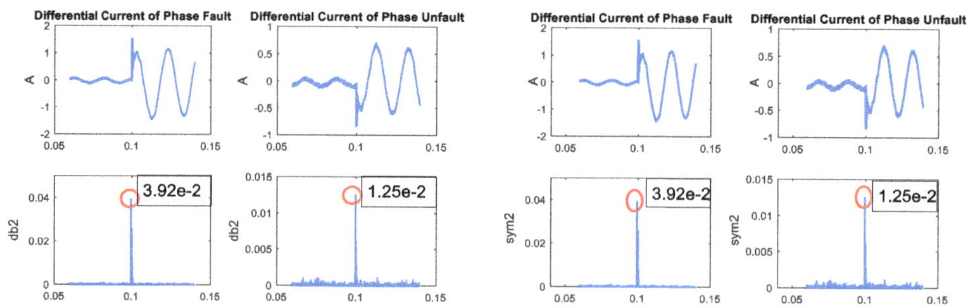

(**a**) (**b**)

Figure 8. *Cont.*

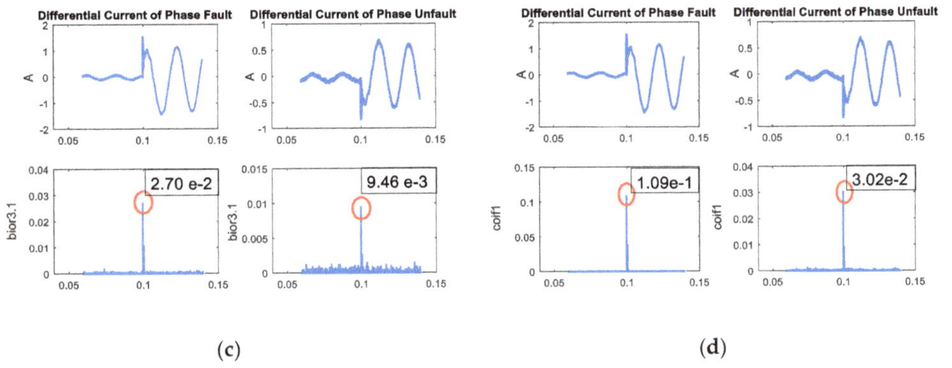

Figure 8. Maximum values of the coefficients from the wavelet transform at scale 2 used in classifying the faults in relay 1: (**a**) Daubechies (db2); (**b**) symlets (sym2); (**c**) biorthogonal (bior3.1); (**d**) Coiflets (coif1).

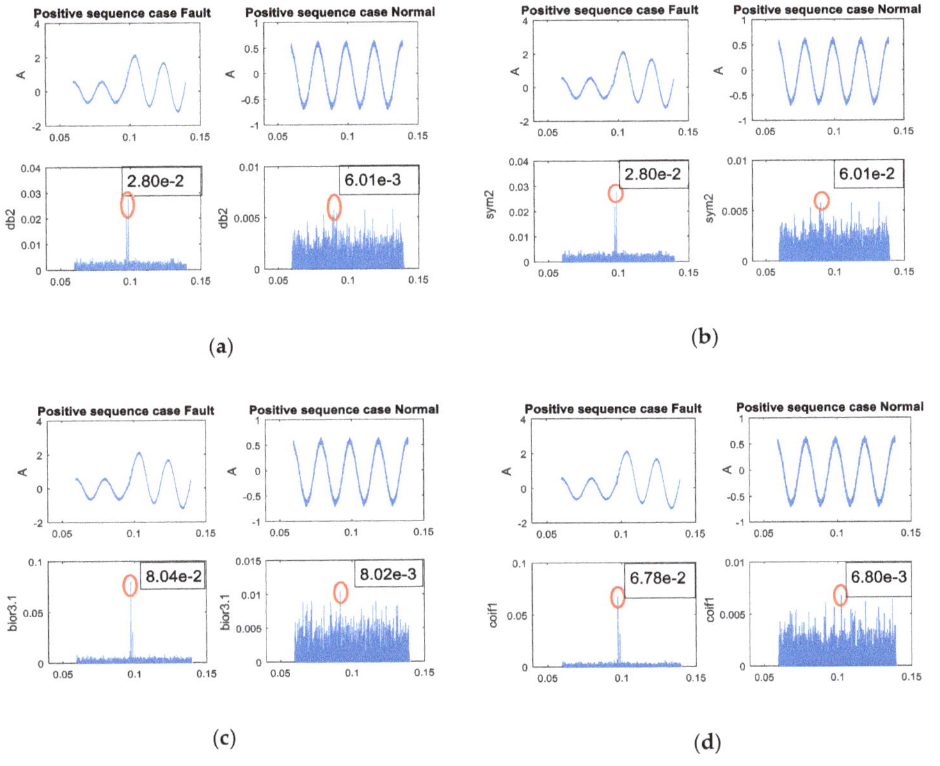

Figure 9. Maximum values of the coefficients from the wavelet transform at scale 2 used in classifying the faults in relay 2: (**a**) Daubechies (db2); (**b**) symlets (sym2); (**c**) biorthogonal (bior3.1); (**d**) Coiflets (coif1).

5. Results

The results from the applied faulty classification algorithm on obtaining a signal from the experimental setup in the case of a fault on the transmission line and transformer is shown in Tables 2 and 3, respectively. From the table, it can be seen that the coefficient values were different when different mother wavelets were used. If the coefficients of the wavelet transform were in accordance with the conditions in Figure 9, the faults inside the transformer would be detected under condition 1, while the faults occurring in the external transformer would be detected under condition 2. For this reason, the first case was an internal fault. By considering the coefficients of relay 1 (Daubechies (db2)), the faults that occurred in phase attained maximum values, and the unfaulted phases also attained maximum values. Both values had coefficients that were larger than 5×10^{-3}, while the coefficients under normal conditions were smaller than 5×10^{-3}, as shown in Figure 7a. Thus, the coefficients changed more than five times, and hence, an internal fault had occurred. The four mother wavelets exhibited the same behavioral characteristics of the coefficients. For the case of external faults, the faulted phases attained maximum values, and the unfaulted phases also attained maximum values. If these coefficients were smaller than 5×10^{-3}, external faults would occur. For fault classification with relay 2, positive sequences (Daubechies (db2)) were detected, which attained maximum values. Its coefficients were larger than 1×10^{-2}, and the pre-fault coefficients were smaller than 1×10^{-2}, as shown in Figure 7a. By considering the coefficients, they changed more than two times, and therefore, there were faults in the electrical systems. Additionally, all four mother wavelets provided similar results. For all data, these algorithms could discriminate the fault types with an average accuracy of 97.75%. The mother wavelet types of db2, sym2, and coif1 provided the highest accuracy at 98.65%, and the mother wavelet type of bior31 had the lowest accuracy at 95.05%, as summarized in Table 3.

Table 2. Coefficients from the DWT of the current signals for fault classification (relay 2).

Mother Wavelet	Case	Coefficient from the DWT of the Current Signal	
		$P_{\max}$	Result
db2	Fault	2.80×10^{-2}	Fault
	Normal	6.01×10^{-3}	Normal
sym2	Fault	2.80×10^{-2}	Fault
	Normal	6.01×10^{-3}	Normal
bior31	Fault	8.04×10^{-2}	Fault
	Normal	8.02×10^{-3}	Normal
coif1	Fault	6.78×10^{-2}	Fault
	Normal	6.80×10^{-3}	Normal

Table 3. Coefficients from the DWT of the current signals for fault classification (relay 1).

Mother Wavelet	Case	Coefficient from the DWT of the Differential Current Signal			
		Phase Fault $Fault^{diff}_{\max}$	Phase Unfault $Unfault^{diff}_{\max}$	$x^{diff}_{\max}$	Result
db2	Internal	3.92×10^{-2}	1.25×10^{-2}	3.92×10^{-1}	Internal
	External	1.30×10^{-3}	1.07×10^{-3}	1.30×10^{-3}	External
sym2	Internal	3.92×10^{-2}	1.25×10^{-2}	3.92×10^{-1}	Internal
	External	1.30×10^{-3}	1.07×10^{-3}	1.30×10^{-3}	External
bior31	Internal	2.70×10^{-2}	9.46×10^{-3}	2.70×10^{-2}	Internal
	External	1.66×10^{-3}	2.10×10^{-3}	2.10×10^{-3}	External
coif1	Internal	1.09×10^{-1}	3.02×10^{-2}	1.09×10^{-1}	Internal
	External	1.27×10^{-3}	0.92×10^{-3}	1.27×10^{-3}	External

6. Conclusions

This paper illustrates the importance of the mother wavelet for fault classification in electrical systems. When faults occur in transmission line systems and transformers, the coefficients in three phases and a positive sequence are detected. The behavior of the coefficients is dependent on the fault type, fault angle, and fault position. Phase faults have higher coefficients than phase unfaults. In addition, fault conditions can be detected by a positive sequence.

The mother wavelets, i.e., Daubechies (db), symlets (sym), biorthogonal (bior), and Coiflets (coif), are used to compare the coefficient values and behaviors. Figures 5–8 show the coefficient values and behaviors. Each mother wavelet has a similar behavior, but its value is not the same. The same behavior can be obtained by different fault classification algorithms in data simulation, as shown in Figure 9. The data are divided into three parts: 1. algorithm design—50% (888 data), 2. data testing—25% (444 data points), and 3. case study—25% (444 data points).

For the case study of the different mother wavelets, the accuracy of the results is summarized in Table 3. The faults in the case study are discriminated by using the proposed algorithm. It is found that there is an average accuracy of internal fault detection (relay 1) of 96.12% and of external fault detection (relay 2) of 99.07%. The mother wavelet types of Daubechies (db2) and symlets (sym2) provide the highest accuracy under condition 1, while the mother wavelet type of biorthogonal (bior3.1) provides the highest accuracy under condition 2, as shown in Figure 10.

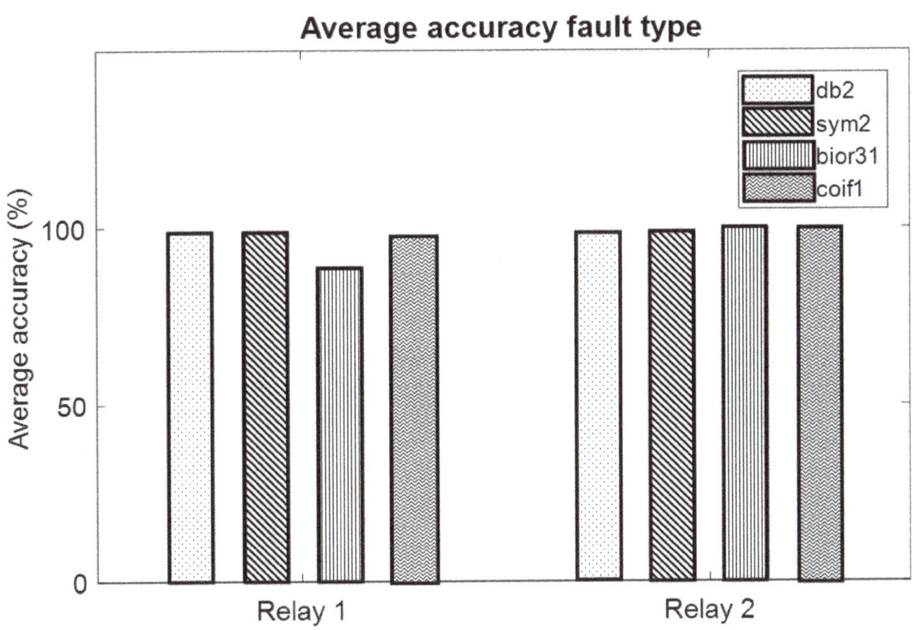

Figure 10. Average accuracy of the mother wavelet in the experimental setup.

The results from the study illustrated the performance of different mother wavelets on the fault classification algorithm. The different mother wavelets provide a different level of accuracy on the fault classification on a transformer, while they do not show a significant impact on fault classification in the transmission line. Thus, this mother wavelet is also one of the factors that must be considered in order to select the suitable mother wavelet for an application. Future work will broaden the suitable mother wavelet selection on other applications and to test the performance of the fault classification algorithm on different power system topologies to verify its application.

Author Contributions: D.A.A. and I.M.Y.N., conceptualization; J.K., formal analysis and design of the experiment. A.N. performed the experiments and investigation; C.P. and A.N. analyzed the data; C.P. and C.J. carried out funding acquisition; D.A.A. and I.M.Y.N. contributed to resources; J.K. and C.J. wrote the paper; C.P and A.N carried out review and editing; A.N., D.A.A., and I.M.Y.N. carried out project administration. All authors have read and agreed to the published version of the manuscript.

Funding: This research was funded by KING MONGKUT'S INSTITUTE OF TECHNOLOGY LADKRABANG RESEARCH FUND, THAILAND, grant number No. KREF156001.

Acknowledgments: The authors wish to gratefully acknowledge financial support for this research (No. KREF156001) from King Mongkut's Institute of Technology Ladkrabang Research fund, Thailand.

Conflicts of Interest: The authors declare no conflict of interest.

References

1. Devi, S.; Swarnkar, N.K.; Ola, S.R.; Mahela, O.P. Detection of transmission line faults using discrete wavelet transform. In Proceedings of the 2016 Conference on Advances in Signal Processing (CASP), Pune, India, 9–11 June 2016; pp. 133–138. [CrossRef]
2. Tripathy, L.N.; Jena, M.K.; Samantaray, S.R.; Dash, D.R. A differential protection scheme for tapped transmission line containing UPFC and wind farm. In Proceedings of the 2014 IEEE Students' Technology Symposium, Kharagpur, India, 28 February–2 March 2014; pp. 319–324. [CrossRef]
3. Adly, A.R.; El Sehiemy, R.A.; Abdelaziz, A.Y.; Ayad, N.M.A. Critical aspects on wavelet transforms based fault identification procedures in HV transmission line. *IET Gener. Transm. Distrib.* **2016**, *10*, 508–517. [CrossRef]
4. Alshawawreh, A. Wavelet transform for single phase fault detection in noisy environment. In Proceedings of the 2014 IEEE 8th International Power Engineering and Optimization Conference (PEOCO2014), Langkawi, Malaysia, 24–25 March 2014; pp. 429–434. [CrossRef]
5. Jin, L.; Jiang, M.; Yang, G. Fault analysis of microgrid and adaptive distance protection based on complex wavelet transform. In Proceedings of the 2014 International Power Electronics and Application Conference and Exposition, Shanghai, China, 5–8 November 2014; pp. 360–364. [CrossRef]
6. Ray, P.; Mishra, D.P.; Mohaptra, S.; Pattnaik, A. Fault classification of a long transmission line using nearest neighbor algorithm and boolean indicators. In Proceedings of the 2016 International Conference on Next Generation Intelligent Systems (ICNGIS), Kottayam, India, 1–3 September 2016; pp. 1–5. [CrossRef]
7. Rathore, B.; Shaik, A.G. Wavelet–alienation based transmission line protection scheme. *IET Gener. Transm. Distrib.* **2017**, *11*, 995–1003. [CrossRef]
8. Singh, S.; Vishwakarma, D.N. Faults classification in series compensated lines based on wavelet entropy and neural network. In Proceedings of the 2016 International Conference on Microelectronics, Computing and Communications (MicroCom), Durgapur, India, 23–25 January 2016; pp. 1–6. [CrossRef]
9. Bhattacharya, S. Fault detection on a ring–main type power system network using artificial neural network and wavelet entropy method. In Proceedings of the International Conference on Computing, Communication & Automation, Noida, India, 15–16 May 2015; pp. 1032–1037. [CrossRef]
10. Swetapadma, A.; Yadav, A. Improved fault location algorithm for multi–location faults, transforming faults and shunt faults in thyristor controlled series capacitor compensated transmission line. *IET Gener. Transm. Distrib.* **2015**, *9*, 1597–1607. [CrossRef]
11. Upendar, J.; Gupta, C.P.; Singh, G.K.; Ramakrishna, G. PSO and ANN–based fault classification for protective relaying. *IET Gener. Transm. Distrib.* **2010**, *4*, 1197–1212. [CrossRef]
12. Singh, S.; Vishwakarma, D.N. Intelligent techniques for fault diagnosis in transmission lines—An overview. In Proceedings of the 2015 International Conference on Recent Developments in Control, Automation and Power Engineering (RDCAPE), Noida, India, 12–13 March 2015; pp. 280–285. [CrossRef]
13. Zhao, M.; Xu, G. Feature extraction of power transformer vibration signals based on empirical wavelet transform and multiscale entropy. *IET Sci. Meas. Technol.* **2018**, *12*, 63–71. [CrossRef]
14. Medeiros, R.P.; Costa, F.B.; Silva, K.M. Power Transformer Differential Protection Using the Boundary Discrete Wavelet Transform. *IEEE Trans. Power Deliv.* **2016**, *31*, 2083–2095. [CrossRef]
15. Tripathy, M.; Nirala, N. Power transformer differential protection algorithm based on dead angle of Wavelet Energy Waveform. Intelligent techniques for fault diagnosis in transmission lines—An overview. In Proceedings of the 2015 International Conference on Recent Developments in Control, Automation and Power Engineering (RDCAPE TENCON 2015–2015 IEEE Region 10 Conference, Macao, China, 1–4 November 2015; pp. 1–2. [CrossRef]
16. Guillén, D.; Esponda, H.; Vázquez, E.; Idárraga–Ospina, G. Algorithm for transformer differential protection based on wavelet correlation modes. *IET Gener. Transm. Distrib.* **2016**, *10*, 2871–2879. [CrossRef]

17. Aktaibi, A.; Rahman, M.A.; Razali, A.M. An Experimental Implementation of the dq–Axis Wavelet Packet Transform Hybrid Technique for Three–Phase Power Transformer Protection. *IEEE Trans. Ind. Appl.* **2013**, *50*, 2919–2927. [CrossRef]
18. Oliveira, M.O.; Bretas, A.S. Application of Discrete Wavelet Transform for differential protection of power transformers. Intelligent techniques for fault diagnosis in transmission lines—An overview. In Proceedings of the 2015 International Conference on Recent Developments in Control, Automation and Power Engineering (RDCAPE 2009 IEEE Bucharest PowerTech), Bucharest, Romania, 28 June–2 July 2009; pp. 1–8. [CrossRef]
19. Ghunem, R.A.; El–Shatshat, R.; Ozgonenel, O. A Novel Selection Algorithm of a Wavelet–Based Transformer Differential Current Features. *IEEE Trans. Power Deliv.* **2014**, *29*, 1120–1126. [CrossRef]
20. AsghariGovar, S.; Seyedi, H. Adaptive CWT–based transmission line differential protection scheme considering cross–country faults and CT saturation. *IET Gener. Transm. Distrib.* **2016**, *10*, 2035–2041. [CrossRef]
21. Suryavanshi, H.; Velandy, J.; Sakthivel, M. Wavelet power ratio signature spectrum analysis for prediction of winding insulation defects in transformer and shunt reactor. *IEEE Trans. Dielectr. Electr. Insul.* **2017**, *24*, 2649–2659. [CrossRef]
22. Yilmaz, S.; Ozgonenel, O.; Karagol, S. Impact of mother wavelet on the performance of wavelet–neural network (WNN) based transformer protection. Intelligent techniques for fault diagnosis in transmission lines—An overview. In Proceedings of the 2015 International Conference on Recent Developments in Control, Automation and Power Engineering (RDCAPE 2016 National Conference on Electrical, Electronics and Biomedical Engineering (ELECO), Bursa, Turkey, 1–3 December 2016; pp. 16–20.
23. Rajaraman, P.; Sundaravaradan, N.A.; Meyur, R.; Reddy, M.J.B.; Mohanta, D.K. Fault Classification in Transmission Lines Using Wavelet Multiresolution Analysis. *IEEE Potentials* **2016**, *35*, 38–44. [CrossRef]
24. Misiti, M.; Misiti, Y.; Oppenheim, G.; Poggi, J.M. *Wavelet Toolbox 4 User's Guide*; The MathWorks, Inc.: Natick, MA, USA, 2007.

applied sciences

MDPI

Article

Optimal Design and Control of MMC STATCOM for Improving Power Quality Indicators

Ahmed A. Zaki Diab [2,3,*], Terad Ebraheem [4], Raseel Aljendy [5,*], Hamdy M. Sultan [2,5] and Ziad M. Ali [1,6]

1 College of Engineering at Wadi Addawaser, Prince Sattam bin Abdulaziz University, Wadi Aldawaser 11991, Saudi Arabia; dr.ziad.elhalwany@aswu.edu.eg
2 Electrical Engineering Department, Faculty of Engineering, Minia University, Minia 61111, Egypt; hamdy.soltan@mu.edu.eg
3 Department of Electrical Engineering, Kyushu University, Fukuoka 819-0395, Japan
4 Electrical Engineering Department, Faculty of Engineering, Tishreen University, Latakia 2237, Syria; terad84ebraheem@gmail.com
5 Electrical Power Systems Department, Moscow Power Engineering Institute "MPEI", Moscow 111250, Russia
6 Electrical Engineering Department, Aswan Faculty of Engineering, Aswan University, Aswan 81542, Egypt
* Correspondence: ahmed.zaki.diab@mu.edu.eg (A.A.Z.D.); aljendy_raseel@yahoo.com (R.A.); Tel.: +2-01021777925 (A.A.Z.D.)

Received: 9 March 2020; Accepted: 30 March 2020; Published: 4 April 2020

Abstract: In recent years, modular multilevel converters (MMC) are becoming popular in the distribution and transmission of electrical systems. The multilevel converter suffers from circulating current within the converter that increases the conduction loss of switches and increases the thermal stress on the capacitors and switches' IGBTs. One of the main solutions to control the circulating current is to keep the capacitor voltage balanced in the MMC. In this paper, a new hybrid control algorithm for the cascaded modular multilevel converter is presented. The Harris hawk's optimization (HHO) and Atom search optimization (ASO) are used to optimally design the controller of the hybrid MMC. The proposed structure of modular multilevel inverters allows effective operation, a low level of harmonic distortion in the absence of output voltage filters, a low switching frequency, and excellent flexibility to achieve the requirements of any voltage level. The effectiveness of the proposed controller and the multilevel converter has been verified through testing with the application of the MMC-static synchronous compensator (STATCOM). The stability of the voltage capacitors was monitored with balanced and unbalanced loads on the studied network.

Keywords: modular multilevel converter; STATCOM; optimization; harmonics; Harris Hawk's optimization; Atom search optimization

1. Introduction

Recently, with the increased popularity and application of power electronic devices in modern industries, some large electrical loads such as from an AC traction system (single-phase), electric furnaces and modern technologies based on high-tech microprocessors in renewable energy applications cause increasing power quality (PQ) problems [1,2]. Power quality problems appear in electric utility as increased harmonic distortions, phase imbalances and low power factors. Electric devices based on power electronics present themselves as non-linear loads, which are characterized as sources of harmonics in the power system. The harmonic currents cause an increase in the RMS value of the current and let the neutral current make circulations in the electric distribution system. The capacity of the distribution system and power losses are affected by the presence of harmonics [2,3].

The effects of nonlinear unbalanced loads on the performance of the power system is an important issue for the power system operators and attracts the interest of many researchers. Many research papers have proposed different techniques and optimization algorithms to improve power quality indicators in the presence of nonlinear and/or unbalanced loads [4]. The static synchronous compensator (STATCOM) is one of the FACTS devices that introduces an efficient and flexible solution for power quality improvement and disturbance mitigation [5–7]. The practical circuit of the STATCOM usually includes two-level voltage source converters (VSC), and a line-frequency transformer is adopted for the enhancement of its voltage and current ratings; however, it results in the heavy, unreliable and expensive design of the compensator [5,6].

The multilevel converters-based STATCOM has received considerable attention from researchers due to its higher capacity that makes it reliable for application in medium- or high-voltage high grids in the absence of line-frequency transformers. Since 1975, the concept of the multilevel converter has been introduced [8]. The multilevel term began with the third-level topology [8,9]. Subsequently, many multilevel converter topologies were developed [10–19]. However, the basic concept of the multilevel converter is the use of a series of semiconductor power switches with several low DC voltage sources to conduct energy conversion and obtain higher energy. The topology of the multilevel converters has succeeded in addressing some of the main problems in the traditional converters [10–19]. The most important of these problems are the problems of power quality, especially the voltage and current harmonics [20–23]. As the number of converter levels increases, the output voltage signal will be as good and as close as possible to the desired waveform [9–19]. But of course, advanced control techniques are needed to improve the performance of multilevel converters and optimize the workflow.

Three major multilevel converters have been applied in high- and medium-power applications, such as the neutral-point clamped multilevel converter (NPCMC), the flying capacitor multilevel converter (FCMC) and the cascaded H-bridge multilevel converter (CHBMC) [8,9]. The major disadvantage of the NPCMC is the inherent voltage imbalance problems, which require an extra auxiliary compensation circuit [8,9]. As for FCMC, besides the need for highly expensive flying capacitors, mainly at low carrier frequencies, this type of multilevel converter is not suitable for those applications with 90° leading or lagging currents. Moreover, for both topologies, as the voltage level has to be increased in high power applications, an increase in the size of the capacitors or in the climbing diodes is required, which makes the control circuit and converter structure more complex. Compared with the previously explained two former topologies, CHBMC requires fewer components in the circuit design, which makes it convenient for physical layout and packaging. The major disadvantage of CHBMC is that it cannot perform properly under imbalanced conditions without feeding from isolated DC sources supplied from multi-winding transforms, which gives rise to the same problems associated with the use of the line-frequency transformer [8,9].

To overcome the previously mentioned problems, a new modular multilevel converter-based STATCOM (MMC STATCOM) is recommended for medium and high voltage applications. The MMC STATCOM is characterized by its availability, compact and modular construction, generation of least harmonics, etc. Moreover, the MMC STATCOM is able to exchange both active and reactive power via the common DC-bus. Accordingly, the active power, which is redistributed in the internal circuits, is utilized for the purpose of negative sequence balance. Consequently, the MMC STATCOM can operate under three-phase imbalance without any intermission.

In this paper, an optimization problem is introduced for the optimal design of the multilevel converter and for determining the optimal parameters of the DC voltage controller. The minimization of total harmonic distortion (THD) in the power system under consideration is introduced as the objective function while keeping the ripple voltage within the allowable ranges and obtaining minimum circulating current. Novel optimization algorithms, namely Atom Search Optimization (ASO) and Harris Hawk optimization (HHO) that have not been reported in the literature for tackling the aforementioned problem of optimization of the values of the capacitors of the MMC STATCOM have been applied. Moreover, to ensure the effectiveness of the proposed optimization techniques, the results obtained from the application of these methods have been tested under two different case studies, nonlinear load and three-phase imbalance.

The paper is organized as follows: in Section 2, a brief review of MMC topologies is introduced in detail; Section 3 introduces the Model of the Modular Multilevel Converter (MMC); the Modular Multilevel Converter-based STATCOM (MMC STATCOM) is introduced in Section 4; Section 5 presents the problem formulation; Section 6 presents the Harris hawk's optimization (HHO) and atom algorithm in detail; the result simulation presents in Section 7; and the conclusion is stated in Section 8.

2. A Brief Review of MMC Topologies

There are many topologies of the MMC. A comprehensive comparison among these topologies is presented in Table 1 [24,25].

Table 1. Comprehensive comparison between Modular Multilevel Converter (MMC) topologies.

Topology	Configuration	Sub-Modules (SMs)	Remarks
Single-Star Bridge Cell (SSBC)	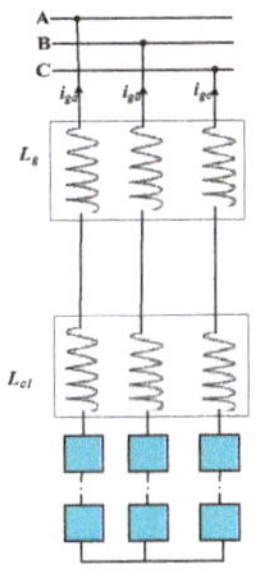	Full-bridge SM 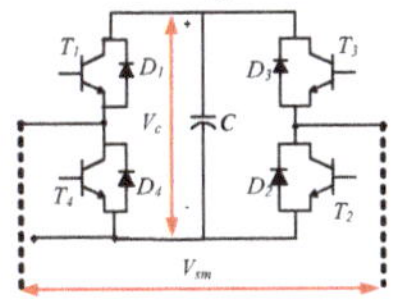	1. No circulating currents 2. The balance of the voltage during abnormal operating conditions is limited by the rating of the devices 3. No communal for STATCOM
Single-Delta Bridge Cell (SDBC)	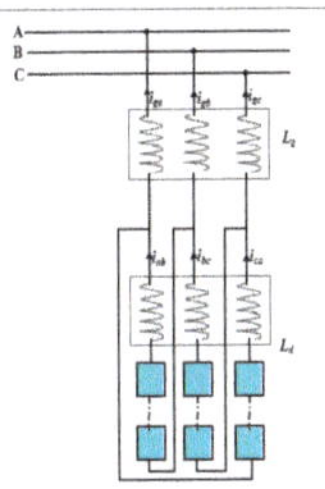	Full-bridge SM 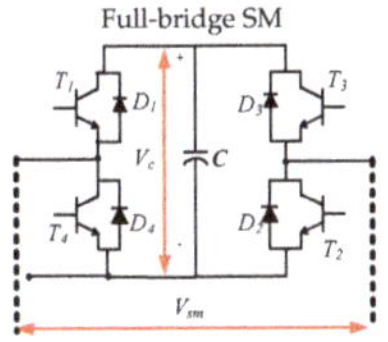	1. Circulating current 2. STATCOM applications 3. Smaller energy storage 4. Lower losses 5. Commonly used for compensation of positive sequence components 6. Susceptible to unbalanced voltage conditions
Double-Star Chopper Cell (DSCC)	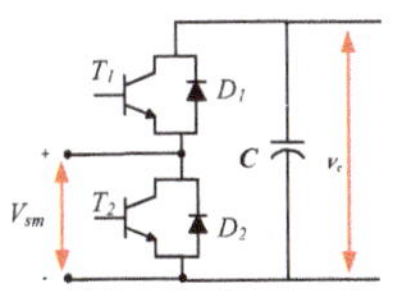	Half-bridge SM 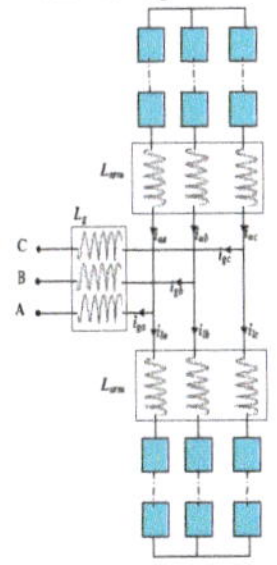	1. Circulating current 2. STATCOM applications 3. Large energy storage 4. Low power losses 5. Excellent presentation for STATCOM
Double-Star Bridge Cell (DSBC)		Full-bridge SM 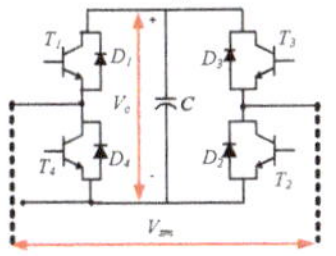	1. Large number of power devices 2. High cost 3. Unsuitable for STATCOM applications

For more explanation to state the motivation of this paper, the level of harmonics in the output voltages and currents of the MMC is very essential and has to be within the allowable limits. The harmonic content could be decreased without the need for any filters with the help of modern control methods [26,27]. In order to obtain the output voltage with an acceptable level of THD in the AC side cascaded configuration, the switching frequency of the two-level converter (main converter) is adjusted to 400 Hz, while the frequency of carriers is taken as 2 kHz. Due to the possible mismatch of synchronization between the main converter and the active filter (H-bridge cascaded model), the fifth and seventh harmonic components may arise, which must be canceled using a number of filters.

As example, reference [26] presented a comparison between the output voltage and the harmonic spectra generated from the single-phase AC cascaded configuration with two H-bridges on the AC side, which are presented in Figures 1 and 2, and the output voltage and the harmonic spectra generated from the single-phase AC cascaded configuration with 4H-bridges on the AC side, which are presented in Figures 3 and 4. It can be seen from Figure 2 that the third harmonic component has a large value and also that the value of THD is extremely high. That is because the method of third harmonic subtraction allows the multilevel converters to work independently of the load power factor and without the problems of capacitor voltage balancing [26]. The data presented in Figure 4 prove the superiority of the last configuration of 4H-bridges, as the profile of the output voltage harmonic is slightly better, but with an increase in the number of the power electronic devices and the cost of the system. Moreover, reference [26] concludes that the increasing of the series sub-modules (SMs) in the case of using two H-bridges or 4H-bridges will lead to a reduction in the harmonics and over-shoot in the output voltages of the MMC.

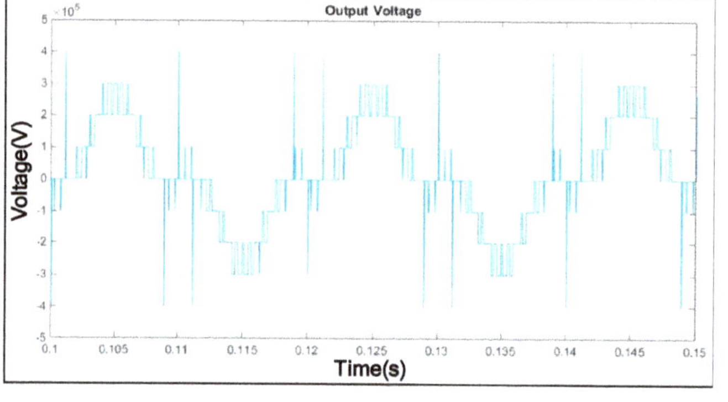

Figure 1. Voltage for a single AC side cascaded configuration including two H-bridges [26].

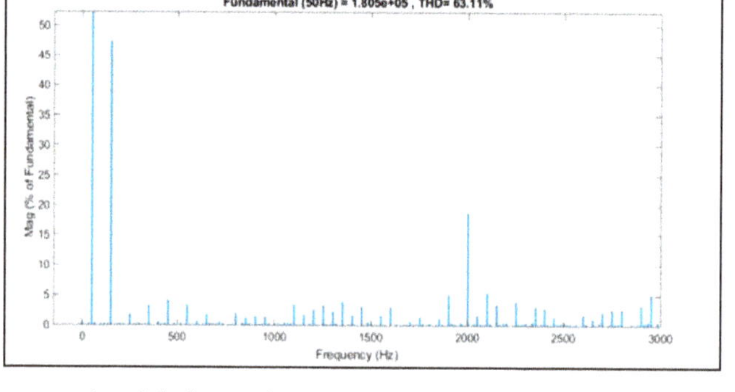

Figure 2. A representation of the harmonic component in the output voltage of the AC side cascaded configuration using two H-bridges [26].

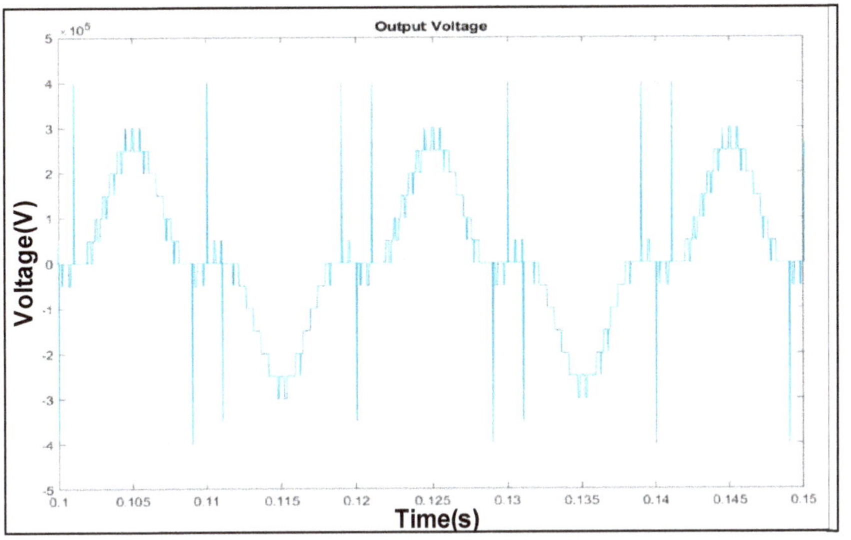

Figure 3. Voltage for a single AC side cascaded configuration including 4H-bridges [26].

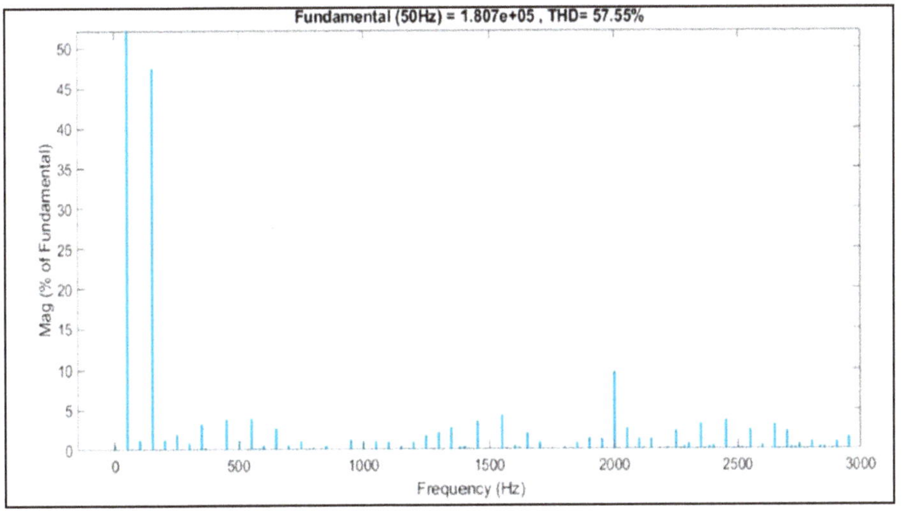

Figure 4. A representation of the harmonic component in the output voltage of the AC side cascaded configuration using 4H-bridges [26].

The introduced discussion leads the authors to present an optimal design of the control system and the optimal capacitor value of the MMC for the STATCOM application, using the half-bridge SMs to decrease the number of devices and the cost of implementation.

3. Model of the Modular Multilevel Converter (MMC)

The modular multilevel converter (MMC) has been introduced firstly by Marquardt in 2001, and is presented as the developed configuration including a number of subsystems for achieving the desired voltage level [15,18,24]. This type of converter is designed for use in medium and high transmission voltages. The overall configuration of the modular multilevel converter is illustrated in Figure 5, where each leg of this converter consists of a lower and an upper arm connected to each other by a DC voltage link.

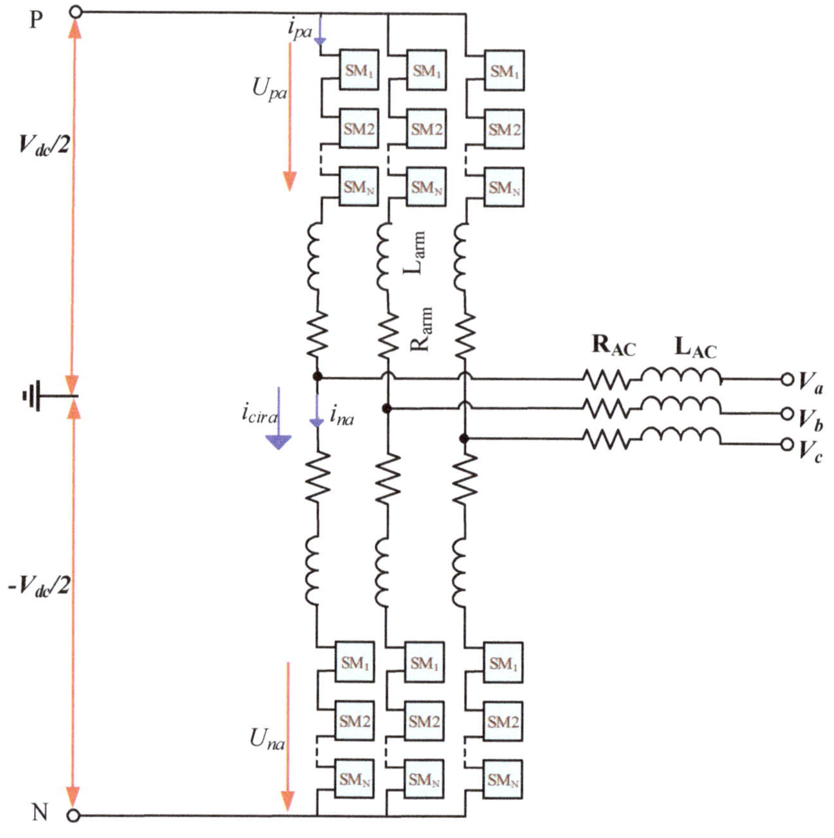

Figure 5. The circuit configuration of the MMC.

Each arm includes a number of SMs, and in turn, every sub-system includes two switches and two reverse diodes connected to each other via a DC capacitor. The switches S1 and S2 are operated in reverse. Each SM could be configured as a half-bridge, full-bridge, H-bridge or clamp-double circuit [18]. Due to its low cost and high efficiency, the MMC based on the half-bridge SMs shown in Figure 6 has been widely used in industrial applications and high voltage DC projects. It is possible to perform a selectable individual control for each SM in the converter leg. Principally, the converter legs provided a controllable Voltage Source Converter (VSC) for each phase. The desired waveform voltage can easily be achieved at the terminal by adjusting the terminal voltage of the SM's leg.

The equivalent circuit presentation of the typical MMC using half-bridge SMs is shown in Figure 7, where u_{vj} is the phase converter output voltage and i_{vj} denotes the line current; i_{pj} and i_{nj} are the upper and lower leg currents, respectively, that are presented as [8,10,27]:

$$\begin{aligned} i_{pj} &= \frac{i_{vj}}{2} + i_{diffj} \\ i_{nj} &= \frac{i_{vj}}{2} - i_{diffj} \end{aligned} \quad (1)$$

where i_{diffj} denotes the phase j inter-circulating current that circulates between the upper and lower legs and is expressed as below:

$$i_{diffj} = \frac{i_{pj} - i_{nj}}{2} \tag{2}$$

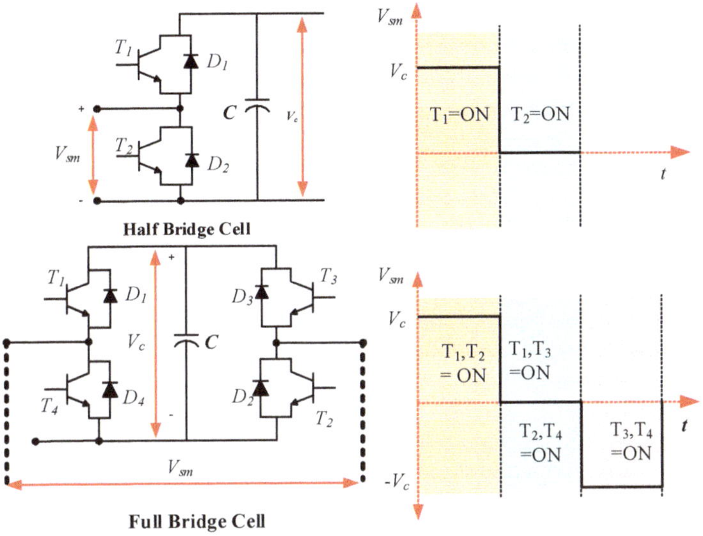

Figure 6. A schematic of the configuration of the MMC based on half-bridge and full-bridge sub-modules (SMs).

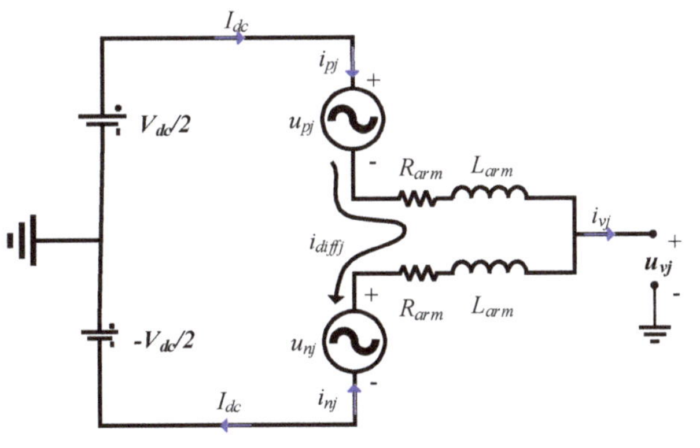

Figure 7. The one phase equivalent circuit.

As discussed in [7], the MMC operates according to the following equations:

$$u_{vj} = e_j - \frac{R_{arm}}{2} \cdot i_{vj} - \frac{L_{arm}}{2} \cdot \frac{di_{vj}}{dt} \; (j = a, b, c) \tag{3}$$

$$R_{arm} \cdot i_{diffj} + L_{arm} \cdot \frac{di_{diffj}}{dt} = \frac{V_{dc}}{2} - \frac{u_{nj} + u_{pj}}{2} \; (j = a, b, c) \tag{4}$$

where e_j denotes the inner *emf* produced in phase *a* and is given as follows:

$$e_j = \frac{u_{nj} - u_{pj}}{2} \quad (5)$$

The internal dynamic performance of the MMC is explained in Equation (4) and can be characterized as follows:

$$u_{diffj} = R_{arm} \cdot i_{diffj} + L_{arm} \cdot \frac{di_{diffj}}{dt} = \frac{V_{dc}}{2} - \frac{u_{nj} + u_{pj}}{2} \ (j = a, b, c) \quad (6)$$

where u_{diffj} denotes the phase j inner unbalance voltage.

The MMC is characterized by its small size and simplicity in installation, and as the number of SMs increases, there is a decrease in harmonics and a matching of the required power quality specifications. However, the multilevel converter suffers from converter inter-circulating current, which increases the switches' conduction loss and increases the thermal stress on the capacitors and switches' IGBTs. Therefore, maintaining the balance of the capacitor voltages is one of the main rules of the design of the MMC.

4. Modular Multilevel Converter-Based STATCOM (MMC STATCOM)

In order to simplify the analysis of the proposed network, the upper and lower arms of the MMC STATCOM are proposed to be equivalent to one power unit, which includes two IGBTs shunted by opposite-connected diodes and a capacitor forming a bidirectional chopper. Figure 8 shows a simplified one-phase equivalent circuit of the proposed MMC STATCOM. Point P in the circuit refers to the positive DC-link, while point N refers to the negative DC-link. In addition, the following assumptions are provided for more analysis,

1. The DC-bus voltage V_d is constant without a ripple.
2. There is a uniform distribution of the DC input current I_d among the three phases.
3. The switching losses in the main components of the circuit are ignored.

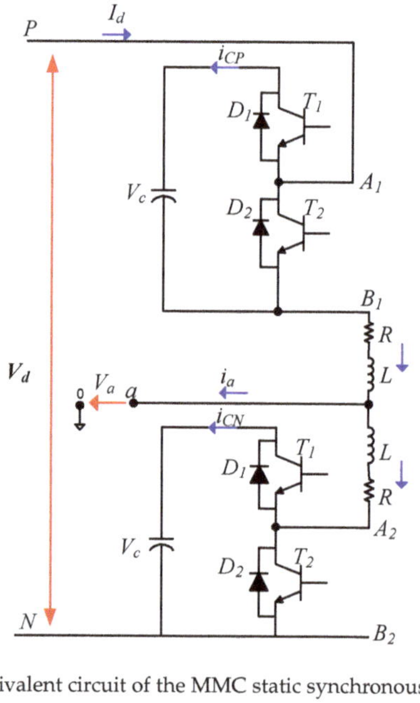

Figure 8. The one phase equivalent circuit of the MMC static synchronous compensator (STATCOM).

The terminal voltage u_a and output current i_a are assumed to be pure sinusoidal:

$$u_a = U\sin(\omega t) \tag{7}$$

$$i_a = I\sin(\omega t + \varphi_i) \tag{8}$$

where U denotes the phase voltage peak value, I denotes the phase current peak value, ω is the angular frequency and φ_i is the phase shift between voltage and current.

For a symmetrical three-phase system, the phase current i_a is divided equally between the two legs. Thus, the resulting share from the upper and lower legs can be identified as follows [27]:

$$i_{Pa} = \frac{1}{2}I[K_I + \sin(\omega t + \varphi_i)] \tag{9}$$

$$i_{Na} = \frac{1}{2}I[K_I - \sin(\omega t + \varphi_i)] \tag{10}$$

where K_I is the current coefficient that is expressed as

$$K_I = \frac{2}{3}\frac{I_d}{I} \tag{11}$$

The differential equations of the upper and lower arms of Figure 8 can be introduced as follows:

$$C\frac{dV_c}{dt} = s_1 i_{Pa} \tag{12}$$

$$C\frac{dV_c}{dt} = s_2 i_{Na} \tag{13}$$

$$L\frac{di_{Pa}}{dt} + Ri_{Pa} = \frac{V_d}{2} - U\sin(\omega t) - s_1 V_c \tag{14}$$

$$L\frac{di_{Na}}{dt} + Ri_{Na} = \frac{V_d}{2} + U\sin(\omega t) - s_2 V_c \tag{15}$$

where V_c is the voltage of the capacitor, s_1 and s_2 are the upper and lower leg switching functions, L denotes the AC inductance of the leg inductor and R is the equivalent resistance of the leg inductor.

4.1. DC-Link Capacitor

In this section, the calculations are only provided on the upper leg, as there is no difference between the upper and lower legs [28]. For more simplification, the calculations are performed taking into account the following assumptions:

1. There is a small ripple in the DC capacitor voltage.
2. Only the fundamental component of the switching function is considered.

In the steady-state condition, the upper leg switching function is defined as:

$$s_1(t) = M\sin(\omega t) \tag{16}$$

where M denotes the modulation index. From Equations (9), (12) and (16), the following formula is produced:

$$C\frac{dV_c}{dt} = \frac{1}{4}MI\cos\varphi_i + \frac{K_I}{2}MI\sin(\omega t) - \frac{1}{4}MI\cos(2\omega t + \varphi_i) \tag{17}$$

After integration, the following equation exists:

$$V_c = V_{DC} - \frac{MIK_I\cos(\omega t)}{2\omega C} - \frac{MI\sin(2\omega t + \varphi_i)}{8\omega C} \tag{18}$$

where V_{DC} denotes the average DC voltage. The ripple in the current of the upper leg is described by the following formula:

$$\Delta V_c = -\frac{MIK_I\cos(\omega t)}{2\omega C} - \frac{MI\sin(2\omega t + \varphi_i)}{8\omega C} \quad (19)$$

By similarity, the ripple in the current of the lower leg is calculated as follows:

$$\Delta V_c = -\frac{MIK_I\cos(\omega t)}{2\omega C} - \frac{MI\sin(2\omega t + \varphi_i)}{8\omega C} \quad (20)$$

When the phase angle between the voltage and current φ_i equals ±90° during the static VAr generating condition and by ignoring the resistance of the inductor, Equation (19) will be converted to:

$$\Delta V_c = -\frac{MIK_I\cos(\omega t)}{2\omega C} + \frac{MI\sin(2\omega t)}{8\omega C} \quad (21)$$

To obtain the maximum value of the peak-to-peak ripple voltage of the DC capacitor, Equation (21) is differentiated, which results in:

$$\Delta V_{cpp} = \frac{MI(K_I + 1)^2}{4\omega C} \quad (22)$$

Accordingly, the capacitance of the DC capacitor can be defined as:

$$C \geq \frac{MI(K_I + 1)^2}{4\omega \Delta V_{cpp}} \quad (23)$$

From Equations (14) and (20) it is clearly shown that the voltage ripple contains the fundamental and the second harmonic, which dramatically increases the rating of the capacitor, resulting in an increase in the total cost of the system. An accurate determination of the capacitance value in each sub-model must be taken into account, as a higher value of the capacitance leads to an increase in the total cost of the MMC converter, and on the other hand, a lower value increases the ripple level in the output voltage.

4.2. Design of SM Capacitance

The value of the capacitor affects the power quality indices such as the level of the THD and the voltage balance of the system. Many methods in literature have been introduced to determine the values of the capacitors. These methods can be divided into two main methods; the first is based on the idea that the average sum capacitor voltage per arm should be kept constant. The second is based on keeping constant the individual average capacitor voltages for the sub-modules for each arm. Both of the two methods are based on determining the required stored energy of the STATCOM-based MMC. The conclusion based on the literature review is that the first method is considered to be a better choice than the second one because using the first method improves the reliability of the system, and it is more convenient for the sub-module capacitor, as well as the semiconductor device, to perform under low voltage [20]. A general description of the first method has been presented.

Based on the energy storage demands of the converter, the capacitance of the SM can be determined. The minimum value of the capacitance of the SM is calculated as follows [29]:

$$C = \frac{2 * N * E}{V_{dc}^2} \quad (24)$$

where E denotes the lowest limit of energy storage for each arm. Due to the similarity between both arms, E for the upper arm can be expressed as reported in [25]:

$$E = \frac{\Delta E}{k_{\max}^2 - \max(\frac{n_u^2 - e_{u,v}k_{\max}^2}{1-e_{u,v}})}, \tag{25}$$

where k_{max} refers to the capacitor voltage upper boundary. Typically, $k_{max} = 1.1$ is commonly used.

Finally, as ΔE and E are in direct dependence on the rated power of the converter, the demands for energy storage of the converter can be given as:

$$W = \frac{6}{S_n} E_{nom} \tag{26}$$

where W denotes the energy storage needed for each MVA rating. According to [24,30,31], the capacitance in each sub-module is selected so that the energy stored in all capacitors of the converter sub-modules is about 30–40 kJ/MVA. As the capacitance of the sub-module is not chosen yet, the ripple level of the sub-module capacitor voltage in the operating point is simply determined. There are two possible ways to obtain an anticipatory value of the voltage ripple. The first way is to assume a ripple level of ±10%, whereas the second way is based on assuming a value for the capacitor in each sub-module in the range, which achieved 30–40kJ/MVA stored energy, then performing simulation in the time domain to determine the level of the voltage ripple.

4.3. Design of Arm Inductance

Typically, the inductors used in the arms of the MMC are dry-type air-cored ones. The value of the inductance of the arm can be determined based on the way that the short circuits on the DC-side will be demonstrated. In addition, the inductance of the converter arm plays an important role in enhancing the characteristics of the inter-circulating current and limiting the fault current. Where bypass thyristors are used in the sub-modules or where they are separated using additional diode-based power modules, the AC sharing of the fault current on the DC-side does not need to be restricted to provide protection for the sub-modules' anti-parallel diodes. The arm inductance helps in the limitation of the high frequency harmonics in the inter-circulating current and to provide a smooth control for the circulating current. A typical value of the arm inductance occurs around 0.05 p.u. [24].

On the other hand, when bypass thyristors are not used in the sub-modules or they are not separated using additional diode-based power modules, protection for the anti-parallel diodes on the DC-side of the converter have to be designed to protect the sub-modules from the high sharing from the AC-side to the fault current in the DC-side. Consequently, as the slow-acting breakers interrupt under this fault current, the arm inductors must restrict the current magnitude to give protection for the anti-parallel diodes, which act as a conduction path for the fault currents. The value of the arm inductance is determined using DC-fault actual time simulation, taking into account the interruption time of the circuit breakers, the time delay until fault detection and the strategy of performing the fault in the control system. Typically, the value of the inductance occurred in a range of 0.10–0.15 p.u.

Moreover, a short circuit between the terminals of the DC-buses is performed in order to present the most critical fault condition. In order to make a limitation on the short circuit current, the value of the arm inductance should be determined according to [29]:

$$L_{arm} = V_{dc}/2\alpha \tag{27}$$

where α (kA/s) denotes the maximum rising rate of the current. Based on (27), when the maximum rising rate is = 0.1(kA/μs) [29], for grid connected converters, the p.u. value of the arm inductance values are restricted at 0.3 pu for Double-Star Chopper Cell (DSCC-MMC), as well as for Single-Delta Bridge Cell (SDBC-MMC) configurations.

Consequently, modern metaheuristic techniques such as Atom Search Optimization (ASO) and Harris hawk's optimization (HHO) have been proposed in this work for the optimization the of the values of capacitors and the coefficients of the DC voltage regulator of the MMC STATCOM, in order to minimize the total harmonic distortion (THD) in the power system under consideration, keeping the ripple voltage in the allowable ranges and obtaining minimum current circulation within the converter.

5. Problem Formulation

Usually, in electric power systems, shunt compensating devices are used for voltage and reactive power control. A MATLAB model for the system under study is shown in Figure 9. The parameters of each component in the system are summarized in Table 2. The system under study compromises the utilization of the MMC STATCOM, which is increasingly used in modern electric networks. The proposed MMC STATCOM is based on full-bridge MMC to form 22 modules per phase power converter. The STATCOM can produce or take reactive power from the grid. The transfer of both types of reactive power is obtained via phase reactance.

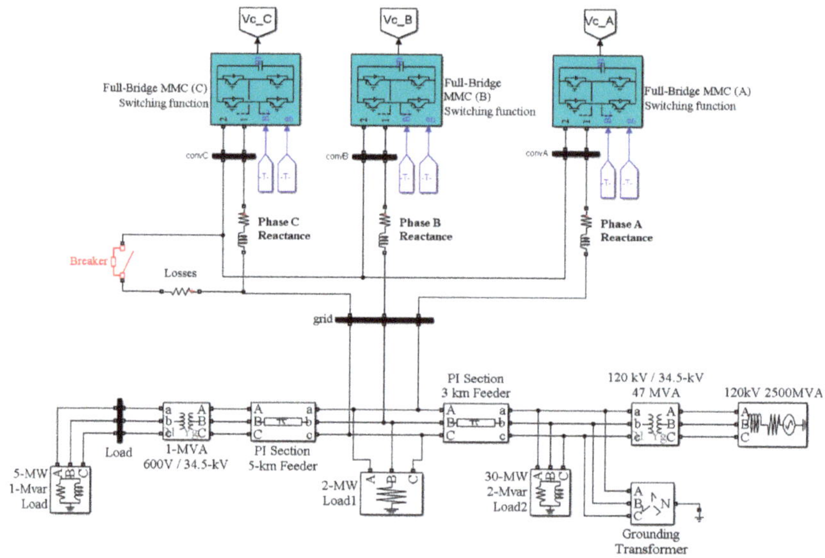

Figure 9. A model of the power system under study.

Table 2. Parameters of the studied network.

STATCOM power capacity	12 MVA	Transformer T1	120 kV/34.5 kV
RMS line to line voltage at PCC	34.5 kV	Transformer T2	34.5 kV/600 V
Number of MMC in each phase	22 modules	Load 1	2 MW
Total stored energy in all submodule capacitors	30 kJ/MVA		
DC link voltage	1600	Load 2	30 MW & 3 MVAr
Line frequency	50 Hz	Transformer X/R ratio	18
Carrier frequency	300 Hz	Load 3	5 MW & 1 MVAr

The proposed MMC generates a three-phase voltage, which has the same phase as that of the grid voltage at the point of coupling. When the amplitude of the voltage generated from the MMC is lower than the grid voltage amplitude, the STATCOM performs like inductance and absorbs reactive power from the grid utility. When the amplitude of the MMC voltage is higher than that of the bus of common coupling, the STATCOM works like a capacitor and reactive power will be injected into the grid. The control circuit of the MMC is shown in Figure 10a. The controller of the high voltage side includes a phase-locked loop (PLL), transformation and measurement blocks, DC-bus voltage regulator and current regulator. Meanwhile, the controller of the low voltage side includes individual and phase balancing controllers of the DC-bus voltage, which are responsible for generating the control signals for producing the pulses applied to the gates of the transistors. Moreover, the optimized controller for phase balancing based on the HHO or ASO optimization techniques is shown in Figure 10b, and the optimized controller of DC voltage based on the HHO or ASO optimization techniques is shown in Figure 10c.

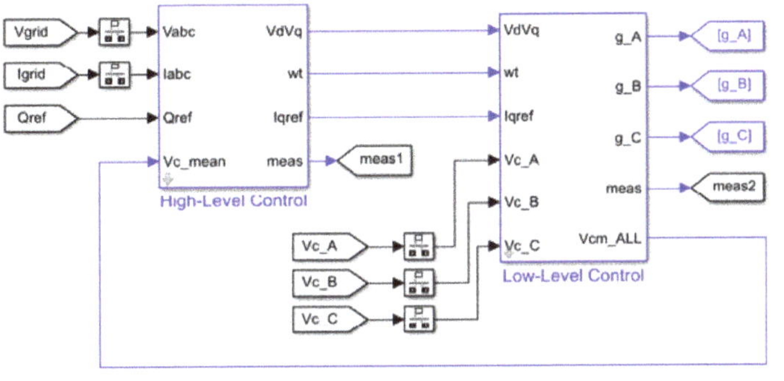

(a)

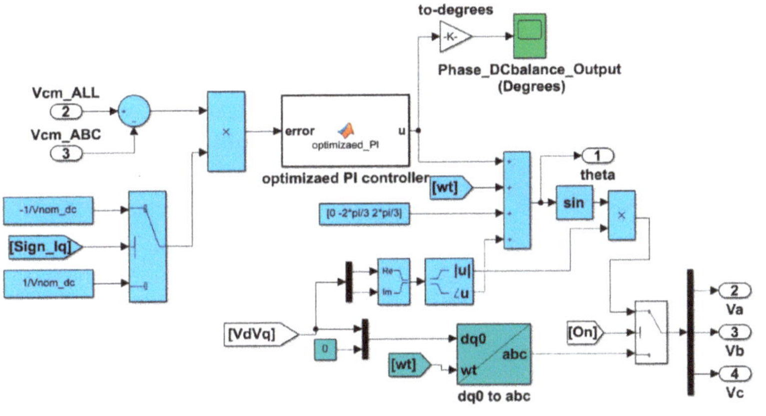

(b)

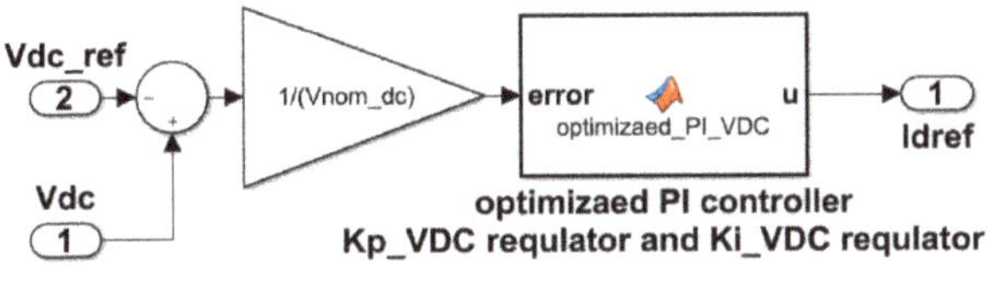

(c)

Figure 10. The controller of the MMC converter: (**a**) Simulink configuration; (**b**) An optimized controller for phase balancing based on the Harris hawk's optimization (HHO) or Atom search optimization (ASO) optimization techniques; (**c**) An optimized controller of DC voltage based on the HHO or ASO optimization techniques.

As discussed in Section 2, the impact of the second component harmonic voltage of the module capacitor on the MMC impedance behavior is expressed. Optimal determination of the value of the module capacitor results in an improvement in the second component of harmonic in the voltage of the module capacitor. Moreover, by reducing the capacitance value, in increase in the second component of harmonic voltage will be observed. Furthermore, it is easy to find that with an increase in the magnitude of the module capacitor voltage second harmonics, an effective increase is observed in the magnitude of the impedance response at low frequencies, but there is no change noticed in positive-sequence. Starting from the introduced explanation, the optimal determination of the module capacitor can be addressed as the design of the impedance of the MMC. However, if a stable performance of the STATCOM is desired, the proposed design should give the minimum demands for energy storage and filtering ripples arising in the capacitance voltage, which shape the constraints of controlling module capacitance to produce an optimal impedance response. Therefore, this paper proposes that the value of the capacitor should be selected to be optimized to reduce the THD using the HHO and ASO algorithm. The objective function to achieve optimal capacitor values leading to the minimum quadratic error of the THD can be calculated as follows:

$$F_1 = \int_0^{\infty} t \cdot THD^2 dt \tag{28}$$

Moreover, to achieve the optimal values of both K_p and K_i for each PI controller, which lead to a minimum quadratic error, the objective function can be calculated as follows:

$$F_2 = \int_0^{\infty} t \cdot e^2 dt \tag{29}$$

where e. is the error signal between the desired and the typical signals and is fed to the PI controller.

The objective function to minimize F can be determined as follows:

$$\text{Minimize } \{F(C,(K_\text{p},K_\text{i})_\text{VDC},(K_\text{p},K_\text{i})_\text{phase balancing}\} \tag{30}$$

where F is the summation of the F_1 and F_2.

6. Proposed Optimization Methods

In the past decades, with the increasing development in the fields of society, economy and industry, numerous complicated and hardly-solved optimization problems have arisen in all fields. Recently, metaheuristic techniques of optimization have been widely implemented to solve complicated engineering problems, as they possess high capabilities of exploration and exploitation to reach global solutions in a short time when compared with traditional optimization methods [31–34].

Taking into account the No Free Lunch Theorem of Optimization [35], the best solution of different optimization problems cannot be fulfilled based on one optimization technique. Based on this theorem, the field of creating and developing better optimization algorithms is still active and full of great achievements. As a result of these achievements, a recent swarm algorithm inspired by physics, called Atom Search Optimization (ASO), is proposed for tackling the optimization problem provided in this study. ASO presented superior performance over conventional and recent optimization techniques when validated on numerous mathematical problems. In addition, the proposed ASO algorithm provided a successful performance when utilized in the optimization problem of hydrogeologic parameter estimation.

In a related context, like the most of existing optimization methods, the searching process in the proposed ASO algorithm is performed in two phases; exploration and exploitation stages [30]. In the first phase, the optimization algorithm should make utilization and promotion of the randomly initialized operators for the deep exploration of the optimal solution within the search space. Thus, the exploratory behavior of a well-developed optimization algorithm should be randomly enriched to generate more random solutions to different areas of the addressed problem formed in the early stages of the research process [36].

6.1. Harris Hawk's Optimization (HHO)

The HHO algorithm was developed in 2019, which in its operation mimics the behavior of the hunting of the Harris hawks, a group of intelligent birds that live in the USA [32,33]. The hunting manner of these intelligent birds depends on surprise. A group of hawks appear to their prey (rabbits) from different sides at the same time to surprise it, at which point the best-fit hawk of the group (the leader) surrounds it. Mathematically, the technique of catching the prey of the hawks can be divided into three stages: (i) the exploration stage, (ii) the stage of transition from exploration to exploitation and finally (iii) the exploitation stage. The first phase of the attack is the most important as it determines the way that the hawks will act. Firstly, all hawks sit randomly in a high place to give them a good chance of observing the environment and wait for their chase. When the prey appears in the surrounding area, the hawks may attack it in several possible manners, and only the leader will decide the way of attack. All hawks might cooperate and surround the prey, or the leader will give the opportunity to one of them, and it will depend on the behavior of the prey to escape. For each of the two possible tactics, the hawks may take their places depending on the location of the neighboring hawk as expressed in (31) or be distributed randomly as described in (32):

$$X(t+1) = \left(X_{best}(t) - X_{avg}(t)\right) - \psi(LB + \tau(UB - LB)) \ \alpha < 0.5 \tag{31}$$

$$X(t+1) = X_{rand}(t) - \beta\left|X_{rand}(t) - 2\varphi X(t)\right| \ \alpha \geq 0.5 \tag{32}$$

where $X(t)$ is the position of the hawks at the present iteration t; $X(t+1)$ is a vector of the new positions of the hawks in the next iteration; $X_{best}(t)$ is the position of the chase (rabbits); and α, β, ψ, φ and τ are randomly distributed numbers in the range from 0 to 1. UB and LB denote the upper and lower limits of the position variables. $X_{rand}(t)$ denotes a hawk, which is randomly selected from the present population, and $X_{avg}(t)$ denotes the mean position of the hawks. The average position of the hawks is determined as follows:

$$X_{avg}(t) = \frac{1}{N}\sum_{i=1}^{N} X_i(t) \tag{33}$$

where $X_i(t)$ denotes the position of each hawk in iteration t and N indicates the number of hawks. Depending on the energy of the prey during the escaping process, HHO can make a transition from exploration to exploitation. The energy of the prey (rabbit) is determined as follows:

$$E = 2 \times E_0\left(1 - \frac{t}{T}\right) \tag{34}$$

where, E_0 denotes the initial energy in each iteration, which is randomly taken from [−1, 1], and T presents the maximum iterations.

As the hawks have detected the prey as explained in the previous stages, they have to make a sudden attack, which depends on the probability of escape. In the mathematical modelling, the probability of escape r will be <0.5 for a successful escape and $r \geq 0.5$ if the prey unsuccessfully gets away. Depending on the way that the prey will take during escape, the hawks will carry out a soft or hard siege. Regardless of the energy of escape, a hard or soft siege will occur. When $|E| \geq 0.5$ and $r \geq 0.5$, the prey has enough energy to make random jumps but fails to escape. During these miserable attempts, the hawks surround the prey until it is exhausted by its force and then suddenly attack it. Thus, the soft siege is described as:

$$X(t+1) = \Delta X(t) - E\left|(J \times X_{best}(t)) - X(t)\right| \tag{35}$$

where $\Delta X(t)$ denotes the difference between the prey's position and the position of the hawks in the present iteration. J is the strength of the prey's random escape.

When $|E| < 0.5$ and $r \geq 0.5$, as the prey has no energy to escape, a hard siege will be performed. Accordingly, the hawks hardly circle the prey to execute a surprise attack. Therefore,

$$X(t+1) = X_{best}(t) - E\left|\Delta X(t)\right| \tag{36}$$

Different stages of the HHO algorithm are represented in Figure 11. Moreover, detailed explanations and mathematical formulations of soft and hard siege manners are provided in [32,33].

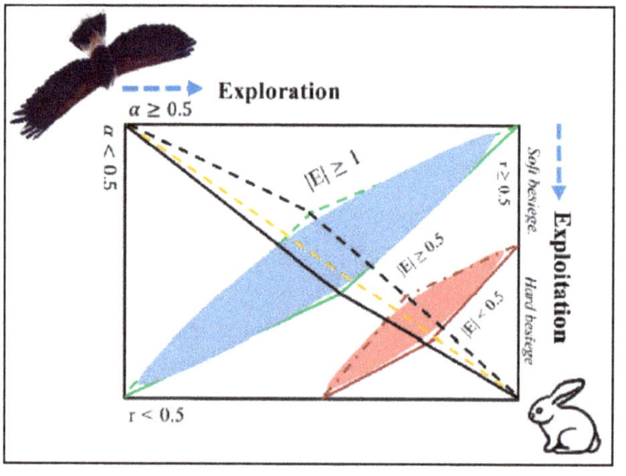

Figure 11. The stages of the HHO algorithm [33].

6.2. Atom Search Optimization (ASO)

Atom search optimization is a novel physics-based method for global optimization problems [34,35]. ASO is a population-based technique, which imitates the motion of atoms subjected to interactions and other constraint forces [33,34]. The total forces of interaction that affect the ith atom in the dth dimension are described as follows:

$$F_i^d(t) = \sum_{j \in Kbest} rand_j F_{ij}^d(t) \tag{37}$$

where $rand_j$ is a number randomly distributed in the range [0, 1] and $Kbest$ is a group of K atoms having the best values fitness function. The value of K has to be decreased gradually as the iterations advance, to let ASO perform more exploitation in the latest iterations, as shown in the following expression:

$$K(t) = N - (N-2) \times \sqrt{\frac{t}{T}} \tag{38}$$

where N denotes the total number of atoms forming an atomic structure, t denotes the present iteration, and T is the total number of iterations. The interaction force F_{ij}^{d} presents the gradient of the potential Lennard-Jones (L-J) and is described in the following formula [29]:

$$F_{ij}^{d} = -\eta(t)\left[2\left(h_{ij}(t)\right)^{-13} - \left(h_{ij}(t)\right)^{-7}\right]\frac{\vec{r}_{ij}}{r_{ij}} \tag{39}$$

where $\eta(t)$ presents the depth function, which controls the regions of attraction and repulsion; $h_{ij}(t)$ is the ratio between the distance between two atoms r_{ij} and the scaled distance between them $\sigma(t)$.

$$h_{ij}(t) = \frac{r_{ij}}{\sigma(t)} \tag{40}$$

$$\eta(t) = \alpha\left(1 - \frac{t-1}{T}\right)^{3} e^{-\frac{20t}{T}} \tag{41}$$

where α denotes the depth weight. The function of the force of interaction under different values of η with respect to a variable scaled distance (h changed from 0.9 to 1) is presented in Figure 12.

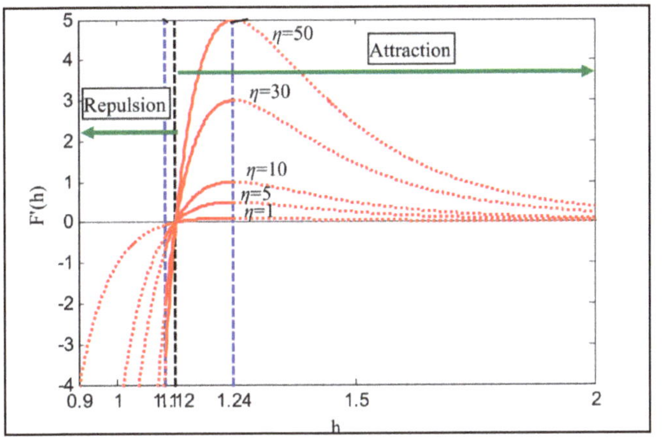

Figure 12. The behavior of the interaction force function with the scaled distance (h) under different depth values [34].

Equation (42) presents the expression of the scaled distance between any two atoms i and j:

$$h_{ij}(t) = \begin{cases} h_{\min} & \frac{r_{ij}(t)}{\sigma(t)} < h_{\min} \\ \frac{r_{ij}(t)}{\sigma(t)} & h_{\min} \le \frac{r_{ij}(t)}{\sigma(t)} \le h_{\max} \\ h_{\max} & \frac{r_{ij}(t)}{\sigma(t)} > h_{\max} \end{cases} \tag{42}$$

where h_{min} and h_{max} describe the upper and lower limits of the distance (h), respectively, and are determined by the following formula:

$$\begin{cases} h_{\min} = g_0 + g(t) \\ h_{\max} = u \end{cases} \tag{43}$$

$$g(t) = 0.1 \times \sin\left(\frac{\pi}{2} \times \frac{t}{T}\right) \tag{44}$$

$$\sigma(t) = \| x_{ij}(t), \frac{\sum\limits_{j \in Kbest} x_{ij}(t)}{K(t)} \|_2 \tag{45}$$

where g_0 and u denote the lower and upper limits, respectively. $g(t)$ denotes a drift factor to give the algorithm the ability to transform from the exploration to the exploitation stage.

Presuming that there is a covalent bond between each atom in ASO and the best one, the constraint force resulting from this combination be expressed as follows:

$$G_i^d(t) = \lambda(t)\left(x_{best}^d(t) - x_i^d(t)\right) \tag{46}$$

where

$$\lambda(t) = \beta \times e^{-\frac{20t}{T}} \tag{47}$$

where $x_{best}^d(t)$ denotes the location of the best atom in the dth dimension, $\lambda(t)$ indicates the Lagrangian multiplier, and β denotes a multiplier weight. After defining the forces of interaction and L-J potential resultant constraint force, the acceleration, to which the ith atom is projected in the dimension d and iteration t, is determined as follows:

$$\begin{aligned} a_i^d(t) &= \frac{F_i^d(t)}{m_i^d(t)} + \frac{G_i^d(t)}{m_i^d(t)} \\ &= \alpha\left(1 - \tfrac{t-1}{T}\right)^3 e^{-\frac{20t}{T}} \\ \times \sum_{j \in Kbest} & \frac{rand_j\left[2\left(h_{ij}(t)\right)^{-13} - \left(h_{ij}(t)\right)^{-7}\right]}{m_i(t)} \\ \cdot \frac{\left(x_j^d(t) - x_i^d(t)\right)}{\|\vec{x}_i(t), \vec{x}_j(t)\|_2} & + \beta e^{-\frac{20t}{T}} \frac{\left(x_{best}^d(t) - x_i^d(t)\right)}{m_i(t)} \end{aligned} \tag{48}$$

where $m_i^d(t)$ denotes the mass of the atom number i in the dimension d of iteration t, and is determined from the value of its fitness function as described below:

$$M_i(t) = e^{-\frac{Fit_i(t) - Fit_{best}(t)}{Fit_{worst}(t) - Fit_{best}(t)}} \tag{49}$$

$$m_i(t) = \frac{M_i(t)}{\sum_{j=1}^{N} M_j(t)} \tag{50}$$

where

$$Fit_{best}(t) = \min_{i \in \{1,2,\ldots,N\}} Fit_i(t) \tag{51}$$

$$Fit_{worst}(t) = \max_{i \in \{1,2,\ldots,N\}} Fit_i(t) \tag{52}$$

Finally, the velocity of the ith atom in the search space and the new position acquired at the next iteration $(t+1)$ can be described by the following formulas:

$$v_i^d(t+1) = rand_i^d \cdot v_i^d(t) + a_i^d(t) \tag{53}$$

$$x_i^d(t+1) = x_i^d(t) + v_i^d(t+1) \tag{54}$$

For a deeper explanation of the atom search optimization technique, the procedure is discussed in detail in [34,35].

7. Results and Discussion

The simulation results have been obtained to validate the effectiveness of the proposed optimization techniques in determining the optimal values of the capacitance and the parameters of the DC voltage regulator. Moreover, to validate the accuracy of the those obtained from the results, these values have been applied to the system under study under the integration of nonlinear loads and the condition of three-phase imbalance. The performance of the MMC STATCOM in the two study cases has been introduced in the simulation results.

7.1. Nonlinear Load Case Study

For the simulation purpose, a dedicated software program for the optimal estimation of the capacitor value, K_p and K_i, is developed in MATLAB/Simulink. The results of the proposed methods are compared with those of the conventional particle swarm optimization (PSO) [37–39]. The design parameters and the limits of the optimized variables are identical for the all optimization techniques proposed in this paper. The end value of iterations has been adjusted at 10 iterations and the search agents are proposed to be 10 agents. The convergence curves of the optimization algorithms are shown in Figure 13. The detailed results of the optimization process for ASO, HHO and PSO are summarized in Table 3. As shown from Figure 13, the HHO algorithm reached the optimum values of the capacitors and coefficients for the PI controllers within two iterations, and ASO reached the best value of the fitness function after four iterations. While the conventional PSO algorithm is the slowest among them, moreover, its finding of the best value of the fitness function (F) is achieved after eight iterations, and it is worse than that with the HHO and ASO. Both of the proposed algorithms were able to obtain convergent results and satisfactory results to improve the voltage and reduce the THD of the current in the studied network, but HHO reached the optimum solution in less time.

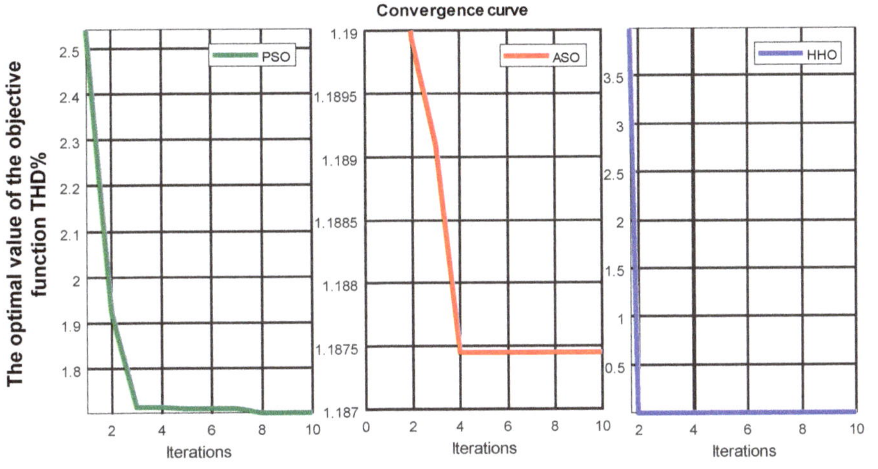

Figure 13. The curves of ASO, HHO, and particle swarm optimization (PSO).

Table 3. Optimization parameters of the proposed optimization methods.

Algorithm	C	Ki_VDC Regulator	Kp_VDC Regulator	Ki_VDC Regulator Phase Balancing	Kp_VDC Regulator Phase Balancing	THD%
HHO	0.0084	800	10	5	0.5	0.60
ASO	0.0047	775	10	5	0.5	1.21
PSO	0.0128	773	8.21	3.24	0.455	2.48

The optimal values for the capacitors of MMC and PI controller constants obtained using HHO and ASO have been applied to the network under study. The profile of the MMC DC voltage for both algorithms is shown in Figure 14, where it can be noticed that stability of the DC voltage is achieved in both cases.

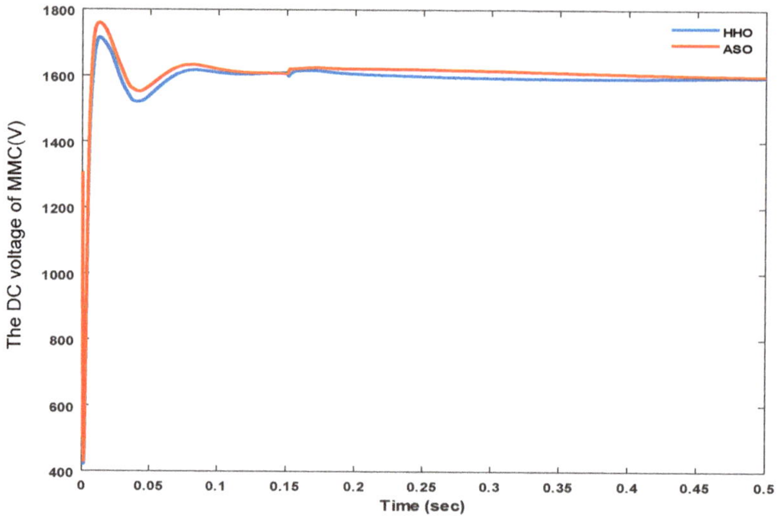

Figure 14. The DC voltage of the MMC using HHO and ASO.

The THD of the current in the original state before optimization was 10.93% in the moment of the connecting MMC STATCOM, as shown in Figure 15, while the THD analysis of the grid current is shown in the Figure 16a, where the value of the THD is decreased to 0.60% with the optimal values of the capacitor converters and the optimal values of the PI controllers as determined using HHO. Moreover, the THD of the current with the optimal values of the capacitor converters and the optimal values of the PI controllers as determined using the ASO algorithm decreased to 1.21%, as shown in Figure 16b. Additionally, Figure 16c shows a comparison between the THD analysis of the current grid with different methods (HHO, ASO and PSO).

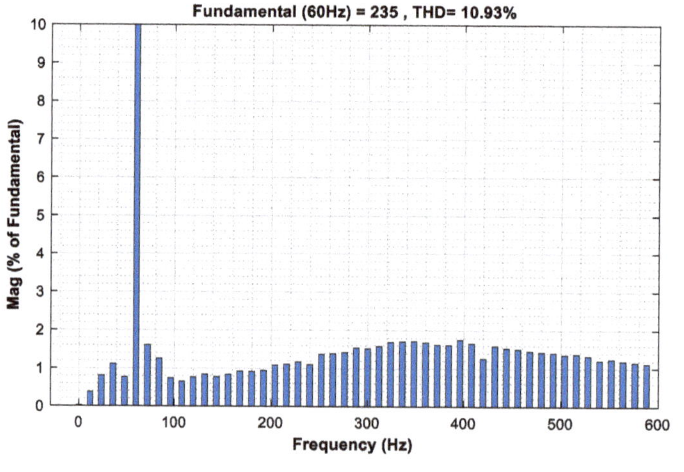

Figure 15. Total harmonic distortion (THD) analysis of the grid current in the original case.

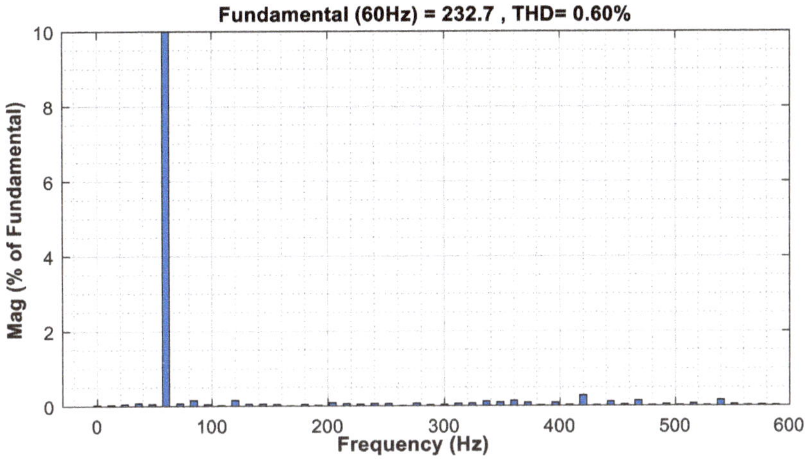

(a)

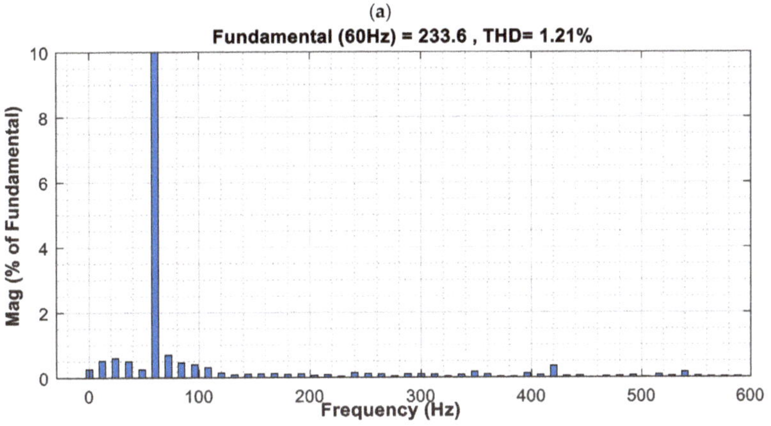

(b)

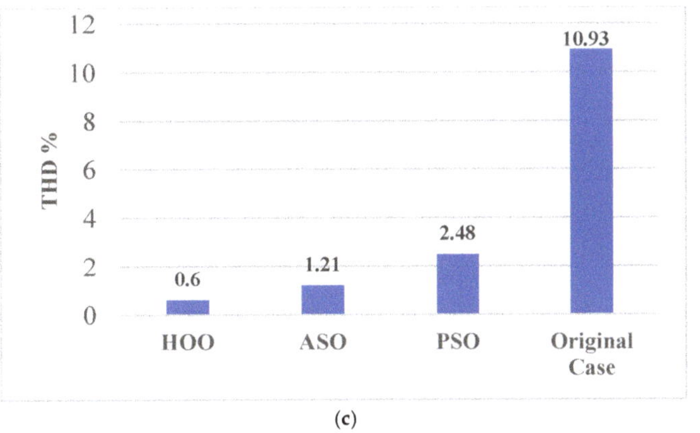

(c)

Figure 16. (a) The THD analysis of the current grid using HHO; (b) The THD analysis of the current grid using ASO; (c) Comparison between the THD analysis of the current grid with different methods.

The improvement of the voltage profile by compensating the reactive power is one of the main prospective results of MMC STATCOM. In order to examine the operation of the MMC STATCOM with the reactive power command, Q_{ref} is changed from (−5 MVAr); in this case the STATCOM operates in the inductive mode, to (+10 MVAr), where the STATCOM operates in the capacitive mode.

The grid side voltage and current of the MMC STATCOM are shown in Figure 17. As shown from Figure 17, when the set-point of the reference reactive power changed at 0.15 s, the current drawn from the grid steps from lagging the grid voltage by 90 degrees to leading the voltage by 90 degrees. The control system of the STATCOM introduces a rapid reaction for modifying the terminal voltage of the inverter in order to produce 10 MVAr of reactive power (capacitive mode). The load current and output voltage are represented in Figure 18.

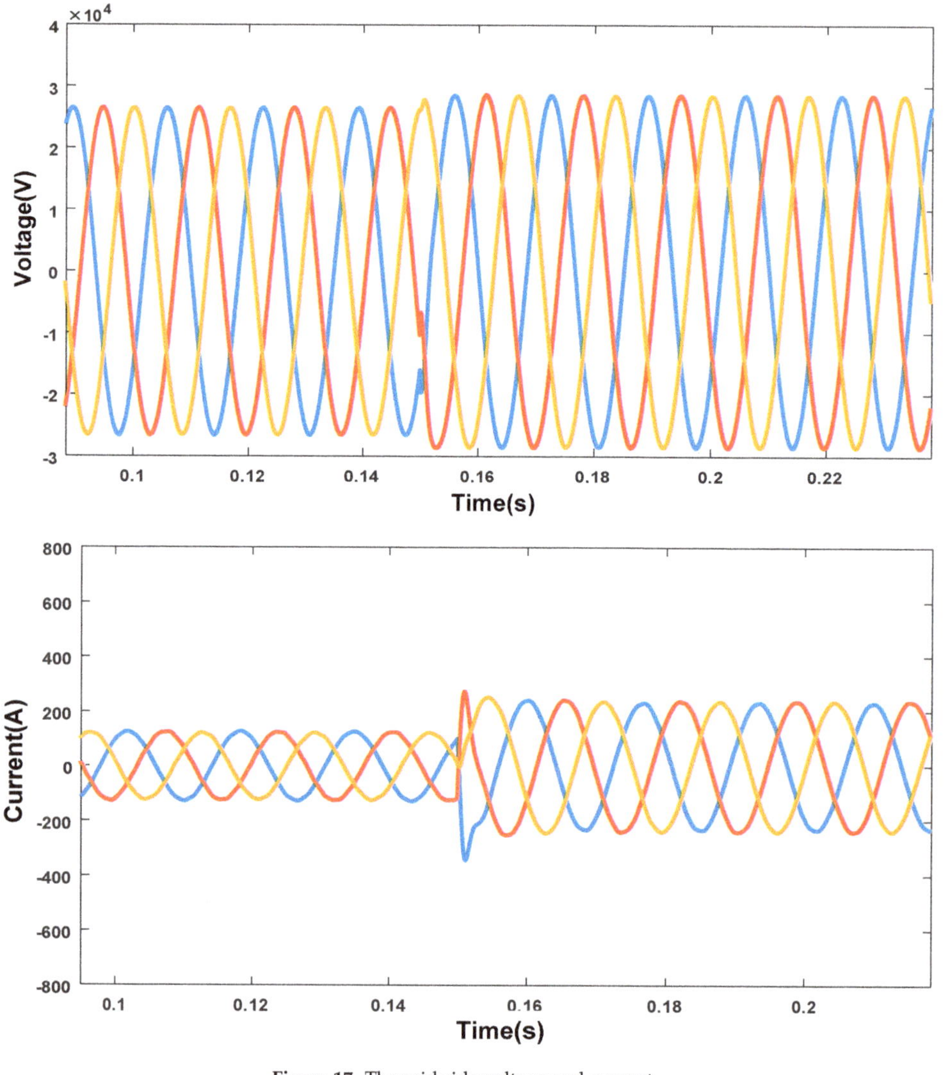

Figure 17. The grid side voltage and current.

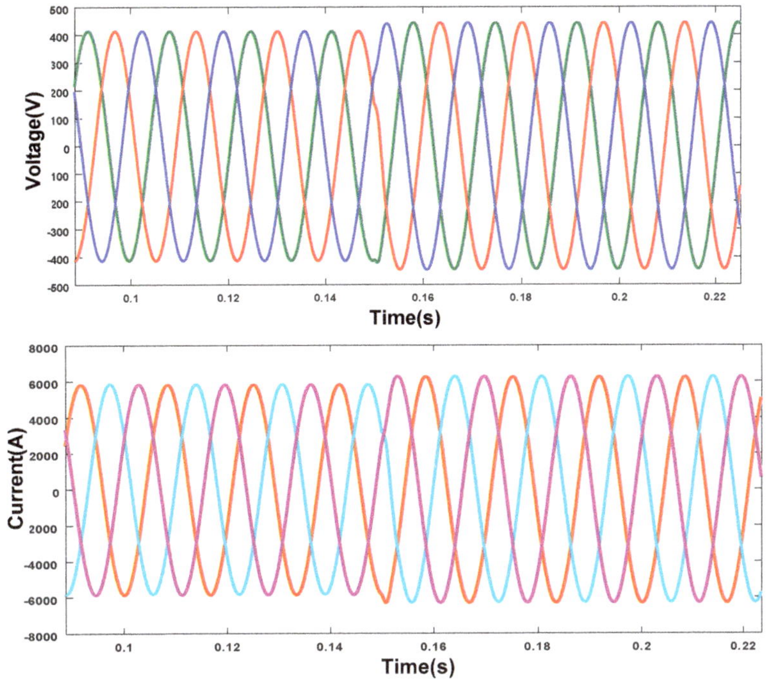

Figure 18. The load side voltage and current.

The active and reactive power with the connection of the MMC STATCOM is presented in Figure 19a,b respectively. From the figure, a high dynamic performance with acceptable raise time has been achieved for the reactive power variation. Moreover, the three-phase voltage of the converters in this case of study is shown in Figure 20.

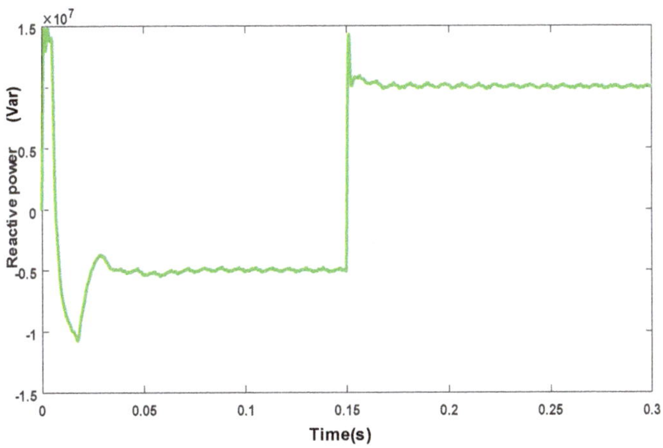

(a)

Figure 19. *Cont.*

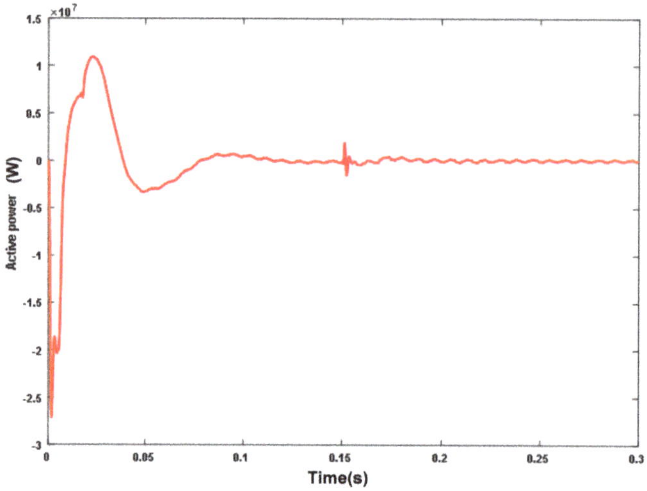

(b)

Figure 19. The reactive and active power with connection to the MMC STATCOM. (**a**) Reactive power; (**b**) Active power.

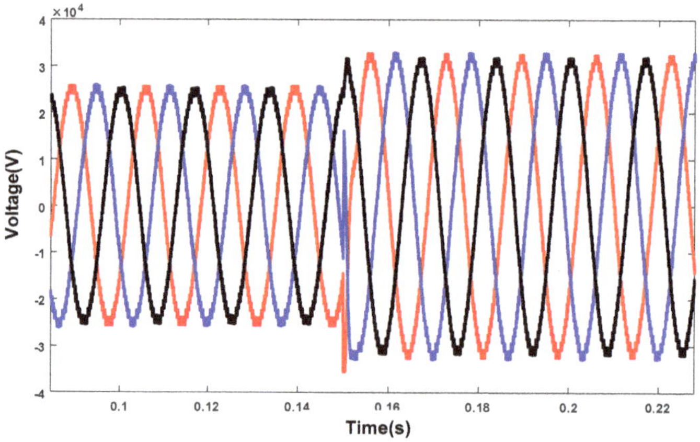

Figure 20. Phase voltage converters.

7.2. Unbalanced Load Case Study

For more validation of the proposed configuration, the system has been tested under the case of unbalanced load. In this case of study, the load on phase C has been assumed to be changed at 1 s by 3 s, and the current at the grid side with imbalanced load is shown in Figure 21. The effectiveness of the proposed control scheme based on the optimization technique of HHO is examined. The voltages and currents at the load side are shown in Figure 22. Figure 23 shows the grid side voltages and currents, which indicate a balance in the grid currents and voltages. Moreover, Figure 24 shows the DC link voltage of the MMC for this case of study. As observed from the figures, in the case of imbalanced load, the proposed MMC STATCOM with the optimal design based on the optimization algorithms was able to maintain the voltage and currents as balanced as possible. Moreover, the maintenance of the voltage in the DC-link, constant without fluctuations, was also desired, and it has been achieved as observed in Figure 24.

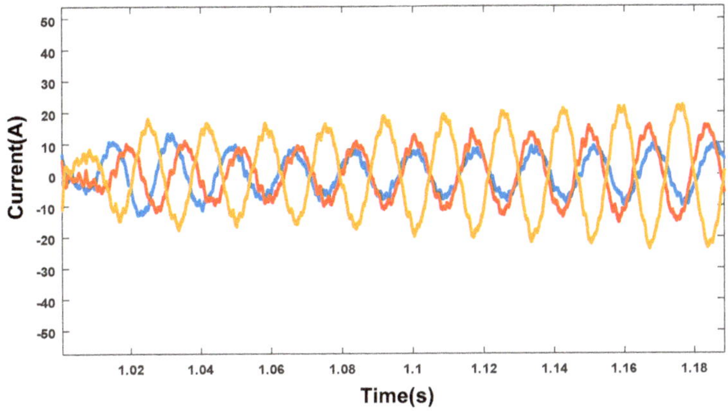

Figure 21. The current at the grid side with an imbalanced load.

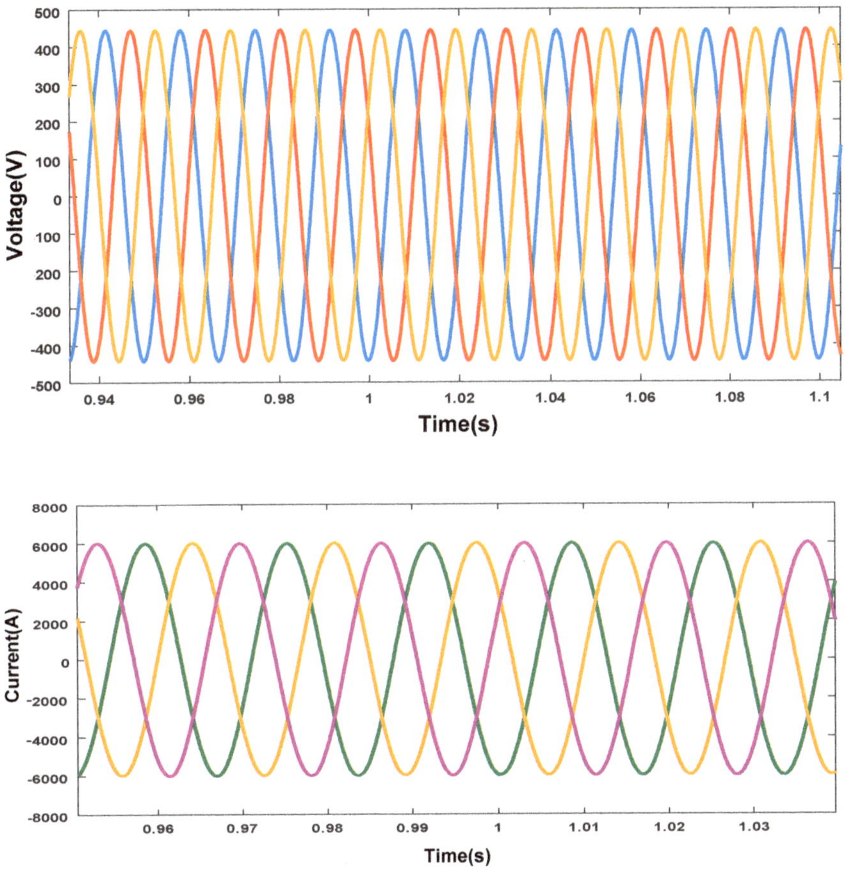

Figure 22. The magnitude voltage and current load with an imbalanced load.

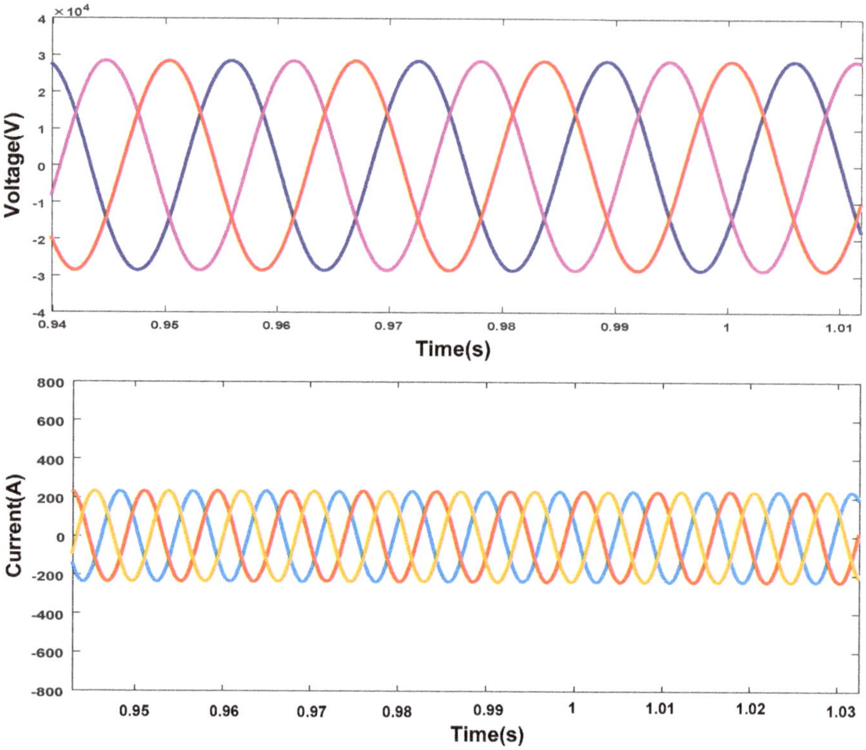

Figure 23. The voltage and current at the grid side with an imbalanced load with MMC STATCOM.

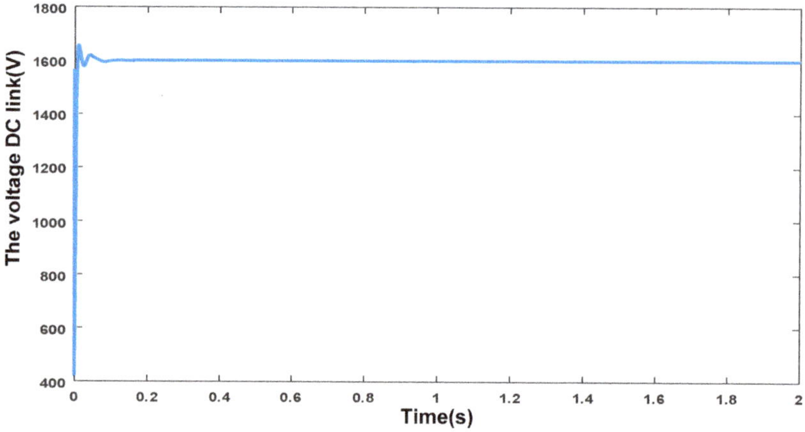

Figure 24. The voltage DC link of the MMC with an imbalanced load.

8. Conclusions

The compensation of harmonics is a major task in the modular multilevel converter (MMC). The Full-Bridge MMC block with 22 modules per phase is presented in this paper. The optimal design of the MMC capacitor and the optimized parameters of the PI controller were determined using HHO and ASO techniques. Moreover, two realistic representative cases of studies were designed and tested:

nonlinear and imbalanced loads. The THD of the current before applying the proposed algorithm was 10.97%, where with the optimal values of the capacitor converters, the THD decreased to 0.60% by using the HHO algorithm, whereas THD decreased to 1.21% using the ASO. The results show that with the application of the MMC STATCOM, the voltage and current load are rapidly adjusted in the case of the imbalanced load. Moreover, the voltage DC-link of the converter has a stability value, and this will reduce the value of the current circulation within the converter. In the two cases of nonlinear and imbalanced loads, the voltage in the DC-link has a stabilized value, and this, in turn, improves the performance of the control system and the power quality indicators in the studied network. In addition, the application of the MMC in grid-connected photovoltaic and wind energy systems and micro grids is also an important issue for our future works. In spite of this, the study of this topic is outside of the framework of this paper, and it can be covered in future works. Finally, the authors would like to remark that, due to the lack of laboratory equipment and apparatus with the high voltage rating, the real time implementation of the proposed algorithm was not performed. However, in future work, one of the aspects of our work will be to experimentally implement the presented control system. Additionally, future work will present the application of other control algorithms to achieve an improvement of the overall system performance.

Author Contributions: Conceptualization, A.A.Z.D. and R.A.; Data curation, H.M.S. and Z.M.A.; Formal analysis, A.A.Z.D. and H.M.S.; Methodology, T.E., A.A.Z.D. and R.A.; Resources, A.A.Z.D., R.A. and Z.M.A.; Software, H.M.S.; Supervision, A.A.Z.D.; Validation, T.E. and A.A.Z.D.; Visualization, A.A.Z.D., H.M.S. and Z.M.A.; Writing—original draft, T.E., A.A.Z.D., R.A. and H.M.S.; Writing—review & editing, A.A.Z.D., R.A. and Z.M.A. All authors together organized and refined the manuscript in the present form. All authors have read and agreed to the published version of the manuscript.

Funding: This research received no external funding.

Conflicts of Interest: The authors declare no conflict of interest.

References

1. Sandeer, K.N. Power Quality Issues and its Mitigation Techniques. Master's Thesis, National Institute of Technology Rourkela, Orissa, India, 2014.
2. Kartashov, I.I.; Tulsky, V.N.; Shamonov, R.G.; Sharov, Y.V.; Nasyrov, R.R. *Control of Power Quality*, 2nd ed.; MPEI: Moscow, Russia, 2017; pp. 347–350.
3. Shahewar, S.; Malthon, R.; Sushmalatha, S. Power Quality Issues and Their Mitigation Technologies. *Int. J. Ind. Electron. Electr. Eng.* **2014**, *2*, 55–61.
4. Tejinder, S.S.; Lakhwinder, S. Comparative Analysis of Custom Power Devices for Power Quality Improvement in Non-linear Loads. In Proceedings of the 2015 2nd International Conference on Recent Advances in Engineering & Computational Sciences (RAECS), Chandigarh, India, 21–22 December 2015. [CrossRef]
5. Wankhede, M.V.; Chandrakar, V.K. Static Synchronous compensator (STATCOM) for the improvement of Electrical System Performance with Non-linear load. *Int. J. Power Syst.* **2016**, *1*, 27–32.
6. Devendra, M.H.; Vinod, K.C. Harmonic domain modeling of Space vector based STATCOM. *Energy Power Eng.* **2016**, *8*, 195–203.
7. Martinek, R.; Bilik, P.; Baros, J.; Brablik, J.; Kahankova, R.; Jaros, R.; Danys, L.; Rzidky, J.; Wen, H. Design of a Measuring System for Electricity Quality Monitoring within the SMART Street Lighting Test Polygon: Pilot Study on Adaptive Current Control Strategy for Three-Phase Shunt Active Power Filters. *Sensors* **2020**, *20*, 1718. [CrossRef] [PubMed]
8. Leon, M.T.; Xiaojie, S. Multilevel Power Converters. In *Power Electronics Handbook*, 2nd ed.; Rashid, M.H., Ed.; Elsevier: San Diego, CA, USA, 2018; pp. 385–416. [CrossRef]
9. Corzine, K.; Familiant, Y. A new cascaded multilevel H-bridge drive. *IEEE Trans. Power Electron.* **2002**, *17*, 125–131. [CrossRef]
10. Kolb, J.; Kammerer, F.; Braum, M. Dimensioning and design of a modular multilevel converter for drive applications. In Proceedings of the IEEE 15th International Power Electronics and Motion Control Conference, Novi sad, Serbia, 4–6 September 2012.

11. Dekka, A.; Bin, W.; Zargari, N.R.; Fuentes, R.L. A space-vector PWM-based voltage-balancing approach with reduced current sensors for modular multilevel converter. *IEEE Trans. Ind. Electron.* **2016**, *63*, 2734–2745. [CrossRef]
12. Hammond, P.W. Four-quadrant AC-AC Drive and Method. U.S. Patent 6,166,513, 26 December 2000.
13. Aiello, M.F.; Hammond, P.W.; Rastogi, M. Modular Multi-Level Adjustable Supply with Series Connected Active Inputs. U.S. Patent 6,236,580, 22 May 2001.
14. Aiello, M.F.; Hammond, P.W.; Rastogi, M. Modular Multilevel Adjustable Supply with Parallel Connected Active Inputs. U.S. Patent 6,301,130, 9 October 2001.
15. Marquardt, R. Stromrichterschaltungen Mit Verteilten Energiespeichern. German Patent DE 10,103,031, 24 January 2001.
16. Shi, X.; Tolbert, L.; Wang, F. Modular multilevel converters with integrated arm inductors for high quality current waveforms. In Proceedings of the IEEE Energy Conversion Congress and Exposition, Mile High City, CO, USA, 15–19 September 2013.
17. Quraan, M.; Yeo, T.; Tricoli, P. Design and control of modular multilevel converters for battery electric vehicles. *IEEE Trans. Power Electr.* **2015**, *31*, 507–517. [CrossRef]
18. Thitichaiworakorn, N.; Hagiwara, M.; Akagi, H. Experimental verification of a modular multilevel cascade inverter based on double-star bridge-cells (MMCI-DSBC). *IEEE Trans. Ind. Appl.* **2014**, *50*, 509–519. [CrossRef]
19. Nor Shahida, H.; Norzanah, R.; Dygku, A.A.; Osman, A.H. Reviews on multilevel converter and modulation techniques. *Renew. Sustain. Energy Rev.* **2017**, *80*, 163–174.
20. Nasyrov, R.R.; Aljendy, R.I.; Abdelaziz, A.L.; Diab, A.A.Z. Enhancement of Power Quality with Hybrid Distributed Generation and FACTS Device. *IETE J. Res.* **2019**, 1–12. [CrossRef]
21. Ufa, R.; Gusev, A.; Diab, A.A.Z.; Suvorov, A.; Ruban, N.; Andreev, M.; Askarov, A.; Rudnik, V.; Abdalla, O.; Ali, Z.M.; et al. Analysis of application of back-to-back HVDC system in Tomsk electric power system. *Energy Rep.* **2020**, *6*, 438–444. [CrossRef]
22. Mohamed, I.S.; Rovetta, S.; Do, T.D.; Dragicević, T.; Diab, A.A.Z. A neural-network-based model predictive control of three-phase inverter with an output LC filter. *IEEE Access* **2019**, *7*, 124737–124749. [CrossRef]
23. Nasyrov, R.R.; Aljendy, R.I.; Diab, A.A.Z. Adaptive PI controller of active power filter for compensation of harmonics and voltage fluctuation based on particle swarm optimization (PSO). In Proceedings of the 2018 IEEE Conference of Russian Young Researchers in Electrical and Electronic Engineering (EIConRus), Moscow, Russia, 29 January–1 February 2018; pp. 719–724.
24. Marquardt, R. Modular multilevel converter topologies with dc-short circuit current limitation. In Proceedings of the IEEE 8th International Conference on Power Electronics and ECCE Asia (ICPE & ECCE), Jeju, Korea, 29 may–2 June 2011.
25. Sharifabadi, K.; Harnefors, L.; Nee, H.P.; Norrga, S.; Teodorescu, R. *Design, Control, and Application of Modular Multilevel Converters for HVDC Transmission Systems*; John Wiley & Sons: Hoboken, NJ, USA, 2016. [CrossRef]
26. Cupertino, A.; Farias, J.; Pereira, H.; Seleme, S.; Teodorescu, R. Comparison of DSCC and SDBC Modular Multilevel Converters for STATCOM Application During Negative Sequence Compensation. *Ieee Trans. Ind. Electron.* **2019**, *66*, 2302–2312. [CrossRef]
27. Sajedi, S.; Farrell, M.; Basu, M. DC side and AC side cascaded multilevel inverter topologies: A comparative study due to variation in design features. *Electr. Power Energy Syst.* **2019**, *113*, 56–70. [CrossRef]
28. Siemaszko, D.; Antonopoulos, A.; Ilves, K.; Vasiladiotis, M.; Angquist, L.; Nee, H.P. Evaluation of control and modulation methods for modular multilevel converters. In Proceedings of the IEEE International Power Electronics Conference, Sapporo, Japan, 21–24 June 2010.
29. Yang, X.; Li, J.; Fan, W.; Wang, X.; Zheng, T. Research on Modular Multilevel Converter Based STATCOM. In Proceedings of the 2011 6th IEEE Conference on Industrial Electronics and Applications, Beijing, China, 21–23 June 2011; pp. 2569–2574.
30. Ilves, K.; Norrga, S.; Harnefors, L.; Nee, P. On energy storage requirements in modular multilevel converters. *Ieee Trans. Power Electron.* **2014**, *29*, 77–88. [CrossRef]
31. Jacobson, B.; Karlsson, P.; Asplund, G.; Harnefors, L.; Jonsson, T. VSC-HVDC transmission with cascaded two-level converters. In *Cigré Session*; CIGRE: Paris, France, 2010; pp. B4–B110.
32. Bairathi, D.; Gopalani, D. A Novel Swarm Intelligence Based Optimization Method: Harris' Hawk Optimization. In Proceedings of the International Conference on Intelligent Systems Design and Applications, Vellori, India, 6–8 December 2018.

33. Heidari, A.A.; Mirjalili, S.; Faris, H.; Aljarah, I.; Mafarja, M.; Chen, H. Harris hawks optimization: Algorithm and applications. *Future Gener. Comput. Syst.* **2019**, *97*, 849–872. [CrossRef]
34. Zhao, W.; Wang, L.; Zhang, Z. A novel atom search optimization for dispersion coefficient estimation in groundwater. *Future Gener. Comput. Syst.* **2019**, *91*, 601–610. [CrossRef]
35. Zhao, W.; Wang, L.; Zhang, Z. Atom search optimization and its application to solve a hydrogeologic parameter estimation problem. *Knowl.-Based Syst.* **2019**, *163*, 283–304. [CrossRef]
36. Wolpert, H.; Macready, G. No free lunch theorems for optimization. *IEEE Trans. Evol. Comput.* **1997**, *1*, 67–82. [CrossRef]
37. Yang, S. Review of meta-heuristics and generalised evolutionary walk algorithm. *Int. J. Bio-Inspired Comput.* **2011**, *3*, 77–84. [CrossRef]
38. Gaing, Z.L. A particle swarm optimization approach for optimum design of PID controller in AVR system. *IEEE Trans. Energy Convers.* **2004**, *19*, 384–391. [CrossRef]
39. Shojaei, A.; Joos, G. An improved modulation scheme for harmonic distortion reduction in modular multilevel converter. In Proceedings of the 2012 IEEE power and energy society general meeting, San Diego, CA, USA, 22–26 July 2012.

applied sciences

MDPI

Article

A High Gain DC-DC Converter with Grey Wolf Optimizer Based MPPT Algorithm for PV Fed BLDC Motor Drive

A. Darcy Gnana Jegha [1], M. S. P. Subathra [1,*], Nallapaneni Manoj Kumar [2,*], Umashankar Subramaniam [3] and Sanjeevikumar Padmanaban [4,*]

1 Department of Electrical and Electronics Engineering, Karunya Institute of Technology and Sciences, Coimbatore 641114, India; darcyjegha@gmail.com
2 School of Energy and Environment, City University of Hong Kong, Kowloon, Hong Kong
3 Renewable Energy Lab, College of Engineering, Prince Sultan University, Riyadh 12435, Saudi Arabia; usubramaniam@psu.edu.sa
4 Department of Energy Technology, Aalborg University, 6700 Esbjerg, Denmark
* Correspondence: subathra@karunya.edu (M.S.P.S.); nallapanenichow@gmail.com (N.M.K.); san@et.aau.dk (S.P.)

Received: 22 February 2020; Accepted: 15 April 2020; Published: 17 April 2020

Abstract: Photovoltaic (PV) water pumping systems are becoming popular these days. In PV water pumping, the role of the converter is most important, especially in the renewable energy-based PV systems case. This study focuses on one such application. In this proposed work, direct current (DC) based intermediate DC-DC power converter, i.e., a modified LUO (M-LUO) converter is used to extricate the availability of power in the high range from the PV array. The M-LUO converter is controlled efficiently by utilizing the Grey Wolf Optimizer (GWO)-based maximum power point tracking algorithm, which aids the smooth starting of a brushless DC (BLDC) motor. The voltage source inverter's (VSI) fundamental switching frequency is achieved in the BLDC motor by electronic commutation. Hence, the occurrence of VSI losses due to a high switching frequency is eliminated. The GWO optimized algorithm is compared with the perturb and observe (P&O) and fuzzy logic based maximum power point tracking (MPPT) algorithms. However, by sensing the position of the rotor and comparing the reference speed with the actual speed, the speed of the BLDC motor is controlled by the proportional-integral (PI) controller. The recent advancement in motor drives based on distributed sources generates more demand for highly efficient permanent magnet (PM) motor drives, and this was the beginning of interest in BLDC motors. Thus, in this paper, the design of a high-gain boost converter optimized by a GWO algorithm is proposed to drive the BLDC-based pumping motor. The proposed work is simulated in MATLAB-SIMULINK, and the experimental results are verified using the dsPIC30F2010 controller.

Keywords: PV water pumping; high gain DC-DC converter; modified LUO converter; hybrid MPPT algorithm; grey wolf optimizer

1. Introduction

The brushless DC (BLDC) motor is commonly used for low and medium power applications because of its high torque/inertia ratio, high efficiency, low electromagnetic interference and lower maintenance. The BLDC motor is also widely used in aerospace, medical, servo appliances, electric vehicles and robotic applications [1,2]. These motors need electronic commutators, which require information about the rotor position to achieve commutation using a voltage source inverter. The speed-torque characteristics of BLDC motors are very significant, and there is no need for the placement of brushes, which in turn reduces the copper and eddy current losses. Due to its permanent magnet excitation, no excitation losses occur, and this enhances the efficiency of the BLDC motor to a sufficiently high value. Photovoltaic (PV) systems can be used to supply power to the BLDC motor for the pumping application [3]. In the conventional method, the motor is driven utilizing a single-phase controlled bridge rectifier. At the output side of the rectifier, a smoothening direct current-link (DC-link) capacitor is used [4]. Owing to the unbalance in the charging and discharging of the DC link capacitor, a higher peak current compared to the fundamental input current occurs, distorting the input alternating current (AC) current. Due to this very high current, the total harmonic distortion (THD) also increases in the range of 65%–70%. Thus, power quality problems occur, in addition to the notching effect and crest factor [5,6].

The BLDC motor drives typically work in a two-phase conduction mode, and during the commutation period it switches over to the three-phase conduction mode. This process is known as overlapped commutation mode. Due to this, there will be an increase in the overlapping current, which in turn produces commutation torque ripples in the BLDC motor drive [7]. Direct current-direct current (DC-DC) converters are classified as non-isolated and isolated converters. The isolated DC-DC converters are used for high voltage conversion where the conduction loss may occur owing to an energy transfer from secondary to the primary winding. However, the non-isolated DC-DC converters are quite reasonable as they are only used for low voltage conversion [8]. The non-isolated DC to DC converters has many salient features when used with the PV system [9]. Overall, the non-isolated DC to DC converters has an excellent performance in the PV system [10]. When the converters operate in a Continuous Conduction Mode (CCM), they offer reduced stress on the switches of the converters, but still, the need for three sensors, a power supply and dc-link voltages increase the cost of converters. When the converter operates in a discontinuous mode, it offers more stress on the converter switches, which requires a mono voltage sensor for controlling the dc-link voltage [11].

The selection of DC-DC converters plays a vital role in deciding the optimum performance of BLDC drives. Research in the field of renewable energy sources shows that general integration has been done with the DC-DC converter topology. From various literature reviews, it is shown that buck converters have been used in speed drive regulation and the reduction of torque ripple by pulse width modulation (PWM) chopping techniques [12]. Richter, 2017 [13] and Singh & Bist 2015, [14] discussed several configurations related to converter-fed BLDC drives. In most cases, the boost converter configuration has been given much importance despite its disadvantage of switching losses that occur in VSI due to the high-frequency PWM signal. For overcome this, low-frequency switching signal based electronic commutators is put in use to control the speed of BLDC drives through voltage source inverter's (VSI) [15]. Variable voltages of the Cuk converter VSI have been used for controlling the speed and VSI functions at the fundamental frequency, thereby reducing switching losses. The main drawback of this system is that, in CCM mode, three sensors are required to limit the voltage and control the speed of the BLDC drive. It also provides an inverting output, which increases the system complexity and slows down the system response, which would, therefore, not match with the system's requirement [16]. Kumar & Singh 2014, [17] used the buck-boost converter for water pumping application and for soft-starting the BLDC by proper control. Buck-boost converters have a discontinuous output current which produces ripples and switching losses in VSI. Kumar & Singh 2016, [18] used a Zeta converter, where PWM pulses have controlled the voltage source inverter's switching sequences. Due to this, a high switching loss occurs on the VSI. Berkovich et al., 2015 [19] used LUO converter with the switched inductor to boost the voltage, and by adding a magnetically coupled inductor, diode and capacitor, the voltage gain is multiplied. A PV system-fed BLDC motor with a Cuk converter-based maximum power point tracking (MPPT) algorithm eliminates the ripple filter, and speed control is achieved through a variable DC link voltage [20]. Depending on the application, different DC-DC converters can be used in renewable energy sources to improve the efficiency of the system [21]. An MPPT based on the Fuzzy and Artificial Bee Colony (ABC) algorithm with a zeta converter was designed, where the sensors and PI controllers in the system often got affected due to the environmental conditions [22]. The closed-loop speed control of the BLDC motor is implemented with the dSPACEDS1103 controller for controlling the speed at a reference value for low to high speeds [22]. Additionally, by self-tuning the fuzzy proportional–integral–derivative (PID) controller, the rotor speed is forced to track the reference speed at all times [23]. The closed-loop speed control of the BLDC motor is obtained from the optimized fuzzy algorithm [24]. A DC-DC boost converter is utilized between the solar PV array and permanent magnet motors, by which the friction between the brushes increases and the torque is lowered [25].

A simple and efficient solar water pumping system using a modified LUO converter has been proposed in this paper. A BLDC motor is attracting much interest, because of its higher efficiency, better performance and simple control when applied in many applications. The induction motors have a nearly 10% lower efficiency than PM motor drives for energy saving applications. The increase in energy demand spurs higher demands for variable speed PM motor drives. A modified LUO converter with an efficient grey wolf optimizer (GWO) MPPT algorithm extracts the maximum power from the PV panel. This converter also provides a low starting current for the BLDC motor, which overcomes the torque ripple problems in the BLDC motor.

The manuscript is structured as follows, in six different sections: in Section 2, the systematic working of the proposed system is presented. In Section 3, the design of the proposed system considering each electrical component is discussed, and detailed mathematical modelling is given. Section 4 presents a brief theory on the perturb and observe (P&O), fuzzy and GWO MPPT algorithms, as well as a detailed flow chart of the GWO algorithm. The simulation and experimental results of the proposed system are presented and discussed in Section 5. Finally, the main conclusions are drawn from the study and views on future works are given in Section 6.

2. Systematic Working of the Proposed System

The configuration of the developed solar-fed BLDC motor using a modified LUO converter is shown in Figure 1. The proposed modified LUO converter operates with a single switch in a continuous conduction mode. This increases the efficiency and suppresses the ripple in the output voltage. The single switch operation reduces the switching stress of the converter. The continuous conduction mode reduces the heat produced in the power semiconductor devices.

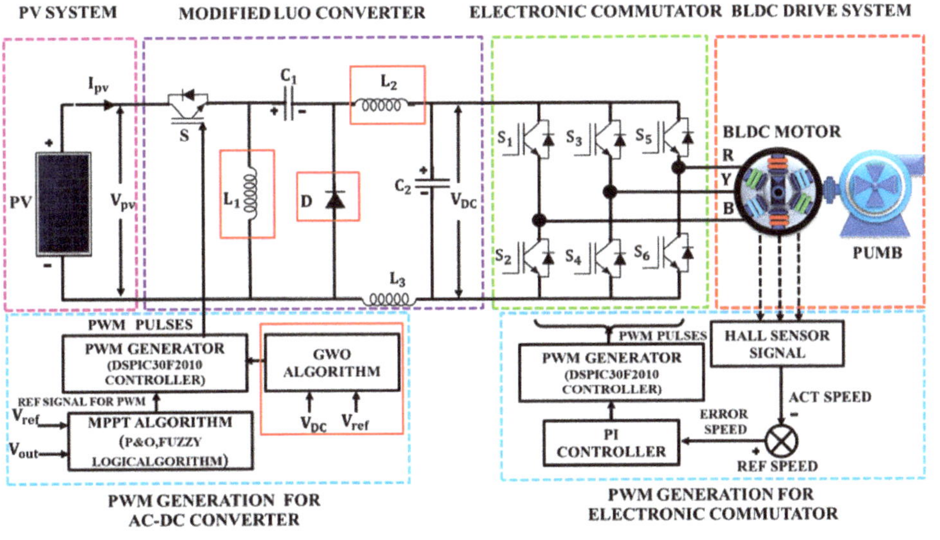

Figure 1. Proposed solar-based system for motor pumping.

The proposed GWO-optimized algorithm reduces the maximum power tracking time and makes the output voltage nearly constant. This constant voltage is fed to the three-phase electronic commutator, which converts the DC into three-phase AC voltage. For find the rotor position, three hall sensors are set up in the rotor displaced by 60° mechanicals. The input current is continuous because of the modified LUO converter, and this makes the BLDC motor make a soft starting. The proportional-integral (PI)-based motor speed control achieves a constant speed of operation [26]. This makes the proposed work highly efficient in a way that could easily be carried out in a low-cost budget. The proposed system is developed using the dsPIC30F2010 microcontroller by Microchip Technology Inc., USA, and the experimental results are verified using the MATLAB Simulink tool by MathWorks, USA.

This system consists of a 1000 W solar panel, insulated-gate bipolar transistor (IGBT)-based modified LUO (M-LUO) converter with an active snubber circuit and IGBT-based VSI with a 3 phase, 1 horsepower (HP) BLDC motor employed with a Hall Effect sensor. Additionally, a dsPIC30F2010 controller is used as a pulse generator, which is controlled by the GWO algorithm. The driver circuit drives the IGBT switch. The system functions when the input from the solar array voltage is given to the modified, LUO converter for maintaining constant voltage, with less ripple content delivered to the inverter system.

The output of the LUO converter is used as an input for the GWO algorithm. For tracking the maximum power, the proposed GWO algorithm is advantageous. This algorithm produces PWM pulses based on the variation of the PV panel voltage, thereby adjusting the duty cycle of the M-LUO converter. In the M-LUO converter, the elementary LUO converter is added with one extra inductor. This converter acts as a second-order low pass filter, and therefore it does not require another filter circuit. This converter with the GWO algorithm provides a constant voltage to all the three phases of the electronic commutator. Switching is done with the help of Hall Effect voltage sensors. The three-phase voltage source inverter converts the DC voltage into AC voltage to provide power to the BLDC motor. The inverter system works at a fundamental frequency to eliminate the high-frequency noise in the inverter output, and high efficiency is thus attained.

3. Proposed System Design

The modelling steps are explained appropriately to make a compelling design of the proposed system, capable of operating in certain determined conditions. A BLDC motor with 1 HP power and a solar panel of the maximum power capacity of 800 W under standard test conditions (STC) are deliberated. Figure 2 shows a schematic representation of the solar cell. In Figure 2a,b, the simplified diagram and the equivalent diagram of the solar cell are represented, respectively. The current I_{ph} represents the current generated due to photons. I_{ph} mainly depends on the incident radiant energy emitted by the sun and its temperature. One single solar cell is designed by using one photon source current, one diode and one set of series and shunt resistors grouped in a series and parallel fashion, forming a photovoltaic cell.

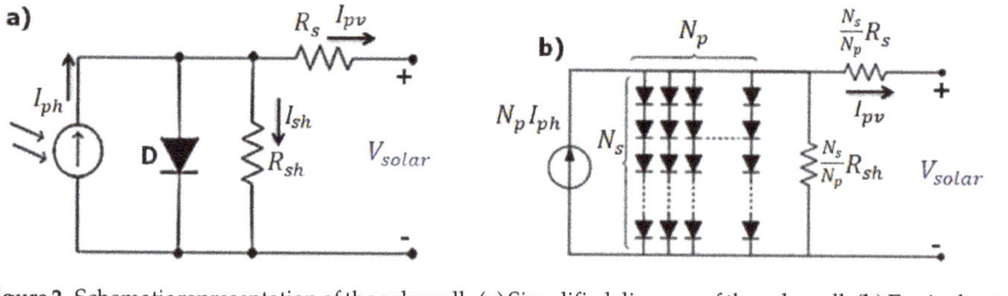

Figure 2. Schematic representation of the solar cell. (**a**) Simplified diagram of the solar cell; (**b**) Equivalent diagram of the solar cell.

3.1. Modelling of the Solar Cell

The output current from the solar cell is:

$$I = N_P \times I_{ph} - N_P \times I_0 \times \left[e^{\left(\frac{\frac{V}{N_S} + I \times \frac{R_S}{N_P}}{n * V_t} \right) - 1} \right] - I_{sh} \quad (1)$$

The diode thermal voltage is given as:

$$V_t = \frac{k \times T}{q} \quad (2)$$

The shunt current is given as:

$$I_{sh} = \frac{V \times \frac{N_P}{N_s} + I \times R_s}{R_{sh}} \quad (3)$$

where N_P is the number of modules connected in parallel; R_s is the series resistance; R_{sh} is the shunt resistance; and the diode thermal voltage is V_t.

With the available physical and mathematical model, various parameters of the solar cell can be estimated. These include the estimated power, output voltage and maximum short circuit current. In the proposed study, a solar panel with the specifications shown in Table 1 is used.

Table 1. Specification of the solar panel.

Parameter	Value with Units
Peak power (P_{mp})	100 W
Open Circuit Voltage (V_{oc})	22.68 V
Short circuit current (I_{sc})	5.86 A
Peak power voltage (V_{mp})	18.75 V
Peak power current (I_{mp})	5.42 A
Number of series-connected cells (N_s)	36

3.2. Design of the Proposed System

The design of the M-LUO converter-fed BLDC for the water pumping application is discussed in this section. The components, such as the input inductor, capacitors L_1 and C_1 respectively, output inductors L_2, L_3 and output dc-link capacitor C_2 work in a continuous conduction mode with low stress on the devices and its elements. Depending on the duty cycle, the mode of operation can be changed.

$$\text{Duty cycle} = V_{out}/(V_{out} + V_{solar}) = 300/(300 + 56) = 0.84 \quad (4)$$

where V_{out} is the average value of the output voltage of the LUO converter, which is equal to the BLDC motor's input voltage.

The estimated output current can be written as:

$$I_0 = P_{out}/V_{out} \quad (5)$$

The BLDC motor rating is taken as $P_{out} = 746$ W:

$$I_0 = 746/300 = 2.48 \text{ A} \quad (6)$$

Then, the estimated maximum permitted ripple values of L_1 and C_1 are:

$$\begin{aligned} \Delta I_L &= I_o \times \frac{V_0}{V_{in(\min)}} \\ &= 2.48\,(300/56) + 30\% \text{ ripples} \\ &= 13.28 + 3.98 \\ &= 17.26 \text{ A} \end{aligned} \tag{7}$$

$$\Delta I_L = \Delta I_{l1} = \Delta I_{l2}$$

The inductor L_1 in the modified LUO converter is calculated by the following equation:

$$\begin{aligned} L_1 &= \mathrm{D}V_{solar(in)}/f_{sw}\Delta I_{l1} \\ &= 0.84 \times 56/10 \times 10^3 \times 17.26 \\ L_1 &= 0.273 \text{ mH} \end{aligned} \tag{8}$$

The inductances L_2 & L_3 in the modified LUO converter are calculated by the following equation:

$$\begin{aligned} L_2 &= (1-\mathrm{D})V_{solar(in)}/f_{sw}\Delta I_{l2} \\ &= (1-0.84) \times 56/10 \times 10^3 \times 17.26 \\ L_2 &= 0.052 \text{ mH} = L_3 \end{aligned} \tag{9}$$

The change in the capacitor voltage across C_1 is given by:

$$\begin{aligned} \Delta V_{cs} &= I_0 \times D/C_s\, f_{sw} \\ &= 2.48 \times 17.26/10 \times 10^{-6} \times 10 \times 10^3 \\ \Delta V_{cs} &= 428.04 \text{ V} \end{aligned} \tag{10}$$

The following equation calculates the capacitor C_1 in the modified LUO converter:

$$\begin{aligned} C_1 &= I_0 \times D/\Delta V_{cs} f_{sw} \\ &= 2.48 \times 0.84/428.04 \times 10 \times 10^3 \\ C_1 &= 486 \text{ nF} \end{aligned} \tag{11}$$

3.3. Modelling of DC-Link Capacitor

For the estimate, the VSI DC-link capacitor, the harmonic content present in the AC supply is reflected on the DC side as a prior harmonic in the three-phase supply system. The minimum and the maximum speeds for the BLDC motor can be calculated through the fundamental frequencies of the VSI. These two frequencies ω_{max} and ω_{min} are further used to calculate the dc link capacitance value which would satisfy the operation of the solar panel when working below the reduced level of irradiance. The highest capacitor value is chosen for designing a DC-link capacitor.

The fundamental output frequency of VSI corresponding to the maximum and minimum pumping speed requirement is:

$$\begin{aligned} \omega_{max} &= 2\pi f_{max} = \frac{2\pi N_{max} P}{120} \\ &= 2\pi \times 1500 \times \frac{6}{120} = 471.20 \text{ rad/s} \end{aligned} \tag{12}$$

$$\begin{aligned} \omega_{min} &= 2\pi f_{min} = \frac{2\pi N_{min} P}{120} \\ &= 2\pi \times 1000 \times \frac{6}{120} = 314.15 \text{ rad/s} \end{aligned} \tag{13}$$

where f_{max} is the maximum fundamental frequency in Hz, f_{min} is the minimum fundamental frequency in Hz, and P is the number of poles of the BLDC motor.

The corresponding capacitor can be calculated as:

$$\begin{aligned} C_{0\ max} &= \frac{I_0}{6} \times \omega_{max} \times \Delta V_{out} \\ &= \frac{2.5}{6} \times 471.2 \times 24 \times 0.05 = 736.8\ \mu\text{F} \end{aligned} \quad (14)$$

$$\begin{aligned} C_{0\ min} &= \frac{I_0}{6} \times \omega_{max} \times \Delta V_{out} \\ &= \frac{2.5}{6} \times 314.15 \times 24 \times 0.05 = 1105.27\ \mu\text{F} \end{aligned} \quad (15)$$

where ΔV_{out} is the amount of the maximum permissible ripple voltage. Finally, $C_{0\ min}$ is chosen to construct the DC link capacitor.

3.4. Water Pump Design

The water pump design relating to the power and speed for the proportionality constant has to be calculated:

$$\begin{aligned} K = \frac{P}{\omega_{max}^3} &= \frac{60}{\left(2\pi \times \frac{1500}{60}\right)^3} \\ &= 1.52 \times 10^{-5} \end{aligned} \quad (16)$$

where P is the BLDC motor's rated power, and ω_{max} is the motor's rated speed in rad/s.

3.5. BLDC Motor Controlled by Electronic Commutation

The BLDC motor is an electronically commutated device and requires data regarding the rotor position for the proper commutation of the stator current. The BLDC motor needs a quasi-rectangular shaped input current drawn into the machine. The back emf of each phase is trapezoidal and 120 electrical degrees-displaced. The BLDC motor is rotated in a sequence to energize the stator winding. Hall sensors sense the rotor position, and it is embedded into the stator. Whenever the magnetic poles of the rotor are nearer to the hall sensors, the sensors feed a low or high signal. The complete commutation sequence can be found by integrating the three-hall sensor output. Another important key factor of electronic commutation is high-frequency switching loss, which can be eliminated by providing the fundamental switching frequency of VSI. The switching sequences for the electronic commutation of the BLDC motor are the same as those employed in [18].

4. Maximum Power Point Tracking (MPPT) Algorithm

An MPPT algorithm is implemented to keep track of the changes caused by the environment and to extricate high power from the solar array. In a system with MPPT, voltage collapse is shunned when Maximum Power Point (MPP) is near to the operating point. The benefits of using the MPPT algorithm lie in its robustness and simplicity in implementation. In this paper, P&O, fuzzy logic and GWO algorithms are compared to extricate high power from the PV system. There are different control algorithms available to monitor the maximum power point. Among these, P&O and Fuzzy logic are better choices of control algorithms. The main aim of this control algorithm is to adjust the variation of the voltage, current and power at the maximum power point obtained under the specific values of irradiance and temperature.

4.1. P&O Algorithm

In this algorithm based on the previous value of the power, the value of the reference voltage and current keeps on changing (i.e., increasing or decreasing) until it attains the maximum power point [27,28]. This perturbation causes changes in the power of the solar module. If the power increases due to the perturbation, it moves the operating point of the PV array to MPP and continues to disturb the PV voltage in the same direction. After reaching the maximum power, the next instant power reduces and consequently reverses the perturbation. The size of the perturbation kept is minimal in order to maintain the power variation. For set the reference perturbation voltage, the additional Proportional Integral controller was utilized to calibrate the duty cycle ratio of the MPPT-based converter. It was observed that there was some power loss due to this perturbation and that it sometimes also failed to track the power under rapidly varying atmospheric conditions. However, this algorithm is well known for its simplicity, showing better stability in the response under slow transient conditions and at a very high voltage oscillation.

4.2. FUZZY Algorithm

The Fuzzy algorithm overcomes the disadvantages of the P&O algorithm. Fuzzy is one of the most optimal and attractive technical tools in all control fields as it deals with indefinite and uncertain information and makes decisions like humans. Thus, fuzzy control is applied to non-linear systems. The solar panel is a powerful, non-linear system. Various control algorithms have been worked out under the maximum power point tracking algorithm [29,30]. The fuzzy-based control algorithm is efficient because it can automatically adjust the output deviation to a constant value based on specific fuzzy rules. The control rules are framed, and the elements are determined based on the theory that in the transient state significant errors need coarse control which requires coarse input/output, while small errors need fine control which requires excellent input/output variables. Based on this, the elements of the rule table are obtained.

4.3. GWO Algorithm

The grey wolf optimization algorithm is being employed to achieve a better MPPT [9]. Mirjalili first proposed the Grey wolf algorithm in 2014. It is a mathematical model which follows the social behaviour of the grey wolf pack when hunting prey. The wolves in the pack are divided into different categories based on their fitness and strength. The different categories are alpha (α), beta (β), delta (δ) and omega (ω). The fittest wolf in the pack is alpha (α), while the wolves with a fitness close to alpha are beta (β) and delta (δ), and the rest of the wolves constitute omega (ω). The same structure is then used by the pack to hunt the prey by tracking, encircling and attacking. Here, the alpha (α), beta (β) and delta (δ) wolves play the primary role.

The same phenomenon is considered in the GWO optimization model. The model divides the available solutions for a given problem into different categories as per their closeness to the desired output and also moves the solutions towards the desired values, just like the wolf pack tracking the prey, encircling the solution and finally achieving the desired solution. The solution which falls into alpha (α) is used by the mathematical model to reach the optimal solution. The equation used by the GWO to achieve this is given below:

$$\vec{X}(t+1) = \vec{X_p}(t) - \vec{A} \times \vec{D} \tag{17}$$

$$\vec{D} = \left| \vec{C} \times \vec{X_p}(t) - \vec{X}\,(t) \right| \tag{18}$$

Here, it indicates the current iteration. $\vec{X_p}(t)$ represents the position of the desired output, and $\vec{X}(t+1)$ represents the position of the current solution. $\vec{A}$, $\vec{C}$ *and* $\vec{D}$ represent the vector coefficients, and these vectors are calculated using the below formula:

$$\vec{A} = 2\vec{a}\,r_1 - \vec{a} \tag{19}$$

$$\vec{C} = 2\vec{r_2} \tag{20}$$

where the values of the vector a are linearly decreased from 2 to 0 as the iterations proceed, and r_1, r_2 are randomly chosen vectors in [0, 1].

In this paper, the GWO math model is used to achieve the maximum power point tracking with the help of a modified LUO converter whose switching pulse's duty cycle can be varied in an optimal way to achieve the maximum power from the PV array. The GWO performs the desired calculations to achieve MPP. The flow chart for the GWO-based MPPT is shown in Figure 3.

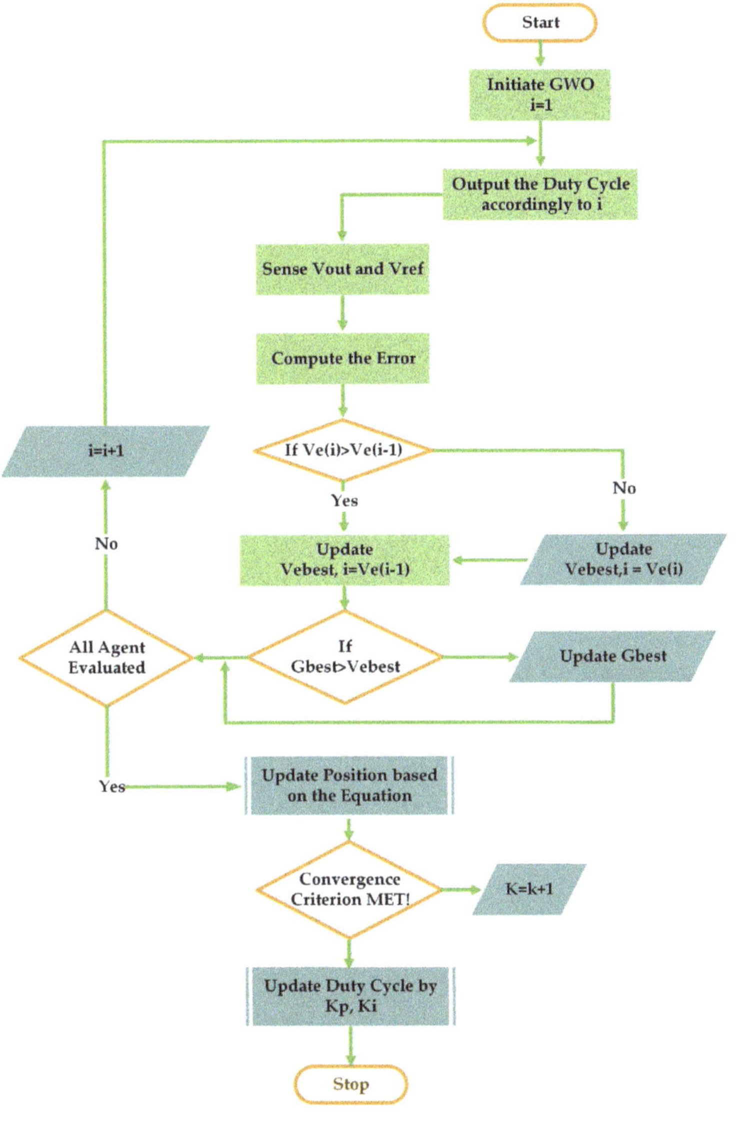

Figure 3. Flow chart of the Grey Wolf Optimizer (GWO)-based Maximum Power Point Tracking (MPPT) algorithm.

The duty cycle (D) of the switching pulse of the modified Luo converter represents the Grey wolf. Therefore, we can rewrite Equation (17) as:

$$D_i(k+1) = D_i(k) - A \times D \tag{21}$$

The objective function of the proposed GWO algorithm is given by:

$$V_{act}(D_i(k)) > V_{ref}(D_i(k-1)) \quad (22)$$

where V indicates the output voltage, d is the duty cycle, 'i' is the number of current Grey wolves and k is the number of iterations. For the MPPT tracking presented in this paper, the duty cycle of the DC-DC converter is referred to like the positions of the wolves in each state, and during iteration, the best value is considered as the K_P, K_i values of the PI controller.

5. Simulation and Experimental Results of the Proposed System

The proposed method is simulated using MATLAB. Here, in the DC-DC converters, IGBTs are utilized to reduce the switching power losses. The MPPT algorithm regulates the control signal produced by the PI controller, PWM pulses for the LUO converter and the input to the three-phase inverter. The proposed method is developed, modeled and simulated under starting and steady-state conditions.

5.1. Simulation Results

5.1.1. Performance of the Converter

The solar panel is consistently operated at the minimum and maximum power level utilizing the proposed GWO optimized algorithm. The tracking of the maximum power is purposely increased at the start in order to achieve the smooth start of the BLDC drive. The perturbation size is appropriately selected in order to avoid oscillation during the peak powerpoint. The DC link voltage across the capacitor is supplied to the VSI for the smooth start of the BLDC drive. The solar panel is operated under the conditions of dynamic variation due to changes in the irradiation level. Using the GWO algorithm, the proportional and integral gain is varied as it is mainly designed to obtain an excellent dynamic response under conditions of uncertainty and external disturbances. The GWO-based closed-loop control algorithm tracks the maximum power from the system. The output voltage (see in Figure 4a) and current (see in Figure 4b) waveforms of the solar PV panel are shown in Figure 4.

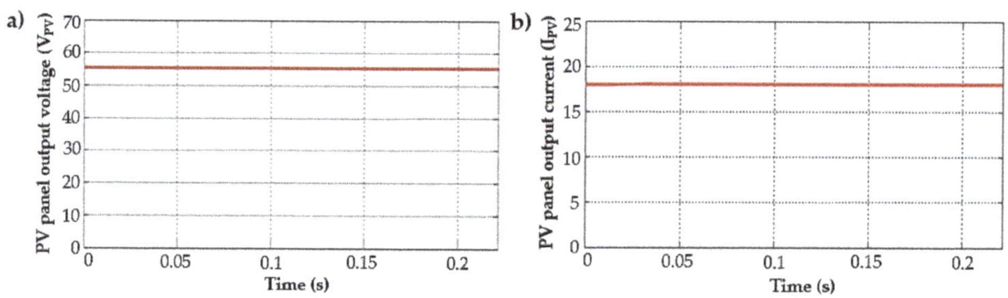

Figure 4. Solar Photovoltaic (PV) panel output. (**a**) Output voltage in Volts (V); (**b**) Output current in Amps (A).

From Figure 4, it is observed that the output voltage from the PV panel is approximately 55 Volts, and the output current generated by the PV panel is 17 Amps. Here, the PV panel is integrated with the MPPT technique. For prove the effectiveness of the proposed controller in MPPT tracking, a comparative study was done using various MPPT algorithms.

In Figure 5, the output waveform of the modified LUO converter using the GWO algorithm is shown. Initially, a peak is observed, and later at the time > 0 s the voltage value settles to a value of approximately 300 Volts. The output voltage produced by the proposed modified LUO converter is compared with the P&O and Fuzzy logic algorithms.

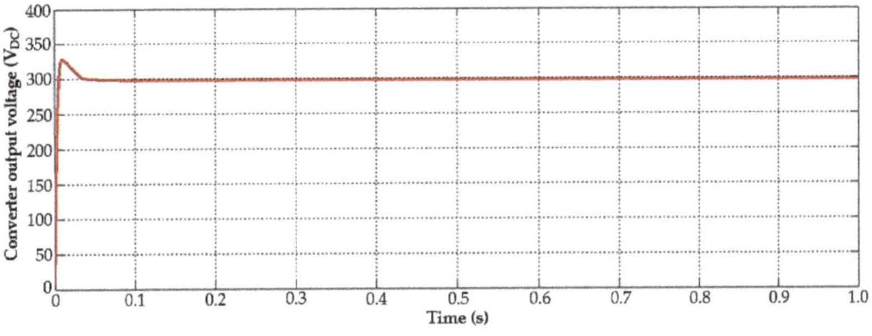

Figure 5. The output voltage waveform of the modified LUO converter using the GWO algorithm.

The input and output currents drawn by the proposed system are shown in Figure 6a,b, respectively. Here, the current ripples are maintained within the permissible limit. The three-phase output voltage waveform of the voltage source inverter is shown in Figure 7.

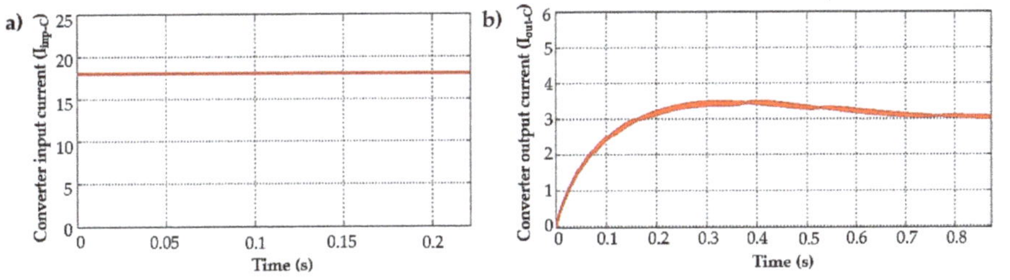

Figure 6. Current waveforms of the modified LUO converter using the GWO algorithm. (**a**) Converter input current; (**b**) Converter output current.

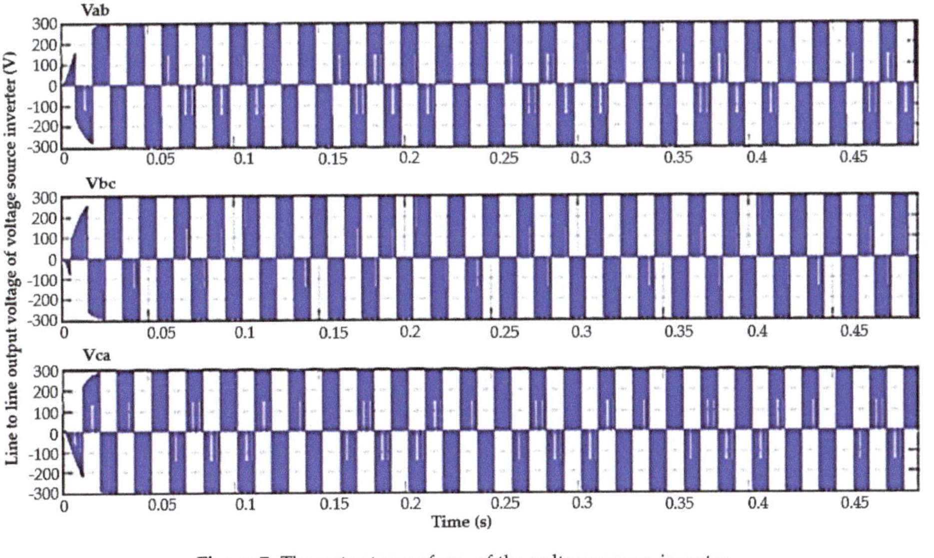

Figure 7. The output waveform of the voltage source inverter.

5.1.2. BLDC Motor Pump Performance

Here, the starting and steady-state performances of the BLDC motor pump are exhibited. If there is any change in the level of irradiance, all the parameters in the BLDC motor pump vary according to the irradiance level. Usually, the BLDC runs above 1100 rpm. However, the required speed to pump water is below this speed at a low irradiance level of a minimum of 200 W/m^2. The proposed work is not intended to find the one among the best optimization algorithms, rather than to prove that the optimization technique can outperform the competitive fuzzy and elementary P&O algorithms. The Luo converter and its modified versions are considered to measure the comparative performance of the BLDC motor. Figure 8 shows the output current waveforms of the BLDC motor for phases A, B and C, respectively. In Figure 9, the back emf of the BLDC motor is shown. The back emf cuts the magnetic flux and opposes the current flowing through the conductor. The starting value of the back emf is zero, and the values depend on the speed of rotation of the armature conductor.

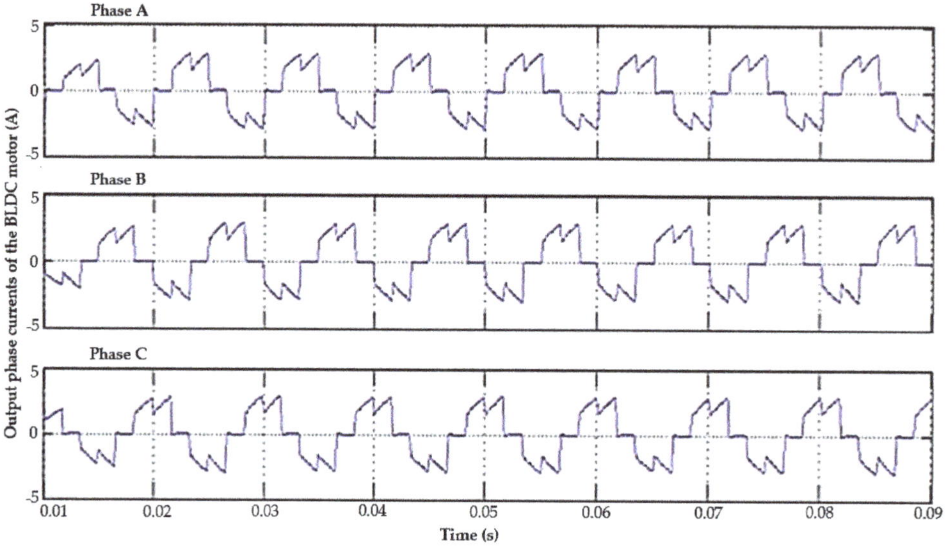

Figure 8. Output current waveform of the brushless DC (BLDC) motor.

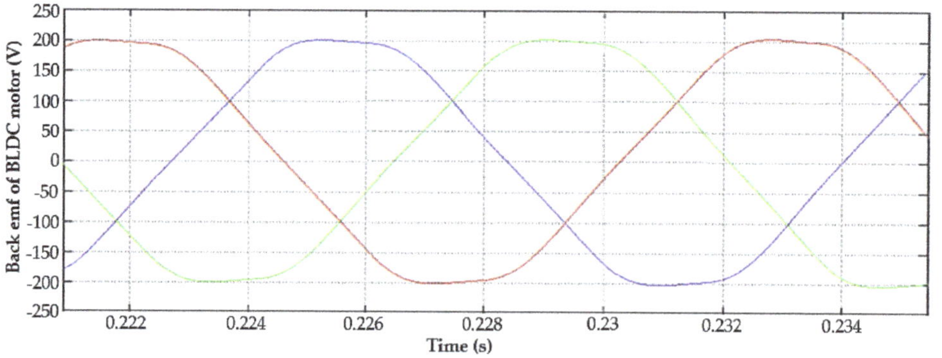

Figure 9. The back emf of the BLDC motor.

Figure 10 shows the starting and steady-state behaviour of the BLDC motor. The speed and electromagnetic torque reach their steady-state value when the maximum power is tracked by the GWO algorithm.

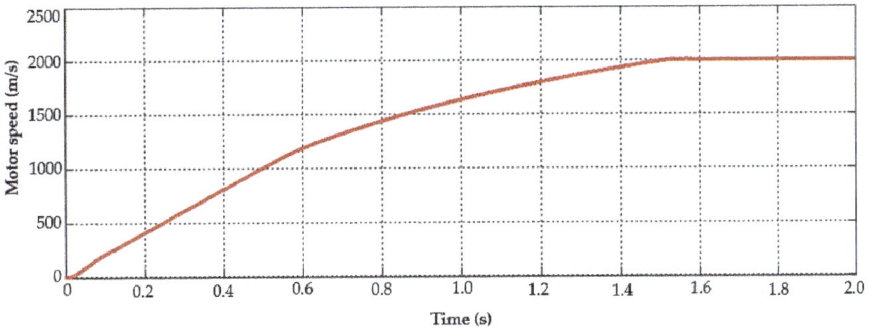

Figure 10. Steady-state response of the BLDC motor speed.

Figure 11 shows the torque waveform of the BLDC motor before using the LUO converter. Here, the starting torque is high and is equal to 6 Nm, and the steady-state torque occurs at 0.12 s with a ripple of 5.5 Nm. Figure 12 shows the torque characteristics of the BLDC motor after using the LUO converter, as a result of which the ripples are reduced.

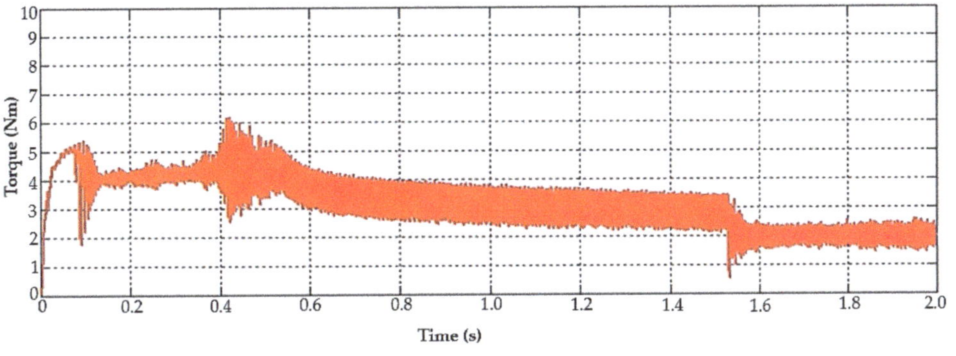

Figure 11. Output load torque characteristics of the BLDC motor before using the LUO converter.

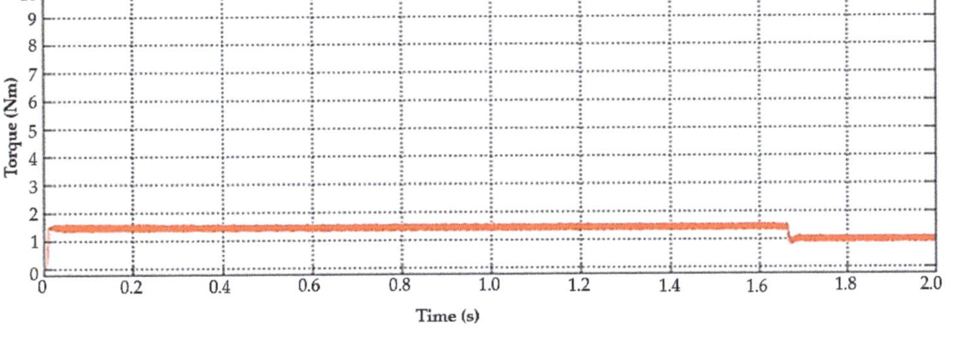

Figure 12. Output load torque characteristics of the BLDC motor after using the LUO converter.

The comparative results are given in Table 2, where it is seen that the proposed system is more efficient when compared to other MPPT algorithms and converter topologies, with an efficiency of 97.8%. The settling time and the output voltage gain are also better when compared to other methods. Usually, the optimization technique consumes significantly more time, which is, however, less than the manual tuning of the PI controller. Table 2 mentions the time taken by the system to settle after the tuning process is completed. The literature also shows that for a particular PI controller, it takes 10 s, whereas it is reduced to 4 s when using an optimization technique.

Table 2. Comparison of the MPPT methods for the elementary and modified LUO converters with $V_{in} = 56$ V.

MPPT Methods /Converter.		Steady-State Settling Time (s)	Output Voltage (V)	Efficiency (%)
P&O	Elementary LUO Converter	0.02	280	86
	Modified LUO Converter	0.019	283	88
Fuzzy	Elementary LUO Converter	0.018	285	92
	Modified LUO Converter	0.016	289	94
GWO	Elementary LUO Converter	0.01	293	95.6
	Modified LUO Converter	0.009	298	97.8

Figure 13 shows the output voltage comparison of different MPPT methods. As the progression reaches the proposed method, the performance (higher output voltage) of the developed method (modified LUO converter with GWO) shows more improvement than the existing approaches. Figure 14 shows the efficiency comparison of the different MPPT methods. As the progression reaches the proposed method, the performance (higher efficiency) of the developed method (modified LUO converter with GWO) shows more improvement than the existing approaches.

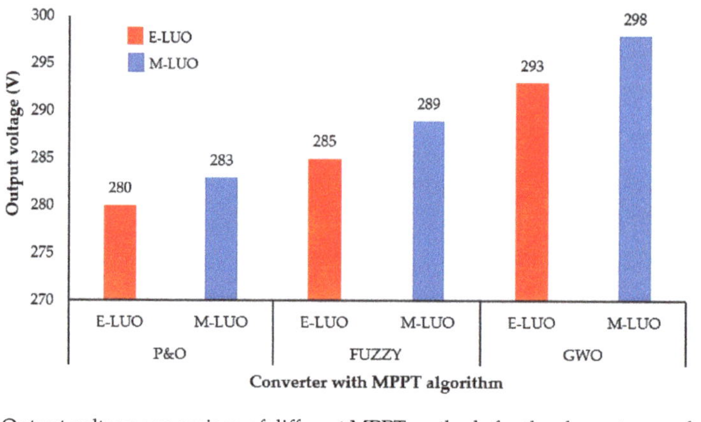

Figure 13. Output voltage comparison of different MPPT methods for the elementary and modified LUO converters (E-LUO and M-LUO).

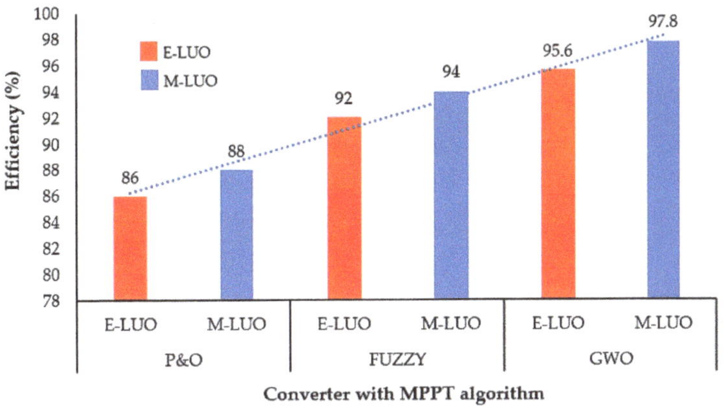

Figure 14. Efficiency comparison of the different MPPT methods for the elementary and modified LUO converters (E-LUO and M-LUO).

5.2. Experimental Results

The proposed method was verified experimentally by conducting a real-time test study. A practical setup was developed: first, the LUO converter and controller circuit designs were developed as demanded by the simulation requirements. The experimental setup shown in Figure 15 includes this designed controller and converter, along with the few additional components like the voltage source inverter, autotransformer, rheostat and BLDC motor. In Figure 16a, the output voltage of the PV panel is shown. This voltage consists of higher-order ripple contents which are oscillatory. For reduce this problem, the voltage is fed to the modified LUO converter. Figure 16b shows the modified LUO converter output voltage. This constant and ripple-free voltage is achieved using the LUO converter and its optimized MPPT algorithm. This constant voltage is fed to the three-phase inverter which converts the DC voltage into AC voltage.

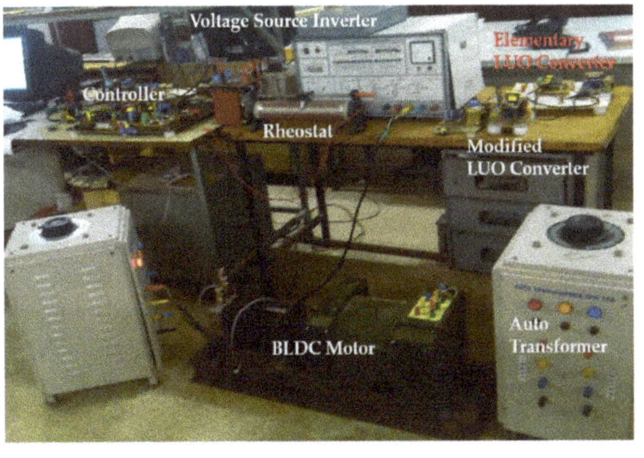

Figure 15. Experimental setup of the proposed system.

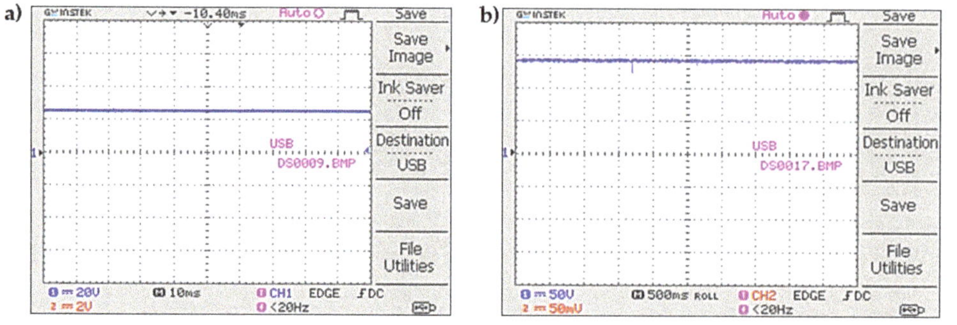

Figure 16. (**a**) Solar panel output voltage waveform; (**b**) Modified LUO converter output voltage waveform.

Figure 17 shows the PWM pulses fed to the switching devices of the voltage source inverter. The PI controller manages the pulse width for speed compensation. The dsPIC produces PWM pulses with a switching frequency of 10 kHz.

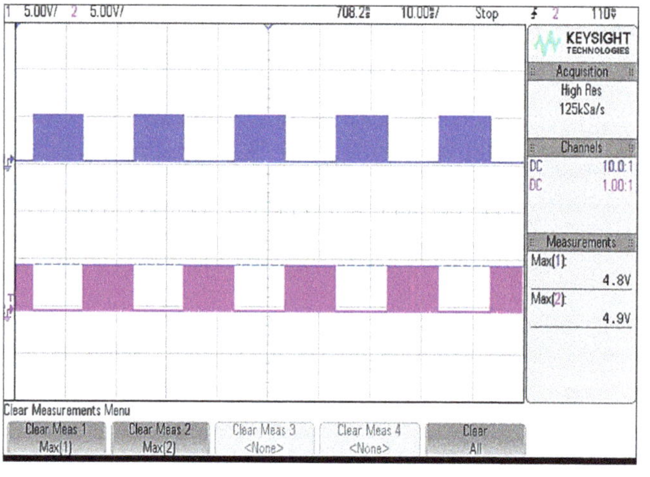

Figure 17. Voltage source inverter Pulse Width Modulation (PWM) pulses.

Figure 18 shows the position of the rotor signal from the three hall effect sensors which are mounted on the stator. Based on the rotor position signals, the electronic commutator switches are energised. Figure 19 shows the output voltage from the three-phase voltage source inverter, which also includes back emf. Due to the presence of this back emf, the supply voltage is trapezoidal.

Figure 20 shows the BLDC motor output current waveform. The modified LUO converter maintains the current within a specific limit. Figure 21 illustrates the speed waveform of the BLDC motor using the closed-loop PI control technique. Here, the reference speed and actual speed are compared and shown in the graph. The PI controller makes the speed constant in terms of the load.

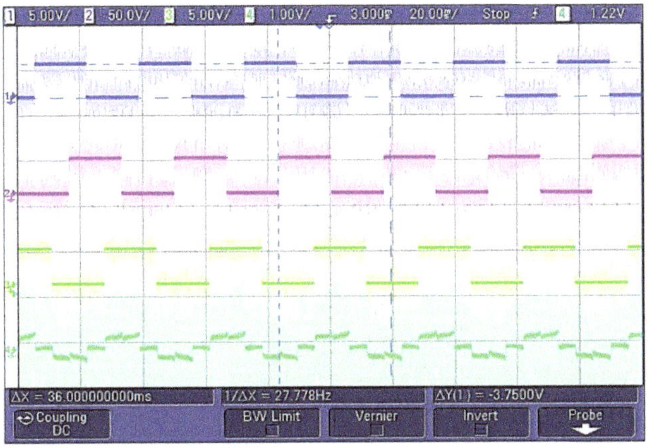

Figure 18. Hall sensor signals and stator current waveform of the BLDC motor.

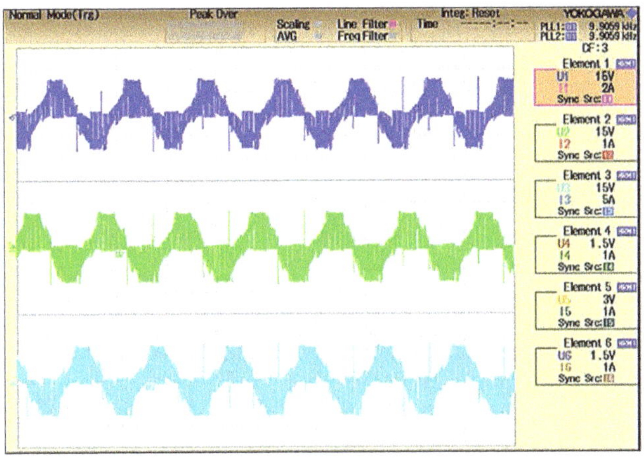

Figure 19. The output voltage waveform of the three-phase inverter.

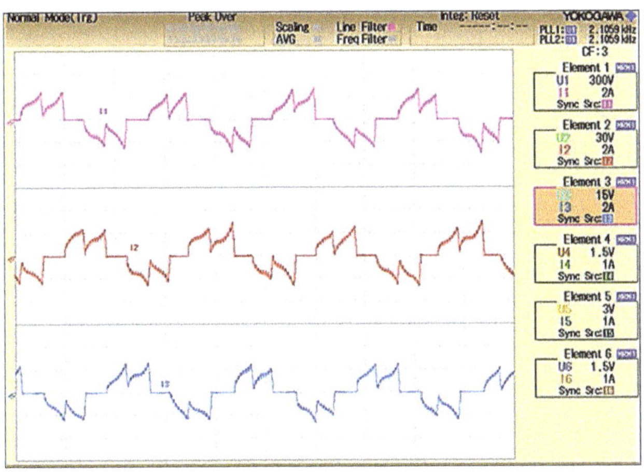

Figure 20. Voltage source inverter-fed BLDC motor output current waveform.

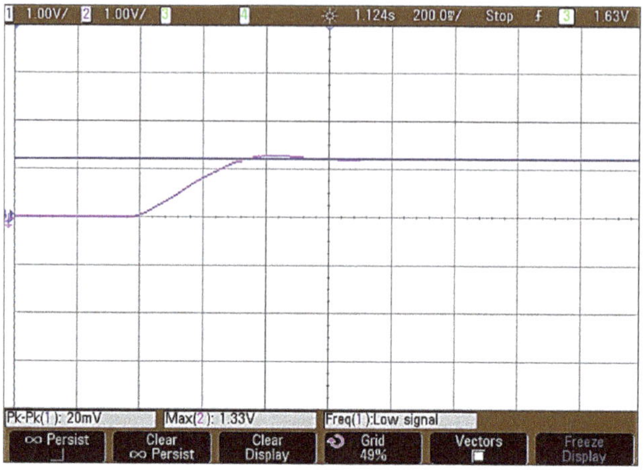

Figure 21. Speed waveform of the BLDC motor.

6. Conclusions

The modified LUO converter-fed VSI-BLDC motor with a solar PV array for a water pumping application was developed using MATLAB/Simulink and the Sim-power system toolbox, whose suitability was exhibited by simulated results. Initially, to meet the various desired objectives, the developed system was logically designed.

The proposed method was simulated and modelled to evaluate the performance under initial and steady-state conditions. The integration of the modified LUO converter was justified by evaluating its performance with the SPV array-based water pumping BLDC motor drive. The developed method availed many desired operations such as the absorption of MPP from the SPV array, BLDC motor soft starting, the fundamental switching frequency of the VSI (resulting in low switching losses), a reduction in the switch stress and the continuous conduction mode of the modified LUO converter for stable operation. Even at a very low solar irradiance, the developed method was operated successfully.

The GWO-optimized algorithm optimally extracted maximum power from the PV system when compared to the P&O and Fuzzy logic MPPT algorithms. The scaling-up of the power rating of the motor and converter did not affect the performance of the pumping system unless the incoming torque of the motor was not affected by the converter. Hence, the proper design of the converter ensured the overall performance of the system. The system may be economical if it is operating with better performance; as stated earlier, the incoming torque has a direct effect on the performance, which affects the overall economic value. While scaling up the system, considering the performance based on the systems' service life is essential, and this will allow us to make decisions on when to change the converter if any operating issues occur. The regular maintenance of the converters may ensure economic gains from the system.

Furthermore, from the results, it was seen that the proposed system was more efficient when compared to other MPPT algorithms and converter topologies, with a maximum efficiency of 97.8%. In comparison, the average performance shown by the Fuzzy optimized modified Luo converter was 94%, and the lowest performance, about 86%, as shown by the conventional P&O algorithm. The settling time was also better when compared to other methods, with a minimum settling time of 0.009 s for GWO and of 0.02 s for the P&O algorithm. Similarly, the voltage gain of the proposed method was higher, with a voltage of 298 V, while for Fuzzy and P&O it was 289 V and 280 V. The experimental outcomes were validated with the simulated results.

Furthermore, this work plans to extend the performance analysis of the BLDC pumping motor for a higher-rated three-phase power system with the proposed converter.

Author Contributions: Data curation, A.D.G.J., and M.S.P.S.; Formal analysis, A.D.G.J., M.S.P.S., and N.M.K.; Funding acquisition, U.S., and S.P.; Methodology, A.D.G.J., and M.S.P.S.; Resources, M.S.P.S., and N.M.K.; Software, A.D.G.J.; Supervision, M.S.P.S.; Visualization, N.M.K., and S.P.; Writing—original draft, A.D.G.J.; Writing—review & editing, M.S.P.S., N.M.K., U.S., and S.P. All authors have read and agreed to the published version of the manuscript.

Funding: This research received no external funding.

Acknowledgments: The authors like to express their sincere gratitude to the Department of Energy Technology, Aalborg University, Esbjerg, Denmark, and Renewable Energy research lab, College of Engineering, Prince Sultan University, Riyadh, Saudi Arabia, for providing technical inputs.

Conflicts of Interest: The authors declare no conflict of interest.

References

1. Bist, V. An Adjustable Speed PFC Bridgeless-SEPIC. In Proceedings of the 2015 IEEE Energy Conversion Congress and Exposition (ECCE), Montreal, QC, Canada, 20–24 September 2015; pp. 4886–4893.
2. Hong, D.K.; Woo, B.C.; Koo, D.H.; Seo, U.J. A single-phase brushless DC motor with improved high efficiency for water cooling pump systems. *IEEE Trans. Magn.* **2011**, *47*, 4250–4253. [CrossRef]
3. Chand, V.; Kalamkar, V.R. Solar photovoltaic water pumping system—A comprehensive review. *Renew. Sustain. Energy Rev.* **2016**, *59*, 1038–1067.
4. Singh, S.; Singh, B. Power factor correction in permanent magnet brushless DC motor drive using single-phase Cuk converter. *J. Eng. Sci. Technol.* **2010**, *5*, 412–425.
5. Shi, T.; Guo, Y.; Song, P.; Xia, C. A New Approach of Minimizing Commutation Torque Ripple for Brushless DC Motor Based on DC–DC Converter. *IEEE Trans. Ind. Electron.* **2010**, *57*, 3483–3490. [CrossRef]
6. Hendershot, J.R.; Miller, T.J.E. *Design of Brushless PM Motors*; Magna Publishing: Hillsborough, OH, USA, 1994.
7. Mandel, Y.; Weiss, G. Reduction of torque ripple in brushless DC motor drives. *IFAC Proc. Vol.* **2009**, *42*, 84–89. [CrossRef]
8. Bahari, N.B.; Jidin, A.B.; Abdullah, A.R.B.; Othman, M.N.B.; Manap, M.B. Modeling and simulation of torque hysteresis controller for brushless DC motor drives. In Proceedings of the 2012 IEEE Symposium on Industrial Electronics and Applications, Bandung, Indonesia, 23–26 September 2012; pp. 152–155.
9. Mohanty, S.; Subudhi, B.; Member, S.; Ray, P.K. A New MPPT Design Using Grey Wolf Optimization Technique for Photovoltaic System Under Partial Shading Conditions. *IEEE Trans. Sustain. Energy* **2015**, *7*, 1–8. [CrossRef]
10. G, D.; Singh, S.N. Selection of non-isolated DC–DC converters for solar photovoltaic system. *Renew. Sustain. Energy Rev.* **2017**, *76*, 1230–1247. [CrossRef]
11. Singh, B.; Singh, S.; Chandra, A.; Al-Haddad, K. Comprehensive study of single-phase AC-DC power factor corrected converters with high-frequency isolation. *IEEE Trans. Ind. Inform.* **2011**, *7*, 540–556. [CrossRef]
12. Zhang, X.; Lu, Z. A new BLDC motor drives method based on BUCK converter for torque ripple reduction. In Proceedings of the 2006 CES/IEEE 5th International Power Electronics and Motion Control Conference, Shanghai, China, 14–16 August 2006; Volume 3, pp. 1631–1635.
13. Richter, M. Zeitverschwendung. *Dtsch. Arztebl. Int.* **2017**, *114*, A2358.
14. Singh, B.; Bist, V. A BL-CSC converter-fed BLDC motor drive with power factor correction. *IEEE Trans. Ind. Electron.* **2015**, *62*, 172–183. [CrossRef]
15. Singh, P.K.; Singh, B.; Bist, V. PFC converter based power quality improvement and ripple current minimization in BLDC motor drive. In Proceedings of the 2016 IEEE 6th International Conference on Power Systems (ICPS), New Delhi, India, 4–6 March 2016; pp. 1–6.
16. Bist, V.; Singh, B. PFC Cuk converter-fed BLDC motor drive. *IEEE Trans. Power Electron.* **2015**, *30*, 871–887. [CrossRef]
17. Kumar, R.; Singh, B. Buck-Boost Converter Fed BLDC Motor Drive for Solar PV Array Based Water Pumping. In Proceedings of the 2014 IEEE International Conference on Power Electronics, Drives and Energy Systems (PEDES), Mumbai, India, 16–19 December 2014; p. 4.
18. Kumar, R.; Singh, B. BLDC Motor-Driven Solar PV Array-Fed Water Pumping System Employing Zeta Converter. *IEEE Trans. Ind. Appl.* **2016**, *52*, 2315–2322. [CrossRef]

19. Berkovich, Y.; Madar, R.; Axelrod, B.; Twina, A. Improved Luo converter modifications with increasing voltage ratio. *IET Power Electron.* **2015**, *8*, 202–212. [CrossRef]
20. Kumar, R.; Singh, B. Solar PV array fed cuk converter-VSI controlled BLDC motor drive for water pumping. In Proceedings of the 2014 6th IEEE Power India International Conference (PIICON), Delhi, India, 5–7 December 2014; pp. 1–31.
21. Sivakumar, S.; Sathik, M.J.; Manoj, P.S.; Sundararajan, G. An assessment on performance of DC-DC converters for renewable energy applications. *Renew. Sustain. Energy Rev.* **2016**, *58*, 1475–1485. [CrossRef]
22. Oshaba, A.S.; Ali, E.S.; Elazim, S.M.A. MPPT Control Design of PV Generator Powered DC Motor-Pump System based on Artificial Bee Colony Algorithm. *WSEAS Trans. POWER Syst.* **2016**, *11*, 190–198.
23. El-samahy, A.A.; Shamseldin, M.A. Brushless DC motor tracking control using self-tuning fuzzy PID control and model reference adaptive control. *Ain Shams Eng. J.* **2018**, *9*, 341–352. [CrossRef]
24. Premkumar, K.; Manikandan, B.V. Engineering Science and Technology, an International Journal Bat algorithm optimized fuzzy PD based speed controller for brushless direct current motor. *Eng. Sci. Technol. Int. J.* **2016**, *19*, 818–840.
25. Depuru, S.R.; Mahankali, M. Boost converter Fed High Performance BLDC Drive for Solar PV array Powered Air Cooling System. *Power Eng. Electr. Eng.* **2017**, *15*, 154–168. [CrossRef]
26. Akkaya, R.; Kulaksiz, A.A.; Aydoğdu, Ö. DSP implementation of a PV system with GA-MLP-NN based MPPT controller supplying BLDC motor drive. *Energy Convers. Manag.* **2007**, *48*, 210–218. [CrossRef]
27. Femia, N.; Petrone, G.; Spagnuolo, G.; Vitelli, M. Optimization of perturb and observe maximum power point tracking method. *IEEE Trans. Power Electron.* **2005**, *20*, 963–973. [CrossRef]
28. Killi, M.; Samanta, S. Modified perturb and observe MPPT algorithm for drift avoidance in photovoltaic systems. *IEEE Trans. Ind. Electron.* **2015**, *62*, 5549–5559. [CrossRef]
29. Hohm, D.P.; Ropp, M.E. Comparative study of maximum power point tracking algorithms using an experimental, programmable, maximum power point tracking test bed. In Proceedings of the Conference Record of the Twenty-Eighth IEEE Photovoltaic Specialists Conference—2000, Anchorage, AK, USA, 15–22 September 2000; pp. 1699–1702.
30. Terki, A.; Moussi, A.; Betka, A.; Terki, N. An improved efficiency of fuzzy logic control of PMBLDC for PV pumping system. *Appl. Math. Model.* **2012**, *36*, 934–944. [CrossRef]

applied sciences

MDPI

Case Report

Analysis of Salt Mixture Contamination on Insulators via Laser-Induced Breakdown Spectroscopy

Shan Lu [1], Xinwei Wang [1], Tianzheng Wang [1], Xinran Qin [2], Xilin Wang [2,*] and Zhidong Jia [2]

[1] Shanxi Electric Power Research Institute, Taiyuan 030000, China; ls8760033@163.com (S.L.); wxw7912@163.com (X.W.); wtz2000@163.com (T.W.)

[2] Engineering Laboratory of Power Equipment Reliability in Complicated Coastal Environments, Tsinghua Shenzhen International Graduate School, Shenzhen 518055, China; txr19@mails.tsinghua.edu.cn (X.Q.); jiazd@sz.tsinghua.edu.cn (Z.J.)

* Correspondence: wang.xilin@sz.tsinghua.edu.cn

Received: 4 March 2020; Accepted: 5 April 2020; Published: 10 April 2020

Abstract: The composition of contamination deposited on transmission line insulators can affect their surface flashover voltage. Currently, there is no rapid on-line method to detect this contamination composition in power grids. In this paper, we applied laser-induced breakdown spectroscopy (LIBS) to analyze contamination on insulator surfaces. Usually, Na and Ca salts are found in contamination along with various sulfate, carbonate, and chloride compounds. As an element's detection method, LIBS can only measure a certain element content, for example, Ca. The mixture of various compounds with the same cations can influence the LIBS signal. The influence of mixing ratios on the calibration curves and relative spectral intensity was studied via LIBS. Na_2CO_3, $NaHCO_3$, $CaSO_4$, and $CaCO_3$ samples containing different proportions of Na and Ca were prepared. The linear correlation coefficients (R2) for the Na and Ca calibration curves generated using various mixing ratios were analyzed. The results showed that the mixture ratio did not dramatically affect the linear calibration curves for mixtures containing the same cations. This finding may significantly reduce the difficulty of applying LIBS analysis for complex contamination on insulators. The laser energy density had effects on the spectral characteristics of the measured elements. The partial least-square regression (PLSR) model can improve the accuracy of Na and Ca prediction.

Keywords: laser-induced breakdown spectroscopy; contamination; insulators; salt; calibration curves

1. Introduction

Insulators are key components on transmission lines because they provide sufficient insulation space between the conductor line and the ground. Usually, insulators are made of glass, ceramics, and polymers [1]. During prolonged outdoor service, contamination or pollution is inevitably deposited on insulator surfaces and consists of dust, particles, and other airborne substances. This contamination cannot be avoided because transmission lines are the main energy transport channel for every grid. Under dry conditions, this contamination is relatively safe. Still, in heavy fog or rainy weather, the soluble components in the contamination can dissolve in the water and form conductive paths on the insulator surfaces, thus reducing the flashover voltage and causing discharge and even accidental flashover. In China, there were several power outage incidents in the 1990s due to transmission flashover caused by contamination deposited on insulators in heavy fog weather [2]. This phenomenon was deemed surface flashover in high-voltage engineering and has become a topic of interest in this field. Thus, the detection of contamination composition has become an important task in the regular operation of the State Grid.

The main factors affecting the surface discharge activity and flashover process of pollution insulators are the equivalent salt deposit density (ESDD) level and relativity air humidity (RH). Higher

ESDD or higher RH will lead to more intense discharges and lower flashover voltage [3,4]. Contaminant composition, especially soluble salts (including common salts such as NaCl, $NaNO_3$, Na_2SO_4, $MgC1_2$, $Mg(NO_3)_2$, $MgSO_4$, $CaCl_2$, $Ca(NO_3)_2$, and $CaSO_4$.), and material properties can affect the flashover process, which may result in an excess or lack of insulation during insulation design. A previous study showed that the tendency for an insulator to flashover is dependent on the type of contaminant, as well as on the equivalent salt density [5]. At present, the only way to analyze these salt compositions is by collecting the contamination during a power outage and taking the sample back to the lab for analysis with chemistry equipment. It needs a new method that can be used for on-site and on-line detection of contamination in insulators.

In recent years, laser-induced breakdown spectroscopy (LIBS) has developed rapidly in power engineering because of its advantages, such as no sample preparation, harmless sampling, and fast detection speed. LIBS has been widely applied as a tool for mineral analysis, archaeology, biomedical analysis, aerospace exploration, etc. [6–8]. The basic experimental process involves exciting plasma with a high-energy laser and collecting its characteristic spectral information to obtain the elemental composition and surface condition of the target material.

In recent years, LIBS has also played an important role in the online detection of power equipment status in high-voltage engineering. Huan et al. [9] applied LIBS technology to detect the vacuum degree of vacuum breakers. Based on the fact that the intensity of characteristic spectral lines of different elements, including Cu, O, N, and H, will change with the vacuum degree of the vacuum breaker, the vacuum degree can be predicted. In their study, principal component analysis (PCA) and the artificial neural network (ANN) model were used to optimize the spectral line selection, and the accuracy of the final model reached 96.67% [10,11]. Our previous work proved that the number of laser pulses has a linear relationship with the depth of ablation. Thus, the distribution of elements along with depth on one test point of silicon rubber was obtained, and the thickness of the aging layer could be calculated accordingly [12–14].

A remote laser-induced breakdown spectroscopy technique combined with a photometric device was proposed and demonstrated at the laboratory scale. It was used for the remote sensing and quantification of surface pollutants such as salt deposits on wind turbine blades from different standoff distances [15]. Nearly every element was studied with LIBS to obtain calibration curves for quantitative analysis of soils, rocks, compounds, and other materials [16–18]. However, on the surface of transmission line insulators, the contamination consists of various compounds that may have the same cations. It is necessary to determine the influence of various compound ratios and system parameters on the calibration results to improve the detection accuracy of LIBS for the online monitoring of insulator surface pollution. Calibration curves are crucial in determining the sample contents with LIBS. There have been many studies on the composition analysis of contamination, and we used artificial pollution and compressed it into pellets. In this work, compounds containing various ratios of Na and Ca were designed and analyzed to determine the effects of these mixture ratios on the LIBS calibration results.

2. Experiments

The experimental setup is depicted in Figure 1. The LIBS system consisted of a Q-switched laser (Beamtech Nimma-900), where the laser pulse duration was 10ns, shot-to-shot energy variation (RMS) ≤1%. We used an Avantes optical fiber spectrometer (190 nm to 650 nm) with 6 channels, the spectral resolution was 0.05nm, and a Stanford Research System Delay Generator SRS DG645. The laser was focused on the surface of the sample by a lens with a focal length of 100 mm. It was ablated operating at a wavelength of 1024 nm. The laser energy of the Nimma-900 could be adjusted from 1 J to 900 J. Laser fluence was determined at 1.9 J/cm^2 for 38 mJ, 3.8 J/cm^2 for 76 mJ, and 6.4 J/cm^2 for 128 mJ.

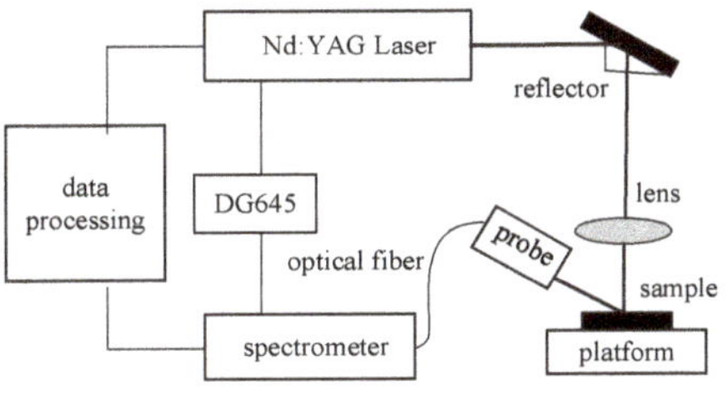

Figure 1. Laser-induced breakdown spectroscopy (LIBS) experimental setup.

All chemicals used in the experiments were acquired from Aladdin and were of analytical reagent grade. Mixed samples of Na_2CO_3 and $NaHCO_3$ were prepared with a series of Na concentration gradients. The total mass of each sample remained constant, and the mass of Na_2CO_3 and $NaHCO_3$ to be mixed was calculated separately according to the total concentration of Na set by the concentration gradient and the various mixing ratios of Na in Na_2CO_3 and $NaHCO_3$. The total concentration of Na for each sample varied from 27.4% to 43.4% because the mass percentage of Na in Na_2CO_3 was 43.4%, and that in $NaHCO_3$ was 27.4%. Similarly, the quantity of $CaSO_4$ and $CaCO_3$ to be mixed in each sample was obtained using the above method, and the Ca mass fraction of the mixtures ranged from 29.4% to 40%, corresponding to the mass percentages of Ca in $CaSO_4$ and $CaCO_3$, respectively. Varying ratios of Na_2CO_3, $NaHCO_3$ and $CaSO_4$, $CaCO_3$ were prepared and mixed, as shown in Table 1. The ratios of the Na and Ca mass fractions of Na_2CO_3:$NaHCO_3$ and $CaSO_4$:$CaCO_3$ are also shown in Tables 1 and 2. Pellets were formed from each sample by using a circle pellet forming press with a diameter of 8 mm under 5 tons of force, which improved the LIBS ablation of the sample.

Table 1. Different ratios of Na_2CO_3 and $NaHCO_3$, $CaSO_4$, and $CaCO_3$.

Sample Number	Na/%	Na_2CO_3/g	$NaHCO_3$/g	Weight/g	Ratio of Na (Na_2CO_3:$NaHCO_3$)
#1-1	27.4	0	1.500	1.5	0:1
#1-2	29.9	0.250	1.250	1.5	0.317:1
#1-3	32.7	0.500	1.000	1.5	0.792:1
#1-4	35.4	0.750	0.750	1.5	1.584:1
#1-5	38.0	1.000	0.500	1.5	3.168:1
#1-6	43.4	1.5	0	1.5	1:0
Sample Number	**Ca/%**	**$CaSO_4$/g**	**$CaCO_3$/g**	**Weight/g**	**Ratio of Ca ($CaSO_4$:$CaCO_3$)**
#2-1	29.4	1.5	0	1.5	1:0
#2-2	31.2	1.250	0.250	1.5	3.675:1
#2-3	32.9	1.000	0.500	1.5	1.470:1
#2-4	34.7	0.750	0.750	1.5	0.735:1
#2-5	36.5	0.500	1.000	1.5	0.368:1
#2-6	40.0	0	1.500	1.5	0:1

Table 2. Wavelength of typical spectral lines for sample by LIBS.

Element	Na I	Na I	Na I	Na I	Na I
Wavelength/nm	568.263	568.859	589.592	588.995	616.075
Element	Ca II	Ca II	Ca II	Ca II	Ca I
Wavelength/nm	315.886	317.933	393.366	396.846	422.672

For the LIBS analysis, the pellets were placed on an optical platform. The vertical distance between the optical platform and convex lens was adjusted, so the laser spot was focused on the sample surface. The delay time of the LIBS system was 3 μs, and the diameter of the laser spot was 0.8 mm. The dark spectrum captured by the spectrometer was acquired 50 times without laser ablation. Then, 5 points were selected randomly for each sample, and the spectral data for each point were obtained after 10 rounds of 10 Hz continuous ablation. The averaged background spectrum was subtracted from the processed spectral data, and the resulting spectra were averaged. The NIST database was used to determine spectral intensities for Ca, C, Na, etc. [19]. The above analyses were repeated for the two groups of samples, #1-1 through #1-6 and #2-1 through #2-6, at a laser fluence of 3.8 J/cm^2 and 6.4 J/cm^2, respectively. Due to the influence of Na matrix effects, LIBS analysis at 1.9 J/cm^2 was added for the analysis of samples #1-1 through #1-6. The elements were measured, as shown in Table 2.

3. Results and Discussion

3.1. Spectral Characteristics for Na_2CO_3 and $NaHCO_3$

Figure 2 shows the whole spectra for samples #1-1 to #1-6 at an energy intensity of 3.8 J/cm^2. The intensity of the spectra increased with increasing concentrations of Na. The Na content in Na_2CO_3 was higher than that in $NaHCO_3$ when the sample had the same amount of substance. Therefore, the spectral intensity of Na_2CO_3 was higher than that of $NaHCO_3$, which was demonstrated in the range of 588 nm to 589 nm. As Na_2CO_3 and $NaHCO_3$ were similar in element composition, except for H, the characteristic spectral wavelengths were very similar in these samples, i.e., either they were pure compounds (#1-1 and #1-6) or a mixture of two compounds (#1-2 through #1-5).

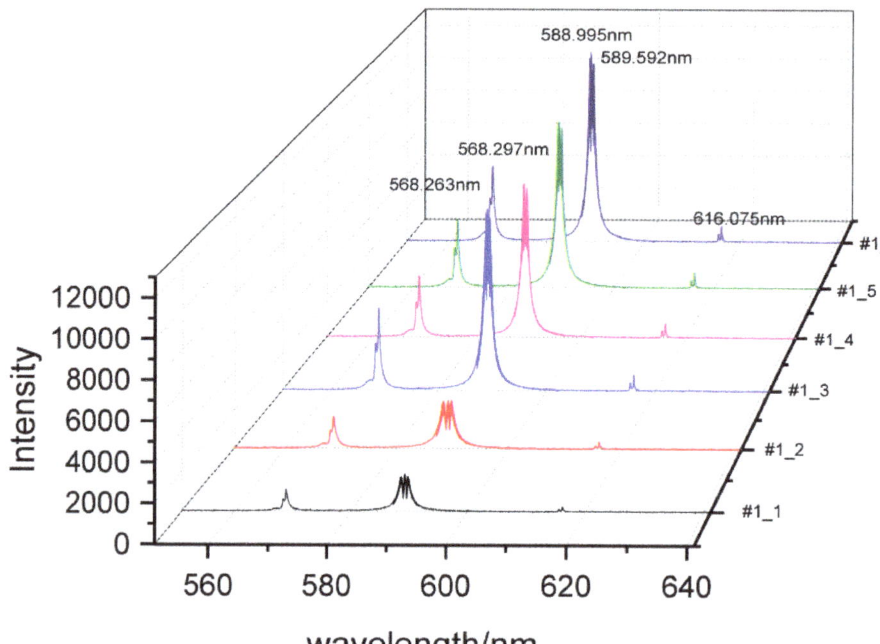

Figure 2. Spectra intensity of samples #1-1 to #1-6 at an energy intensity of 3.8 J/cm^2 shown in black, red, navy blue, green, light blue, and purple, respectively.

In Figure 3, the spectral resonance lines (588.995 nm, 589.592 nm) and nonresonant lines (568.297 nm, 568.859 nm) for Na were selected from Figure 2. Self-absorption was found in the resonance lines (588.995 nm, 589.592 nm) at a laser fluence of 3.8 J/cm^2, as shown by a significant decrease in signal intensity. The variation in the Na concentration in different samples contributed to the self-absorption phenomenon, i.e., the higher the concentration, the more notable the self-absorption. This susceptibility existed because the laser ablated the target surface, and plasma was generated by the trailing edge of the laser pulse and disappeared during condensation. Moreover, the temperature of the entire illuminant was not uniform. After Na atoms are excited by the emission wavelength, the generated photons were absorbed by other Na atoms when passing through the low-temperature region of the plasma, resulting in changes in the intensity and profile of the Na spectrum. The influence of Na concentration on the self-absorption effect could be determined by comparing the contours of the Na resonance lines.

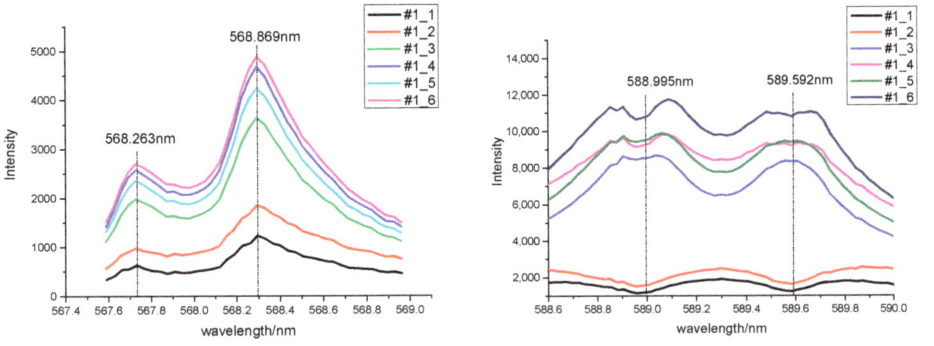

Figure 3. Emission intensity of Na in samples #1-1 through #1-6 measured at different wavelengths at an energy intensity of 3.8 J/cm^2. Spectra show the resonance lines (588.995 nm, 589.592 nm) and nonresonant lines (568.263 nm, 568.859 nm).

The relationship between the Na content, according to the atomic lines (Na I 588.995 nm, Na I 568.859 nm), and the average relative spectral intensity as a function of laser energy intensity is shown in Figure 4. As the laser energy density increased, the spectral intensity corresponding to each spectral line was enhanced significantly. At a laser fluence of 6.4 J/cm^2, as the concentration increased, the intensity first increased and then decreased, which was due to the strong self-absorption effect. This result demonstrated that the range of Na content was limited when using calibration curves to analyze the linear relationship between Na concentration and laser fluence.

However, reducing the laser energy density per pulse and selecting the ion line reduced the atomic emission intensity because of the weak self-absorption effect. It prolonged the dynamic range of the concentration measurements, which allowed the Na atomic lines (Na I 588.995 nm, Na I 568.859 nm) to maintain a good linear relationship at a laser energy of 1.9 J/cm^2. Therefore, when the concentration was held constant, reducing the laser energy, and selecting the proper Na ion line would extend the linear analysis range and improve the accuracy of Na determination using linear calibration curves. For the elements tested, a good linear relationship was the basis of the artificial data methods used to process the LIBS spectral data. This method could also be used as a preprocessing method when employing the artificial intelligence algorithm to determine the relationship between the concentration of other elements and spectral intensity.

The intensity and variation trends of spectral lines were considered to analyze the influence of the Na salt mixture on the calibration results. The linear calibration coefficient (R^2) for Na was obtained with a laser fluence of 3.8 J/cm^2. The R^2 value was greater than 0.7, as shown in Figure 5. Although the middle data point was higher in the spectrum corresponding to 588.995 nm and 589.592 nm, the spectral intensity and concentration corresponding to these data points also had linear relationships. The spectral data presented here represented the original data after subtracting the dark spectrum. The ratio of Na in samples #1-1 through #1-6 varied, but the calibration curves were similar to that of the pure compound. The mixing of various proportions of Na had minimal effects on the linear calibration model, which also indicated that the results of other complex artificial intelligence methods would not be affected. These results provided a convenient LIBS method for the on-line detection of elemental composition from multi-compound mixing.

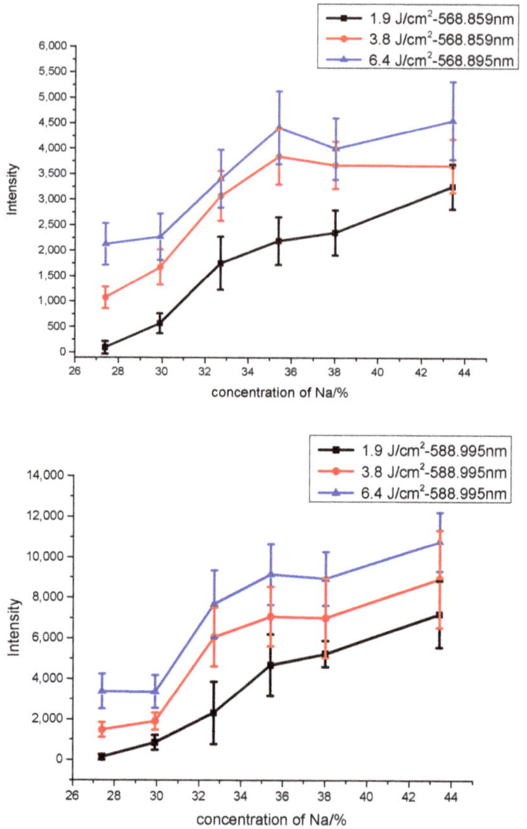

Figure 4. Emission intensity under varying concentrations of Na as a function of laser energy. Signals obtained from the Na I 588.995 nm, Na I 568.859 nm are shown.

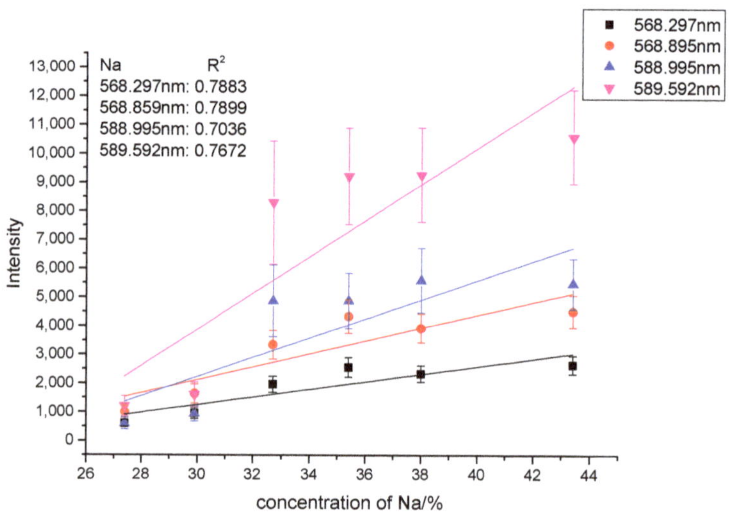

Figure 5. LIBS signal intensity of Na vs. Na concentration at a laser fluence of 3.8 J/cm^2.

PLSR (partial least squares regression) was applied in the LIBS spectroscopy analysis to obtain the calibration model through MATLAB. There were five standard samples in the experiment, four of which were used as the calibration sample set, and their spectral data were used to train the prediction model. Meanwhile, the remaining one is used to predict and verify the model.

To compare the prediction results of the model, we used the Pearson correlation coefficient(R) to measure the calibration effect of the model. The root mean square error (RMSE) was used to describe the prediction accuracy of the model. The closer the R was to 1, the better fitting effect of the calibration curve: the smaller the RMSE, the more accurate the quantitative effect of the model.

As shown in Figure 6, The red diamond data point indicates the predicted value. The degree of fit of the regression line to the observed values was higher than that of the univariate regression model.

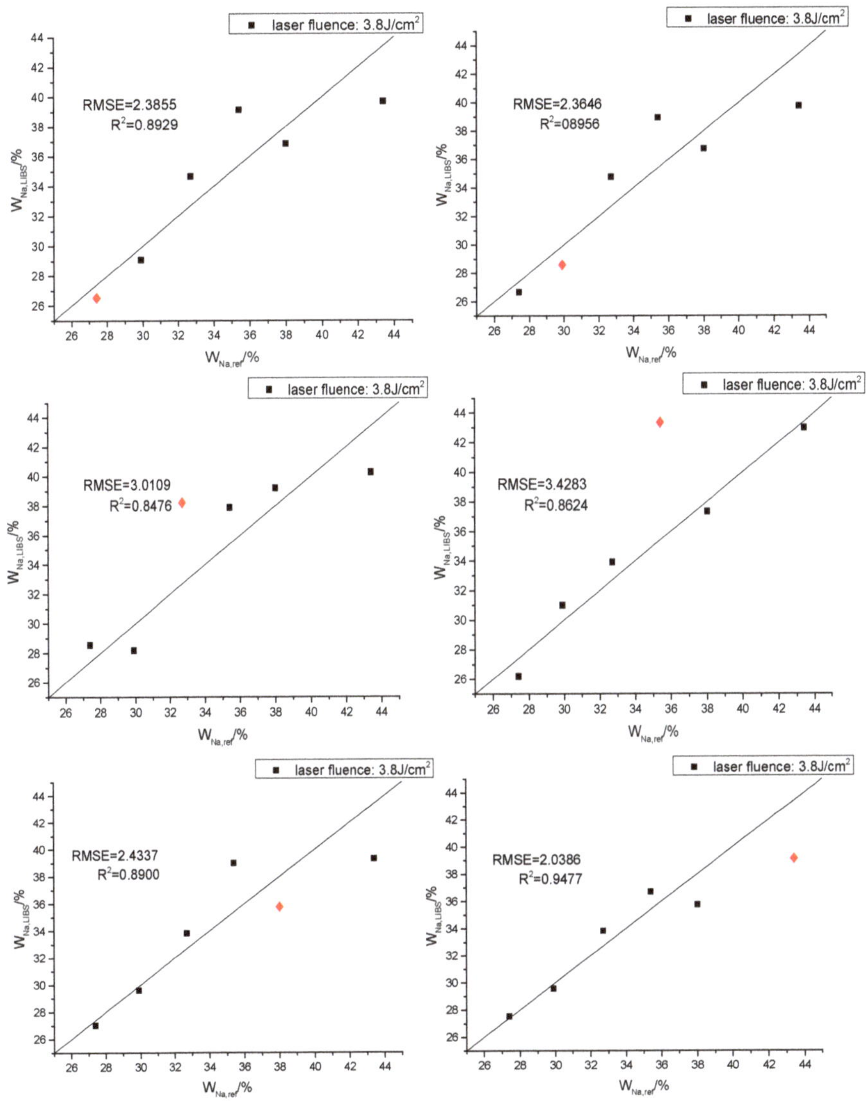

Figure 6. Results of Na data for Partial least squares regression (PLSR) (2 components).

3.2. Characteristic Spectra of $CaCO_3$ and $CaSO_4$

Under local thermal equilibrium conditions, the common method to calculate the plasma electron temperature is the Boltzmann method. The layout number on the atomic bound energy level satisfies the Boltzmann distribution, as shown in the following equation:

$$\ln\left(\frac{\lambda_{mn} I_{mn}}{hcg_m A_{mn}}\right) = -\frac{E_m}{k_B T_e} + \ln\frac{N(T)}{U(T)} \tag{1}$$

where m and n are the upper and lower energy levels of the spectral line transition, λ is transition wavelengths, A is spontaneous transition probability, I is the relative strength of the measured spectral line, Em and gm are the excitation energy and statistical weight of the m level, respectively. h, c, and kB are the Planck constant, the speed of light, and the constant number of Boltzmann, respectively. The electron temperature of the plasma can be derived from the slope of linear fitting [20,21].

In this experiment, three calcium atomic lines were selected. The relevant parameters of these lines are listed in Table 3. Figure 7 shows that as the laser energy fluence increases, the electron temperature increases. The electron temperature of the plasma was 1.38×10^4 K at a laser fluence of 3.8 J/cm^2 and 1.52×10^4 K at a laser fluence of 6.4 J/cm^2.

Table 3. Parameters of calcium atomic emission line.

Species	Wavelength/nm	E_m/cm^{-1}	A_{mn}/10^7s^{-1}	g_m
Ca II	392.065	148515	3.3	9
	504.133	135910	1.6	5
	534.516	156767	1.1	13

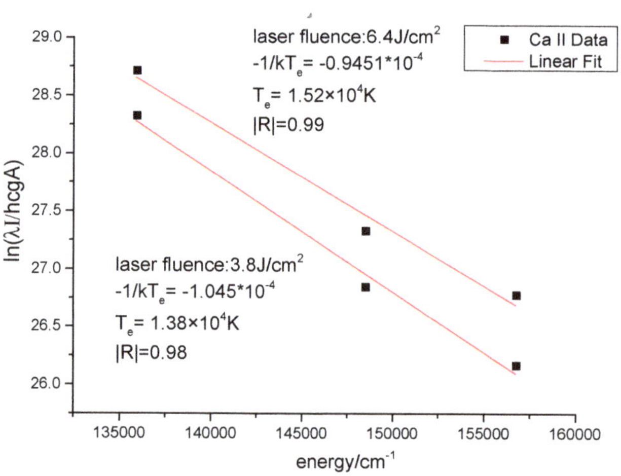

Figure 7. Boltzmann diagram of plasma electron temperature (obtained by 3 calcium atomic lines).

The characteristic spectral wavelengths of Ca in samples #2-1 through #2-6 were searched in the NIST database. According to the average relative spectral intensity of Ca at different wavelengths, linear calibration curves for spectral intensity and concentration were generated, and R^2 values were determined.

The results showed that in the Ca linear calibration curves, at a laser energy fluence of 3.8 J/cm^2, four spectral lines had R^2 values greater than 0.8, and 24 lines had R^2 values greater than 0.9. The maximum R^2 value was 0.972. Furthermore, there were ten calibration curves at the laser fluence of 6.4 J/cm^2 that led to an R^2 value greater than 0.9. The maximum R^2 value was 0.951. In general, the linearity of the laser energy intensity of 3.8 J/cm^2 had better linearity and a better different degree of increase in R^2 at 6.4 J/cm^2. The lines with R^2 values greater than 0.9 at 3.8 J/cm^2 and 6.4 J/cm^2 correspond to the linear calibration curves shown in Figure 8. The distribution of R^2 values greater than 0.9 as they relate to the wavelength at a laser fluence of 3.8 J/cm^2 and 6.4 J/cm^2, respectively, are also shown in Figure 8. The result of the PLSR model is shown in Figure 9. The R^2 values were 0.9813 and 0.9688, which were close to the maximum R^2 value in the Ca linear calibration curves.

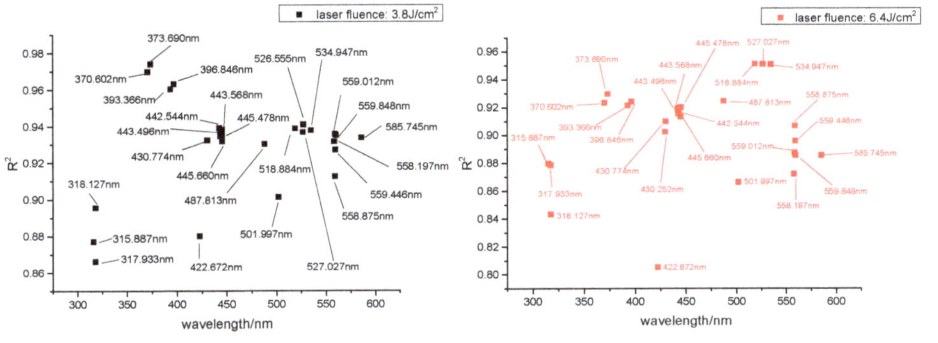

(**a**) R^2 of linear calibration curves at 3.8 J/cm^2. (**b**) R^2 of linear calibration curves at 6.4 J/cm^2.

Figure 8. R^2 values greater than 0.9 at various Ca spectral lines as a function of laser energy intensity. The two laser fluences are drawn in black (3.8 J/cm^2) and red (6.4 J/cm^2).

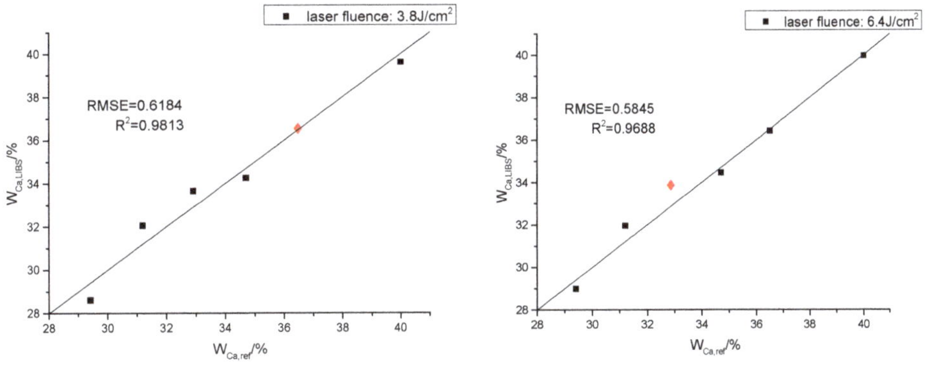

(**a**) R^2 of linear calibration curves at 3.8 J/cm^2. (**b**) R^2 of linear calibration curves at 6.4 J/cm^2.

Figure 9. Results of 5-fold cross-validation of Ca data for PLSR (2 components).

By comparing the linear relationship between the laser energy of 3.8 J/cm^2 and 6.4 J/cm^2, increasing the output energy was not determined to result in a higher number of spectral lines with good linearity. The appropriate laser energy should be selected according to the experimental results.

As shown in Table 1, the mixing ratio of the two Ca ion compounds in samples #2-1 through #2-6 was different, but the linear relationship between the concentration and the intensity was not affected. Increasing the number of compounds containing the same cation did not influence the concentration relationship in the LIBS analysis of Ca.

In the laser-induced plasma generation process, the target material and the ions and atoms excited by the gas molecules in the air together constituted a component in the laser-plasma. The pulsed laser ablated the surface of the sample, and the energy was sufficient to cause the surface temperature to rise, melt, and evaporate to cause ion bond rupture and the further formation of laser-plasma. Both the Ca ions in $CaCO_3$ and $CaSO_4$ were ionized after reaching laser energy sufficient to form an ionic state of Ca. When the ionized Ca acquired enough energy, it transitioned from the ground state to various excited states and then rapidly transitioned back to form atomic emission lines and ion lines. If the concentration of the ionic state was the same, then the atomic emission spectrum was not affected. This disregarded the compound type or mixture of the original state, which was similar to results from Na analysis.

4. Conclusions

In this paper, artificial contamination was used to study the effect of the salt mixture in LIBS signals, especially the calibration curves and relative spectral intensity of different Na and Ca samples. The results showed that the R^2 of the calibration curve between the element content and characteristic wavelength line intensity was affected minimally by the mixing ratio of different compounds with the same cations. Thus, it is beneficial in the LIBS test on-site, that we can get the content of cations regardless of their compounds while the cations would contribute most in the soluble contamination. The laser energy per pulse in the LIBS test would affect the spectral significantly both in line intensity and shape. By reducing the laser energy intensity and selecting the proper emission wavelength of the analytes, the effect of the self-absorption can be weakened, and the detection accuracy can be improved. The PLSR model was used in the data process, which could improve the accuracy of Na and Ca linear analysis.

Author Contributions: Conceptualization and formal analysis, S.L. and X.W.; investigation, X.Q.; resources, S.L.; data curation, S.L. and T.W.; writing—original draft preparation, X.Q.; writing—review and editing, X.W.; supervision, T.W. and Z.J. All authors have read and agreed to the published version of the manuscript.

Funding: This research was funded by the National Natural Science Foundation of China (51607101), Science and technology projects of Shanxi Electric Power Research Institute (SGSXDK00SPJS1900162).

Conflicts of Interest: The authors declare no conflict of interest.

References

1. Liu, Y.; Wu, Y.; Du, B. Dynamic formation mechanism of water droplet and induced surface discharges on silicone rubber composites. *High Volt.* **2019**, *4*, 59–64. [CrossRef]
2. Liang, X.; Wang, S.; Fan, J.; Guan, Z. Development of composite insulators in China. *IEEE Trans. Dielectr. Electr. Insul.* **1999**, *6*, 586–594. [CrossRef]
3. Mcelroy, A.J.; Lyon, W.J.; Phelps, J.D.; Woodson, H.H. Insulators with contaminated surfaces, part I: Field conditions and their laboratory simulation. *IEEE Trans. Power Appar. Syst.* **1970**, *8*, 1848–1858. [CrossRef]
4. Woodson, H.H.; Mcelroy, A.J. Insulators with contaminated surfaces, part II: Modeling of discharge mechanisms. *IEEE Trans. Power Appar. Syst.* **1970**, *8*, 1858–1867. [CrossRef]
5. Williams, L.; Kim, J.; Kim, Y.; Arai, N.; Shimoda, O.; Holte, K. Contaminated insulators-chemical dependence of flashover voltages and salt migration. *IEEE Trans. Power Appar. Syst.* **1974**, *5*, 1572–1580. [CrossRef]
6. Courreges-Lacoste, G.B.; Ahlers, B.; Perez, F.R. Combined Raman spectrometer/laser-induced breakdown spectrometer for the next ESA mission to Mars. *Spectrochim. Acta Part A Mol. Biomol. Spectrosc.* **2007**, *68*, 1023–1028. [CrossRef] [PubMed]
7. Cremers, D.A.; Multari, R.A.; Knight, A.K. *Laser-induced Breakdown Spectroscopy, Encyclopedia of Analytical Chemistry: Applications, Theory and Instrumentation*; John Wiley & Sons, Ltd.: Hoboken, NJ, USA, 2006.
8. Yu, K.-Q.; Zhao, Y.-R.; Liu, F.; He, Y. Laser-induced breakdown spectroscopy coupled with multivariate chemometrics for variety discrimination of soil. *Sci. Rep.* **2016**, *6*, 1–10. [CrossRef] [PubMed]
9. Yuan, H.; Song, L.; Liu, P. Experimental study of vacuum degree online detection of vacuum interrupter based on laser induced breakdown spectroscopy. *High Volt. Appar.* **2017**, *3*, 235–239.
10. Yuan, H.; Gojani, A.B.; Gornushkin, I.B.; Wang, X. Investigation of laser-induced plasma at varying pressure and laser focusing. *Spectrochim. Acta Part B At. Spectrosc.* **2018**, *150*, 33–37. [CrossRef]
11. Wang, X.; Yuan, H.; Liu, D.; Yang, A.; Liu, P.; Gao, L.; Ding, H.; Wang, W.; Rong, M. A pilot study on the vacuum degree online detection of vacuum interrupter using laser-induced breakdown spectroscopy. *J. Phys. D Appl. Phys.* **2016**, *49*, 44LT01. [CrossRef]

12. Wang, N.; Wang, X.; Chen, P.; Jia, Z.; Wang, L.; Huang, R.; Lv, Q. Metal Contamination Distribution Detection in High-Voltage Transmission Line Insulators by Laser-induced Breakdown Spectroscopy (LIBS). *Sensors* **2018**, *18*, 2623. [CrossRef] [PubMed]
13. Wang, X.; Wang, H.; Chen, C.; Jia, Z. Ablation properties and elemental analysis of silicone rubber using laser-induced breakdown spectroscopy. *IEEE Trans. Plasma Sci.* **2016**, *44*, 2766–2771. [CrossRef]
14. Wang, X.; Hong, X.; Chen, P.; Zhao, C.; Jia, Z.; Wang, L.; Lv, Q.; Huang, R.; Liu, S. In-situ and quantitative analysis of aged silicone rubber materials with laser-induced breakdown spectroscopy. *High Volt.* **2018**, *3*, 140–146.
15. Kumar, V.S.; Vasa, N.J.; Sarathi, R. Remote surface pollutant measurement by adopting a variable stand-off distance based laser induced spectroscopy technique. *J. Phys. D Appl. Phys.* **2015**, *48*, 435504. [CrossRef]
16. Hahn, D.W.; Omenetto, N. Laser-induced breakdown spectroscopy (LIBS), part II: Review of instrumental and methodological approaches to material analysis and applications to different fields. *Appl. Spectrosc.* **2012**, *66*, 347–419. [CrossRef] [PubMed]
17. Tognoni, E.; Palleschi, V.; Corsi, M.; Cristoforetti, G. Quantitative micro-analysis by laser-induced breakdown spectroscopy: A review of the experimental approaches. *Spectrochim. Acta Part B At. Spectrosc.* **2002**, *57*, 1115–1130. [CrossRef]
18. Wang, Z.; Yuan, T.-B.; Hou, Z.-Y.; Zhou, W.-D.; Lu, J.-D.; Ding, H.-B.; Zeng, X.-Y. Laser-induced breakdown spectroscopy in China. *Front. Phys.* **2014**, *9*, 419–438. [CrossRef]
19. NIST. National Institute of Standards and Technology, U.S. Department of Technology Database. Available online: http://www.physics.nist.gov (accessed on 8 February 2020).
20. Cremers, D.A.; Radziemski, L.J. *Handbook of Laser-Induced Breakdown Spectroscopy*; John Wilkey & Sons Ltd.: Chichester, UK, 2006.
21. Singh, J.P.; Thakur, S.N. *Laser-Induced Breakdown Spectroscopy*; Elsevier: Amsterdam, The Netherlands, 2007.

MDPI
St. Alban-Anlage 66
4052 Basel
Switzerland
Tel. +41 61 683 77 34
Fax +41 61 302 89 18
www.mdpi.com

Applied Sciences Editorial Office
E-mail: applsci@mdpi.com
www.mdpi.com/journal/applsci

www.ingramcontent.com/pod-product-compliance
Lightning Source LLC
Chambersburg PA
CBHW061459210926
51850CB00008B/1760
* 9 7 8 3 0 3 6 5 1 1 4 0 5 *